BIOLOGICAL MACROMOLECULES AND ASSEMBLIES

Volume 2: Nucleic Acids
and Interactive Proteins

BIOLOGICAL MACROMOLECULES AND ASSEMBLIES

Volume 2: Nucleic Acids
and Interactive Proteins

Editors

Frances A. Jurnak and Alexander McPherson

University of California
Riverside

A Wiley-Interscience Publication

JOHN WILEY & SONS
New York   Chichester   Brisbane   Toronto   Singapore

Library of Congress Cataloging in Publication Data:

Main entry under title:

Nucleic acids and interactive proteins.

(Biological macromolecules and assemblies ; v. 2)
"A Wiley-Interscience publication."
Includes index.
1. Nucleic acids. 2. Proteins. 3. Biochemorphology.
I. Jurnak, Frances A. II. McPherson, Alexander,
1944–   III. Series.
QP801.P64B554 1984 vol. 2 [QP620]   574.19′24s   84-3586
ISBN 0-471-87076-5   [574.87′328]

Printed in the United States of America

10 9 8 7 6 5 4 3 2 1

Contributors

WAYNE F. ANDERSON, MRC Group on Protein Structure and Function, University of Alberta, Edmonton, Alberta, Canada

DAVID M. BLOW, Blackett Laboratory, Imperial College of Science and Technology, London, England

GARY D. BRAYER, University of British Columbia, Vancouver, British Columbia, Canada

PETER BRICK, Blackett Laboratory, Imperial College of Science and Technology, London, England

RICHARD E. DICKERSON, Molecular Biology Institute, University of California, Los Angeles, California

OLGA KENNARD, University Chemical Laboratory, Cambridge, England

MARY L. KOPKA, Molecular Biology Institute, University of California, Los Angeles, California

MITCHELL LEWIS, Smith Kline and French Laboratories, Philadelphia, Pennsylvania

BRIAN W. MATTHEWS, Institute of Molecular Biology, University of Oregon, Eugene, Oregon

ALEXANDER MCPHERSON, University of California, Riverside, California

DOUGLAS H. OHLENDORF, Institute of Molecular Biology, University of Oregon, Eugene, Oregon

CARL PABO, Johns Hopkins Medical School, Baltimore, Maryland

PHILIP PJURA, Molecular Biology Institute, University of California, Los Angeles, California

ALEXANDER RICH, Massachusetts Institute of Technology, Cambridge, Massachusetts

ZIPPORA SHAKKED, The Weizmann Institute of Science, Rehovot, Israel

HENRY M. SOBELL, The University of Rochester School of Medicine and Dentistry, Rochester, New York

THOMAS A. STEITZ, Yale University, New Haven, Connecticut

Y. TAKEDA, University of Maryland, Catonsville, Maryland

ANDREW H.-J. WANG, Massachusetts Institute of Technology, Cambridge, Massachusetts

JIAHUAI WANG, The Institute of Biophysics at Beijing, Academia Sinica, The Peoples Republic of China

IRENE T. WEBER, Yale University, New Haven, Connecticut

ALEXANDER WLODAWER, National Measurement Laboratory, National Bureau of Standards, Washington, D.C. and National Institute of Arthritis, Diabetes, and Digestive and Kidney Diseases, Bethesda, Maryland

Preface

The most exciting and, at the same time, widely anticipated contribution to molecular biology made by X-ray crystallographers in recent years has been the direct visualization at atomic resolution of nucleic acid and a variety of proteins with which it interacts. Indeed, it appears likely that a sound structural basis is now in the making for the interpretation of protein–nucleic acid interactions at the highest level of detail. From this will emerge an understanding in molecular biological and classical chemical terms of the processes by which genetic events are initiated, mediated, and ultimately regulated. At the same time, the multitude of essential cellular processes involving the interplay of these macromolecules can begin to be interpreted in a structurally consistent and rational manner.

Found in this volume are presentations of what may be considered the initial discoveries of a vast wealth of new information, a new level of understanding, that now clearly lies within reach of our skills and methods. But these investigations, each representing many scientist-years of intense effort, patience, and ingenuity, have already provided ample, indeed bounteous, rewards. Fundamental questions of long-standing interest have been answered, a quantum increase in precision has been added to our description of structures, and a new level of insight has been provided into the underlying mechanisms governing the maintenance and expression of the scheme for life.

This moment in time is particularly noteworthy, and this volume is especially appropriate because it marks the convergence of two independent lines of investigation that together have yielded, almost simultaneously, unexpectedly rich results. One line of investigation, the traditional studies of nucleic acid structure grounded in fiber diffraction—supplemented by numerous investigations of mono- and dinucleotides and aided by the new technology of oligonucleotide synthesis—culminated in the crystallization and determination of true examples of nucleic acid molecules. In complementary studies, other crystallographers were experiencing equally good fortune in the determination of the structures of proteins that bound to and interacted with these same nucleic acids. The coincidence of these two lines of investigation has provided the inspiration for this book.

The first three chapters describe the results of research defining the atomic resolution structures of what now appear to be the primary physiological forms of DNA. These are, respectively, the classic A and B forms of DNA and the new, strikingly novel Z form. Within each of these chapters is contained not only the detailed physical description and parameters of the structures but variations and ancillary experiments that begin to define the dynamic potential of the systems. The authors discuss how structure may signal sequence, why the molecular forms exhibit their own particular peculiarities, and why these individual characteristics may be physiologically meaningful. Chapter 4, relying on the observed interactions of small molecules with DNA, suggests some very interesting possibilities for the dynamic range inherent in DNA structure and its functional implications.

Chapters 5 through 7 describe the structures of three proteins that recognize and bind to sequence-defined sites on native double-stranded DNA and in so doing effect the expression of specific genes. Two of these proteins, cro and the lambda repressor fragment, are closely related and share control of a genetic locus on the DNA of bacteriophage lambda. The third, catabolite activator protein (CAP), acts in response to a specific ligand, cyclic AMP, to bind to a unique locus on the DNA of *E. coli* and thereby promotes transcription of a family of genes. A number of striking structural similarities within this set of proteins make evident, virtually by inspection, some of the mechanisms by which proteins and nucleic acids accommodate one another. Subtle differences between these proteins, in turn a reflection of amino acid sequence, at the same time suggest possibilities for the recognition and preference in binding of individual base sequences sequestered within the duplex DNA.

Chapters 8 and 9 describe two proteins, one quite familiar and the other less so, that bind to single-stranded DNA irrespective of sequence. These proteins are pancreatic ribonuclease and the gene 5 DNA unwinding protein from bacteriophage fd. Both can bind either DNA or RNA. These proteins are of particular interest because they are refined structures, in the crystallographic sense, and therefore they allow us to examine at a reasonably high level of detail the chemical interactions that can take place between a protein and a nucleic acid. They also suggest a number of interesting mechanistic possibilities by which enzymes and interactive proteins may manipulate as well as respond to the presence of DNA.

The final chapter discusses proteins that bind exclusively to RNA. The current state of research on the structures of the tRNA synthetases is such that the structure of the substrate, tRNA, is known in exquisite detail and its dynamic properties are under extensive investigation, but the atomic-level details of the interacting proteins have continued to elude the best efforts of the best crystallographers. It is clear from this chapter, however, that even this difficult obstacle is crumbling.

In the not too distant future we may expect to know the full range of protein–nucleic acid interactions. In this volume we find a start, and a very good start it is.

FRANCES A. JURNAK
ALEXANDER MCPHERSON

Riverside, California
October 1984

Contents

The A Form of DNA

1

ZIPPORA SHAKKED
The Weizmann Institute of Science
Rehovot, Israel

OLGA KENNARD
University Chemical Laboratory
Cambridge, England

CONTENTS

1. INTRODUCTION

Structural information on different conformations of DNA helical duplexes has accumulated during the past thirty years from analyses of X-ray diffraction patterns of oriented and sometimes crystalline fibers of native DNAs and of synthetic polynucleotides. In the majority of these systems, two antiparallel right-handed polynucleotide chains have complementary base sequences which are hydrogen bonded in the Watson–Crick manner. Known structures fall into two main conformational classes, called A and B; these differ in the way in which base pairs are stacked and the sugar phosphate backbones are wound around the helix axis, resulting in substantial differences in the relative width and depth of the major and minor grooves (Watson and Crick, 1953; Langridge et al., 1960; Arnott, 1976). External factors such as hydration and salt content have been shown to affect the secondary structure and the transition from one form to another (Wilkins, 1963). However, fiber diffraction studies provide only an averaged conformation of the DNA and hence local

2

variations in the double helix induced by the particular base sequence cannot be detected by this technique.

A knowledge of the detailed structure of DNA became increasingly important as evidence accumulated regarding the recognition of specific sequences by enzymes acting on DNA and by proteins involved in genetic control. The key question to be answered is whether the region of DNA recognized by a protein has a distinctly different structure from the bulk of DNA, which then provides specificity to the DNA–protein interaction.

High resolution studies of DNA became feasible only in the late 1970s, when methods for the synthesis of DNA fragments were sufficiently developed to produce material in the quantity and purity needed for the preparation of single crystals for X-ray analysis. The comparatively few crystallographic studies of self-complementary oligonucleotides so far completed have provided detailed information on both right- and left-handed double helices.

The first of these structures to be analyzed, the tetranucleotide d(pA-T-A-T), contained right-handed double helical fragments two base pairs long. It demonstrated sequence-dependent conformational features and suggested a dinucleotide repeat for the alternating polymer dA-dT (Viswamitra et al., 1978; Klug et al., 1979; Viswamitra et al., 1982). Studies of alternating sequences of cytosine and guanine bases, $d(C-G)_3$ and $d(C-G)_2$, led to the discovery of the left-handed alternating double helix, Z-DNA (Wang et al., 1979; Drew et al., 1980; Crawford et al., 1980).

In addition, four different right-handed DNA fragment structures were determined at near atomic resolution. Of these only the dodecamer d(C-G-C-G-A-A-T-T-C-G-C-G) or CGCGAATTCGCG and its 5-bromocytosine analog adopt the B type of conformation (Wing et al., 1980; Fratini et al., 1982). The three other structures d(G-G-T-A-T-A-C-C) or GGTATACC and its 5-bromouracil-containing analog (Shakked et al., 1981; 1983), d(^{1}C-C-G-G) or CCGG (Conner et al., 1982), and d(G-G-C-C-G-G-C-C) or GGCCGGCC (analyzed at $-8°$ and $-18°$C) (Wang et al., 1982a) are all of the A type. The recently determined RNA-DNA hybrid r(G-C-G)d(T-A-T-A-C-G-C) (Wang et al., 1982b; Fujii et al., 1982) also adopts the A conformation.

With the exception of d(pA-T-A-T) and $d(C-G)_3$, which were determined at atomic resolution, the resolution of the other structures was limited by the diffracting power of the crystals to a range of 1.8 to 2.25

Å. At this resolution crystallographic refinement must be restrained by imposition of prior information about bond lengths and angles (Konnert, 1976; Sussman et al., 1977; Jack and Levitt, 1978). Nevertheless, the main conformational features such as torsion angles about bonds and dihedral angles between base planes can be determined by such procedures with an accuracy of a few degrees even if individual atom positions cannot be resolved.

In this chapter we describe and compare the three A-DNA structures in terms of conformation, crystal packing, and biological implications. Our analysis is based on the published material on CCGG, GGTATACC, and GGCCGGCC and some unpublished data on GGCCGGCC (determined at −8°C) kindly provided to us by A. Wang and A. Rich, and on CCGG by R. E. Dickerson.

2. COMPARISON OF THE STRUCTURAL RESULTS OF THE A-DNA DOUBLE HELIX

2.1. Average Helical Conformations

Four stereoscopic views of each structure are given in Figures 1–4. Crystal structure data are in Table 1. In GGTATACC (I), the asymmetric unit is a double helical octamer. In CCGG (II), two double helical tetramers are stacked one upon the other via the crystallographic twofold axis to simulate a continuous octamer duplex, denoted as CCGG/CCGG (II). In the crystal structure of GGCCGGCC (III), the molecule possesses an intrinsic twofold symmetry, hence the asymmetric unit is a double helical tetramer as for II. The crystal packings of II and III are similar.

The three duplexes have common conformational features typical of the A type of structure as deduced from fiber diffraction (Arnott et al., 1976). The base pairs are invariably positioned away from the helix axis so that the minor grooves are shallow and the major grooves are deep. The base pairs are tilted (rather than perpendicular to the helix axis) in a clockwise sense as seen through the minor groove (Figs. 3 and 6). Superimposed on the common features, each duplex displays a broad spectrum of helical conformations. Evidence that the conformational variability is an intrinsic property of the base sequence rather than a consequence of external factors comes from two observations. First, in

Table 1. Crystal Structure Data

Compound	Space Group	Cell Dimensions (Å)			Z	Resolution (Å)	Current R-Value[b] (%)	Refinement Method
		a	b	c				
d(G-G-T-A-T-A-C-C)	$P6_1$	45.01	45.01	41.55	12[a]	1.8	19.8	Sussman et al., 1977
d(G-G-BrU-A-BrU-A-C-C)	$P6_1$	45.08	45.08	41.72	12	2.25	13.5	Sussman et al., 1977
d(^{I}C-C-G-G)	$P4_32_12$	42.06	42.06	25.17	16	2.1	20.5	Jack & Levitt, 1978
d(G-G-C-C-G-G-C-C) (−8°C)	$P4_32_12$	40.51	40.51	24.67	8	2.25	17.4	Hendrickson & Konnert, 1979
d(G-G-C-C-G-G-C-C) (−18°C)	$P4_32_12$	41.1	41.1	26.7	8	2.25	15.8	Hendrickson & Konnert, 1979

[a] Number of oligomers in the unit cell.
[b] R is defined $\sum |F_o - |F_c|| / \sum F_o$ where F_o and F_c are the observed and calculated structure factors respectively.

5

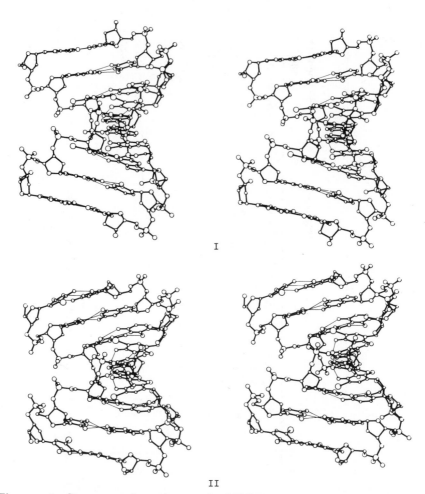

Figure 1. Stereoscopic view of I(GGTATACC), II(CCGG), and III(GGCCGGCC) perpendicular to the helix and twofold axes.

GGTATACC, which does not contain a crystallographic twofold axis, the molecular structure largely retains the internal symmetry of the base sequence, indicating minimal influence of the solvent and of the crystal environment on the molecular conformation. Second, GGCCGGCC and CCGG/CCGG, which crystallize isomorphously and are thus apparently subject to similar crystal forces, display distinctly different conformations.

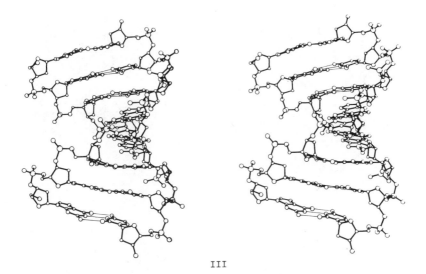

III

Figure 1. (Continued)

The two helices II and III are straight whereas helix I is bent symmetrically about the pseudo twofold axis by 15° over eight base pairs, corresponding to a curvature radius of approximately 80 Å. The bending is to the right in the plane of Figure 1.

The average helical conformations of the three duplexes are significantly different, as reflected in their global helix parameters (Table 2). The variations in average rise per base pair (h) are correlated with the base tilt values. As the tilt decreases the rise per base pair increases, as shown by corresponding values in Table 1. A significantly larger tilt (20°) associated with a smaller h (2.6 Å) is observed for the RNA-DNA hybrid decamer similar to that of the fiber A-DNA form (Arnott et al., 1976).

The helix structure is also characterized by the helical twist or rotation per base pair (t). The average values for the three structures increase in the following order: 32.2° for GGTATACC, 32.6° for GGCCGGCC and 33.9° for CCGG/CCGG. These correspond to 11.2, 11.0, and 10.6 base pairs per turn. The differences in the average periodicities of the three helices are reflected by the views down the helical axes (Fig. 2). The smaller the twist, the larger is the projected gap between the two 5' ends of the octamers.

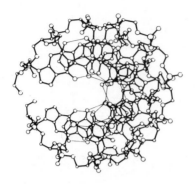
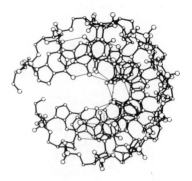

I

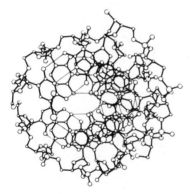
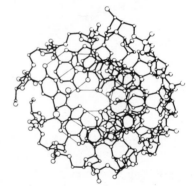

II

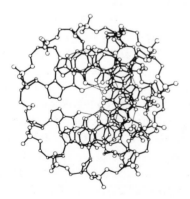
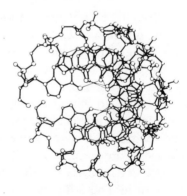

III

Figure 2. Stereoscopic view of I, II, and III down the helix axis.

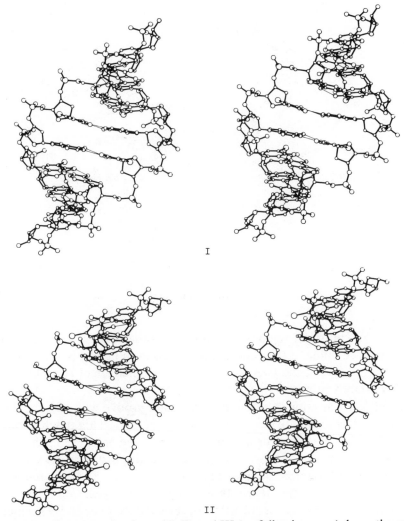

Figure 3.　Stereoscopic view of I, II, and III (on following page) down the minor groove.

The minor grooves of the various A-type helices maintain a fairly constant width whereas the major groove becomes wider as h increases and the base tilt decreases. Groove widths are measured as the shortest P-P separation across the groove less 5.8 Å, the van der Waals diameter of a phosphate group. The average value of the minor groove width (m) in each of the three DNA structures is around 10 Å. It appears that minor

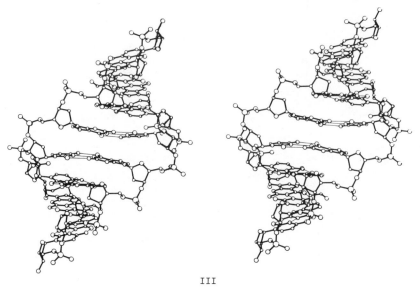

III

Figure 3. (Continued)

groove width at A + T regions is slightly larger than that at G + C regions. The observed value at the central TATA region in helix I is 10.5 Å compared with 9.8 Å at the GG(=CC) ends. A similar effect is observed in the RNA-DNA hybrid.

Significant variations are found in the major groove width for the three structures. Octamers are too short to exhibit the closest approach distance across the major groove for a complete turn of the helix, and hence variations in this parameter are estimated from the distance between the terminal 5'-phosphorus atoms. The largest value (8 Å) is found in GGCCGGCC where the helical translation is the largest (3.0 Å) and the base pair tilt is the smallest (12°). The observed difference (1.5 Å) in major groove width between GGTATACC and CCGG/CCGG is more than what would be expected from the corresponding values for the average rise per residue (2.87, 2.85 Å) and tilt angles (13°, 14°). The widening in the former results from the symmetrical bending of helix I about its approximate twofold axis in a direction that opens up the major groove. The variations in the base tilt and major groove width of the three helices and of an octamer model based on the 11-fold A-DNA form (h = 2.56 Å) deduced from fibers (Arnott and Chandrasekaran, 1982) are illustrated in the space-filling drawings of Figure 5. It is noteworthy that lowering the temperature

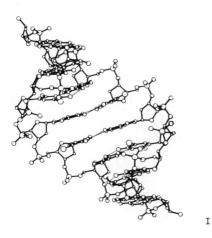
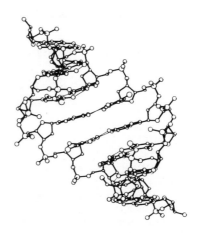

I

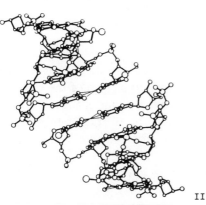
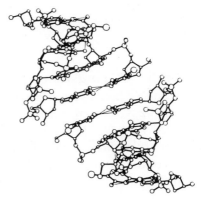

II

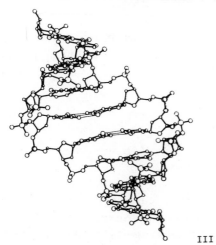
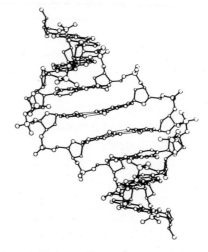

III

Figure 4. Stereoscopic view of I, II, and III down the major groove.

11

Table 2.   Average Helix Parameters[a]

	Twist per Base Pair (°)	Rise per Base Pair (Å)	Base Tilt (°)	Groove Width (Å) Minor	Groove Width (Å) Major[d]	R_P (Å)[c]	D(Å)[b]
GGTATACC	32.2(1)	2.87(1)	13.5	10.2	6.3	9.3	4.0
CCGG/CCGG	33.9(2)	2.85(3)	14.0	10.6	4.8	9.4	3.7
GGCCGGCC	32.6(3)	3.03(4)	12.0	9.6	7.9	9.6	3.6
A-DNA[e]	32.7	2.56	20	10.9	3.7	8.6	4.5

[a] Calculated by a least-squares procedure described by Shakked et al. (1983).
[b] Radial distance of base pairs from helix axis (see definition in Fig. 6a).
[c] Radial distance of P atoms from helix axis.
[d] The major groove width is estimated from the distances between the 5'-phosphorus atoms of the double-stranded octamers less 5.8 Å.
[e] Based on data from Arnott and Chandrasekaran (1982).

of GGCCGGCC crystals from −8 to −18°C leads to significant shrinkage of the double helix, characterized by a closer approach of the base pairs to the helix axis, a greater tilt, and a narrower major groove.

So far we have described and compared the average helical conformations of the three duplexes. We now describe and analyze two aspects of the conformational variability: stacking and relative motion of bases, and the geometry of the sugar-phosphate backbone.

2.2.   Stacking and Relative Motion of Bases

The arrangement of purine and pyrimidine bases in the double helical structures may be conveniently characterized by the following parameters:

1.   Propeller twist.
2.   Helix twist angles.
3.   Roll angles between base pairs.
4.   Stacking patterns of successive base pairs.

Although each has its own characteristic features, the parameters are also interdependent. Propeller twist is the contrarotation of two bases of a pair about their common long axis. A positive twist indicates a clockwise

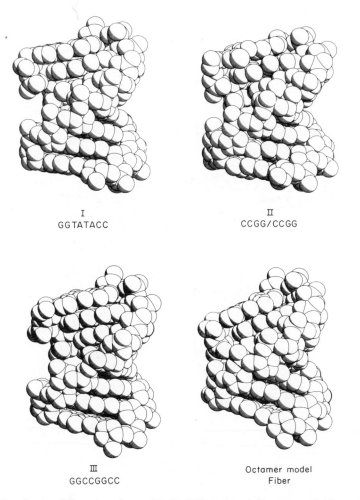

I
GGTATACC

II
CCGG/CCGG

III
GGCCGGCC

Octamer model
Fiber

Figure 5. Space-filling drawings of I, II, III, and an octamer model based on the fiber structure of A-DNA (Arnott and Chandrasekaran, 1982). The view as is for Figure 1. The major groove becomes wider (indicated by the cavity at the left) as the rise per residue increases and the base tilt decreases.

rotation of the nearer base with respect to the farther one when viewed along the long axis [Fig. 6a]. All three structures exhibit considerable positive propeller twists similar to that observed for the B-type dodeca-mers (Dickerson and Drew, 1981; Fratini et al., 1982). The average pro-peller twists for I, II, and III are 12°, 16°, and 9° respectively. It has been

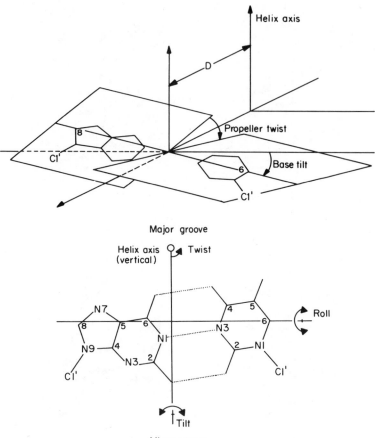

Figure 6. The local axial system used to define the various parameters of base position and orientation in the double helix (see Sections 2 and 3).

suggested that the observed propeller twist results in an improved base stacking along each of the two strands of the double helix (Levitt, 1978). Significant variations in individual propeller twist values are observed in all three duplexes. In particular, the end base pairs in helix I are flattened in comparison with the internal ones, thus allowing efficient intermolecular stacking against the shallow minor grooves of neighboring molecules. This aspect is discussed in Section 2.4.

The variations in helical twist for each structure may be estimated from the angles between successive interstrand C1′–C1′ vectors projected down the global helix axis. The individual values for the three structures are presented in Figure 7. The local twist angles show an alternating pattern of large and small values in all three sequences. Exceptionally large variations are observed for GGCCGGCC where the extreme values are 16° and 44°. Narrower ranges of twist angles are observed for GGTA-TACC (30°–34°) and CCGG/CCGG (31°–37°). In all three cases, pyrimi-

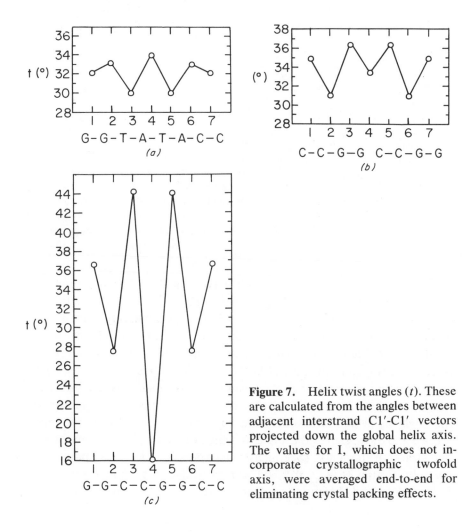

Figure 7. Helix twist angles (t). These are calculated from the angles between adjacent interstrand C1′-C1′ vectors projected down the global helix axis. The values for I, which does not incorporate crystallographic twofold axis, were averaged end-to-end for eliminating crystal packing effects.

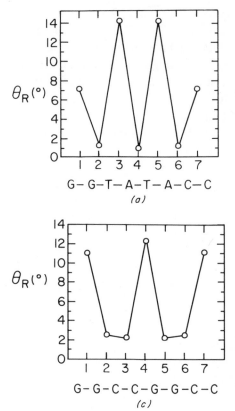

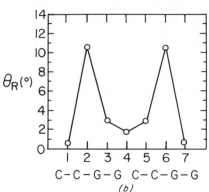

Figure 8. Roll angles between base pairs (θ_R). It measures the extent by which the mean planes through adjacent base pairs open towards the minor groove. See analytical expression for θ_R by Fratini et al., (1982).

dine-purine steps display the smallest twist angles. The homopolymer steps in II and III have the largest twist values. It is interesting that the common CCGG fragments in II and III show qualitatively similar alternating patterns of twist angles, even though the magnitudes of these angles for the two sequences are markedly different. The variations observed in the central fragment of the DNA-RNA hybrid are similar to those of I.

The relative orientation of the mean planes through adjacent base pairs may be analyzed in terms of roll (θ_R) and tilt (θ_T) components. θ_R measures the angle by which two successive base pairs open towards the minor groove and θ_T measures the angle by which the same planes open towards one strand of the helix. (See analytical expressions for these parameters given by Fratini et al., 1982; roll and tilt axes of a base pair are defined in Fig. 6b.) In all structures analyzed so far, the tilt components are negligible. The individual roll angles for the three duplexes are shown schematically in Figure 8. All significant roll angles are positive, meaning that the relevant base pairs open towards the minor groove.

Significant variations in the roll angles are observed for the three structures. Pyrimidine-purine steps exhibit the largest values ($10°-15°$). The corresponding values for purine-pyrimidine steps are almost negligible, indicating that the mean planes through the corresponding base pairs are nearly parallel. Both large and small roll angles are found in homopolymer steps. The common CCGG regions in II and III show similar roll patterns. The roll angles in GGTATACC alternate between smaller and larger values. Similar variations are observed for the DNA-RNA decamer.

The above findings show that the smallest twist angles are associated with the largest roll angles at the pyrimidine-purine steps. In all purine-pyrimidine steps the base pairs are nearly parallel to each other. The conformational features of the homopolymer steps G-G($=$C-C) appear to vary largely as a function of the particular base sequence.

The fourth structural parameter characterizing local changes in the double helix is the stacking pattern of successive base pairs, which reflects the intrastrand and interstrand proximity of the various functional groups of the bases. As a consequence of the different sizes of purine and pyrimidine bases, a uniform double helix has distinctly different stacking patterns for purine-pyrimidine, pyrimidine-purine, and purine-purine (or pyrimidine-pyrimidine) sequences. Local changes in the double helix induced by the particular order of the bases may further modulate these patterns. The stacking mode is strongly affected by the distance of the base pairs from the helix axis and their tilt with respect to that axis (see Fig. 6a). In the A form the base pairs are distant from the helix axis and are highly tilted, whereas in the B form the base pairs are close to the axis and nearly perpendicular to it. As a consequence, distinct patterns characterize the stacking of double helices of the A type and B type.

The various stacking modes observed in the three A duplexes are illustrated in Figure 9. Once again, large variations are observed for each type of dinucleotide sequence. The five distinct G-G($=$C-C) steps observed in the three structures are shown in the left-hand column of Figure 9. The stacking patterns of the two crystallographically nonequivalent homopolymer steps in GGCCGGCC (1b, 1c) are similar even though their corresponding helix twist angles ($37°$, $44°$) and roll angles ($11°$, $2°$) are appreciably different. In contrast, the two nonequivalent homopolymer

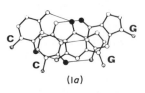
(1a)

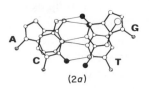
(2a)

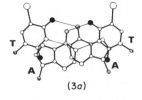
(3a)

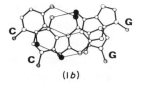
(1b)

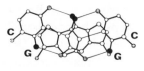
(2b)

(3b)

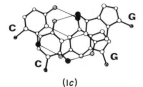
(1c)

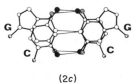
(2c)

(3c)

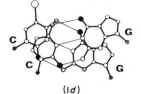
(1d)

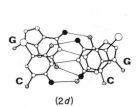
(2d)

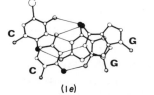
(1e)

Figure 9. Base stacking patterns. The view is perpendicular to the mean plane through the upper base pair. Bases of the right-hand strand are going downward from 5′ to 3′ direction, those of the left-hand strand are going upward from 5′ to 3′ direction. Keto oxygens are black, amino nitrogens are hatched, and C1′ atoms are dotted.

Table 3. Average Torsion Angles (°)[a]

	α	β	γ	δ	ε	ζ	χ
GGTATACC	-62(12)	173(8)	52(14)	88(3)	-152(8)	-78(7)	-160(8)
ᴵCCGG	-73(4)	180(8)	64(10)	80(6)	-161(7)	-67(4)	-161(7)
GGCCGGCC (-8°C)[b]	-75(36)	185(13)	56(22)	91(18)	-166(19)	-75(19)	-149(10)
DNA-RNA decamer[c]	-69(31)	175(14)	55(22)	82(9)	-151(15)	-75(16)	-162(10)
A-DNA[d]	-50	172	41	79	-146	-78	-154
A-RNA[d]	-68	178	54	82	-153	-71	-158
B-Dodecamer[e]	-63(8)	171(14)	54(8)	123(21)	-169(25)	-108(34)	-117(14)

[a] Main chain torsion angles are defined by

$$O3' \overset{\alpha}{\text{—}} P \overset{\beta}{\text{—}} O5' \overset{\gamma}{\text{—}} C5' \overset{\delta}{\text{—}} C4' \overset{}{\text{—}} C3' \overset{\epsilon}{\text{—}} O3' \overset{\zeta}{\text{—}} P \overset{\alpha+1}{\text{—}} O5'$$

Glycosyl torsion angle (χ) is defined as O1'-C1'-N1-C2 for pyrimidines and O1-C1'-N9-C4 for purines. A positive torsion angle indicates a clockwise rotation of the more distant bond.

[b] Omitting the δ values of disordered sugars of G1 and G2.

[c] Crystal structure data of r(GCG)d(TATACGC) by Wang et al. (1982b).

[d] Fiber data of Arnott and Chandrasekaran (1982).

[e] Crystal structure data of d(CGCGAATTCGCG) by Drew et al. (1981).

21

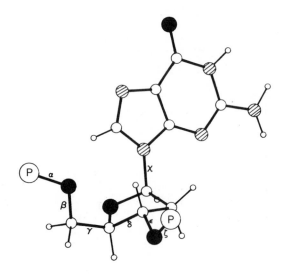

Figure 10. Nucleotide model showing main chain and glycosidic bond torsion angles. Oxygens are shaded and nitrogens are hatched.

The conformation of the sugar is usually defined by its puckering mode. The energetically favorable conformations are those in which four atoms are approximately coplanar and the fifth is displaced from this plane. The ring pucker is defined as X-*endo* or X-*exo* depending whether the out-of-plane atom (X) is on the same side of the plane as C5′ or on the opposite side. The backbone torsion angle δ about the C3′–C4′ bond is closely related to the sugar puckering (Sundaralingam, 1969; Levitt and Warshel, 1978) and permits the assignment of the sugar conformation with reasonable certainty even when, as for the structures discussed here, the resolution of the X-ray analysis is insufficient to locate individual atoms of the furanose ring.

In all the single crystal structures the backbone conformation is similar to that of the 11-fold A-DNA and A-RNA double helices deduced from fiber diffraction studies. The average values for GGTATACC and the DNA-RNA hybrid are closest to A-RNA. Significant variations, however, are observed in individual torsion angles. The most flexible region in the A-type duplexes is found to the 5′-side of the sugar ring. The α and γ torsion angles exhibit the largest variances, as indicated by the corresponding rms deviations from the mean. In contrast, for the B-type do-

decamer the largest variance in the torsion angles is found at the sugar
ring and to the 3'-side of the ring. The dominant sugar conformation in
the present A-type double helices is C3'-*endo*, which is also the one characteristic of the classical A-form geometry (Arnott et al., 1976). The P-
P distances are around 6 Å, which is again typical of the A-type structure.
Significant deviations from this geometry are observed for the structures
of GGCCGGCC, determined at $-8°C$ and $-18°C$. The sugar rings of the
two terminal guanines of the $-8°C$ structure are disordered as indicated
by the unfavorable planar conformation with δ values of nearly 120°. In
both structures the central segment of four base pairs contains sugar residues having alternating conformations with δ angles around 115° and 80°.
The distances between adjacent phosphorus atoms in this region also
alternate between larger values (av. 6.6 Å) and smaller values (av. 5.8
Å). The unusual geometry in this region appears to arise from the specific
stacking patterns as discussed below.

Several structural correlations between the backbone torsion angles
are observed in the present A-type structures. The most significant bi-

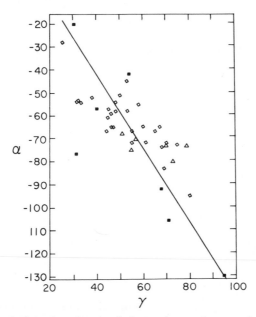

Figure 11.   Correlation plot of main chain torsion angles α and γ. The correlation
coefficient is -0.78. $\Diamond$, I; $\triangle$, II; $\blacksquare$, III.

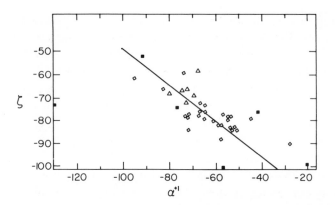

Figure 12. Correlation plot of the torsion angles in the phosphodiester 03'-P-05' linkage, ζ of one residue and α^{+1} of the next residue as defined in Table 3. The correlation coefficient is -0.69. $\diamond$, I; $\triangle$, II; $\blacksquare$, III.

variate relations, are shown with their linear correlation coefficients in Figures 11–15. The values for the terminal residues were omitted from the correlation calculations as these usually suffer from end effects. A negative sign of the correlation coefficient indicates that an increase in one angle is associated with a decrease of the other. All these correlations appear to result from the strong requirement of energetically favorable

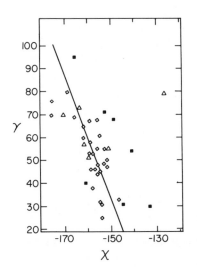

Figure 13. Correlation plot of the glycosyl torsion angle χ against main chain torsion angle γ. The correlation coefficient is -0.65.

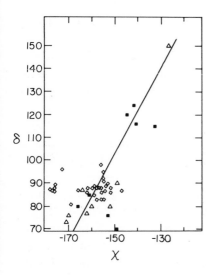

Figure 14. Correlation plot of the glycosyl torsion angle χ against main chain torsion angle δ. The correlation coefficient is 0.62 and drops to 0.0 by omitting the five points whose δ values are greater than 100°.

base stacking. The relative motion of the base pairs is achieved by correlated adjustments in the backbone torsion angles and the rotation about the glycosidic bonds.

The strongest correlation is observed between the torsion angles about P–O5′ and C4′–C5′ bonds, as shown by the α–γ plot (Fig. 11). This relation appears to play a major role in adjusting the relative orientation of the base pairs. The intervening torsional angle β is always *trans*, and hence

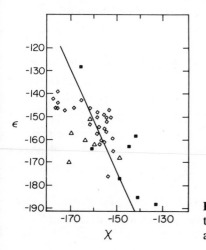

Figure 15. Correlation plot of the glycosyl torsion angle χ against main chain torsion angle ε. The correlation coefficient is −0.66.

nearly parallel motions of successive base pairs may be readily achieved by variations of these two angles in opposite directions.

The torsion angles about the phosphodiester linkage (O3'–P–O5') are also coupled in a negative manner, as shown by the ζ–α^{+1} relationship (Fig. 12). The rotation range about O3'–P ($-100°$ to $-50°$) is much smaller than that about P–O5' ($-130°$ to $-20°$). The larger flexibility at the second bond is facilitated by the concerted variations in γ, as discussed above. The extreme values of these torsion angles are found in GGCCGGCC, whereas the range is significantly narrower for GGTATACC and CCGG/CCGG.

Three structural relations are associated with the glycosidic torsion angle χ (Figs. 13–15). The three successive torsion angles (γ, δ, ϵ) about C5'–C4'–C3'–O3' are coupled with χ through a negative, positive, and negative correlation respectively. The weakest of the three is the χ–δ correlation, which arises from the few residues adopting sugar conformations other than C3'-endo. On omitting these residues the correlation coefficient drops practically to zero.

Relatively small positive changes in χ angles are associated with large negative changes of both γ and ϵ. The χ–γ relationship may be partly ascribed to the weak electrostatic attraction between the aromatic hydrogen (H6 or H8) and the O5' atom. This would cause the rotation of the pyrimidine or purine base about the glycosidic bond to be in the opposite sense to that of the oxygen about the C4'–C5' bond (see Fig. 10; the distance between the corresponding hydrogen and oxygen atoms is approximately 2.5 Å in the observed structures). The observed relation between χ and ϵ seems to serve as a fine tuner for adjusting the relative orientation of each base with respect to its neighbor in the 3' direction.

The most pronounced correlation in the B-dodecamers (Drew et al., 1981; Fratini et al., 1982) is between δ and χ, where a wide spectrum of sugar conformations is associated with a broad range of orientations about the glycosidic bonds. It appears that this type of relation is the principal mechanism for adjusting the relative motions of successive base pairs in the B-type structures. In the A-type duplexes, this relation is weak as the majority of the deoxyribose rings adopt the C3'-endo conformation. The four other structural correlations observed in the A-type structures are absent in the B-dodecamers, indicating that A-type and B-type double helices are subject to distinctly different geometrical constraints.

Sequence-dependent effects are less evident in the conformation of the sugar-phosphate backbone than in the base stacking. In GGTATACC and its isomorphous bromo analog, which do not incorporate crystallographic twofold symmetry, the individual torsion angles do not always follow the internal symmetry of the base sequence. Additionally, there are significant differences between corresponding torsion angles in these two octamers, probably arising from different hydration effects. Several corresponding torsion angles for the two structures of GGCCGGCC determined at $-8°C$ and $-18°C$ also differ greatly, whereas the corresponding base stackings are similar (Wang et al., 1982a). It should be noted that the unusual alternating conformation of the sugar-phosphate backbone in the middle region of these structures is not associated with an alternating purine-pyrimidine sequence. However, it facilitates the observed alternating pattern of helix twist angles with an extremely small value (16°) at the central C-G step and large ones (44°) at the flanking homopolymer steps. In CCGG/CCGG the helical base stacking at the center is preserved even in the absence of connecting phosphates.

All of the above findings indicate that the interaction between the bases is the dominant factor in determining local variability in the conformation of the double helix. Optimal stacking is achieved by correlated adjustments in the backbone torsion angles and the glycosidic orientation.

2.4. Crystal Packing and Intermolecular Interaction

In all three structures the packing of the molecules in the crystal is governed by the stacking of the terminal base pairs on a flat portion of the shallow A-DNA minor groove. In CCGG/CCGG and GGCCGGCC, which are isomorphous, the terminal base pair of one duplex interacts with the minor groove of a neighboring one via a left-handed fourfold screw axis (Fig. 16). In GGTATACC this stacking interaction is utilized in the arrangement of the duplex octamers around the right-handed sixfold and twofold crystallographic axes (Fig. 17).

Close views of the interactions between adjacent duplexes in GGTA-TACC are given in Figures 18a, b, c. Two duplexes related by the twofold screw axis are linked by a pair of hydrogen bonds, indicated by the broken lines in Figure 18a. A second view of the same duplexes given in Figure 18b clearly shows the stacking interaction between the terminal base pair

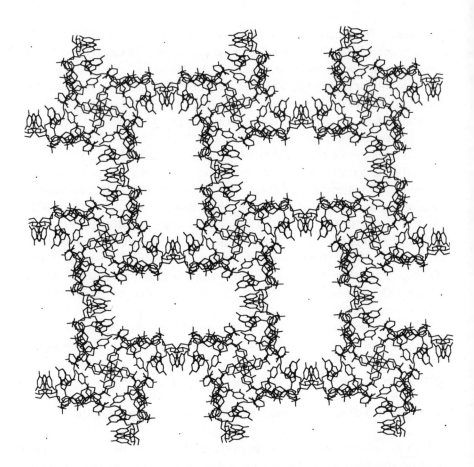

Figure 16. Packing arrangement of GGCCGGCC (space group $P4_32_12$). The view is down the fourfold screw axis. (The figure was kindly provided by A. Wang)

of the upper double helix and the sugar rings at the minor groove of the lower one. Two molecules related by the sixfold screw axis are shown in Figure 18c. Here again the terminal base pair of the upper molecule is stacked against the minor groove of the lower one. Both minor groove–base pair interactions and hydrogen bonding were observed in the DNA-RNA hybrid, but only the stacking interactions were found in GGCCGGCC and in CCGG/CCGG.

The intermolecular interactions observed in the B-type dodecamer also include base-sugar stacking and hydrogen bonds through the minor grooves of neighboring molecules (Wing et al., 1980). However, because the minor groove of the B-DNA double helix is deep and narrow a distinctly different intermolecular geometry is utilized. The minor groove termini of two molecules related by the twofold screw axis are interlocked with base planes tilted 35° to one another so that the end base pair of one helix is parallel to and in van der Waals contact with the sugar-phosphate backbone of the other. The interlocking of the minor grooves leads to the asymmetric bending of the double helix. In the A-type octamers, because

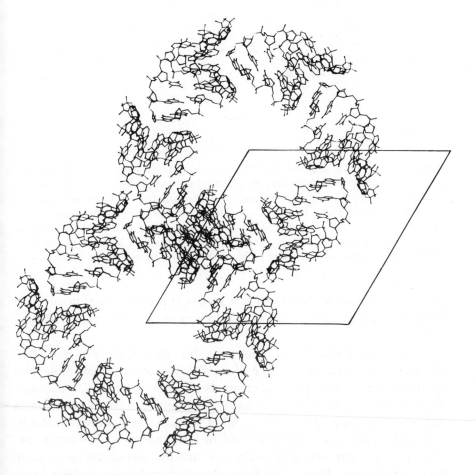

Figure 17.   Packing arrangement of GGTATACC (space group $P6_1$). The view is down the sixfold screw axis.

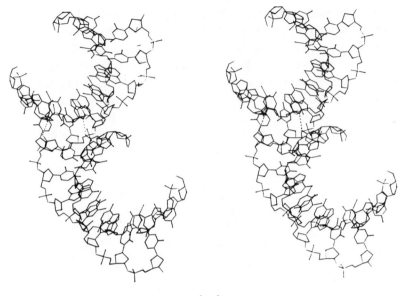

(a)

Figure 18. (*a*) Two adjacent GGTATACC duplexes related by twofold screw axis. The broken lines show intermolecular hydrogen bonds. The view is perpendicular to the 2_1 axis. (*b*) The same duplexes rotated by 90° about a vertical axis with respect to (*a*). It shows the stacking of the terminal base pair of the upper helix on the minor groove of the lower one. (*c*) Two GGTATACC duplexes related by the sixfold screw axis. The view is down this axis. It shows the stacking of the end base pair of the upper helix on the minor groove of the lower one.

of the shallowness of the wide minor groove, the interactions involve the outer surface of the double helix and thus only slightly affect the molecular conformation.

As seen in Figures 16 and 17 for both structures, there are large channels along the unique axes (4_3 or 6_1) in which most of the solvent molecules are found. In GGTATACC the channels are cylindrical, with a diameter of 30 Å. In the other two structures the channels have elliptical cross sections measuring 10 Å by 20 Å. All of these crystals, as well as the B-type dodecamers, contain about 50% of DNA by mass. The other constituents are water, spermine molecules, and the various cations used in the crystallization. Most of the ordered solvent molecules are found near the phosphate oxygens and in the major grooves of the duplexes, whereas

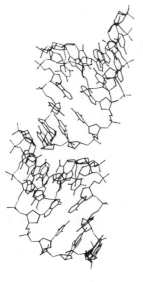
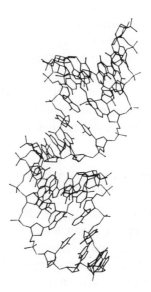

(*b*)

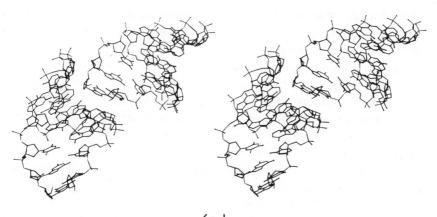

(c)

Figure 18.   (Continued)

the minor grooves are relatively dry. At the present resolution (~2 Å) neither the spermine molecules nor the counter ions could be identified with certainty in any of the A-DNA structures.

The B-type structures show a different pattern of hydration (Drew and Dickerson, 1981). The minor groove has a distinctive and very regular two-layer hydration structure that is believed by the authors to be largely responsible for the stability of the B helix. The disruption of this hydration spine must then be a necessary step in the B-to-A helix transition.

It may well be that the common intermolecular interaction observed in the crystals of I, II, and III, with the end base pairs stacked against the shallow minor grooves of neighboring molecules, helps to stabilize the A-type of double helix with respect to the B-type.

3. BIOLOGICAL IMPLICATIONS

As discussed in the introduction to this chapter, a major objective of the single-crystal X-ray analyses of deoxyoligonucleotides is to explore the influence of the base sequence on the structure of the DNA double helix and to trace sequence-induced conformational variations that might contribute to the specificity of protein–DNA recognition. All of the oligonucleotides reviewed in this chapter contain specific sequences known to interact with proteins.

It is generally assumed that the predominant form of DNA in biological systems is of the B geometry. However, most of the deoxyoligonucleotide structures studied to date have the A type of conformation, suggesting that this form is stable under a relatively wide range of conditions. It may well be that both A and B forms are important in biological events.

The central four-base-pair fragment in GGTATACC is part of a consensus sequence in the promoter regions of both prokaryotic and eukaryotic genes and is involved in RNA–polymerase binding prior to unwinding of the DNA double helix and subsequent transcription (Pribnow, 1979; Siebeslist et al., 1980). Variations in this sequence result in changes in promoter efficiency, indicating that both the overall conformation of the DNA and specific interactions involving the bases might be important. The similarity between the conformations of the T-A-T-A containing DNA octamer, the 11-fold form of RNA (Arnott and Chadrasekaran, 1982), and the hybrid r(G-C-G)d(T-A-T-A-C-G-C) (Wang et al., 1982b) suggests that certain DNA regions are capable of adopting a structure compatible with that of RNA or DNA-RNA hybrid and may be recognized as such by RNA-polymerase. The particular features of the T-A junctions observed both in GGTATACC and in the RNA-DNA hybrid—that is, the relative

destacking of the bases, the small twist angle and the large roll angle—
may facilitate the local unwinding and melting of double-stranded DNA
at promoter regions prior to transcription.

The two other sequences, GGCCGGCC and CCGG/CCGG, contain
the recognition sites CCGG and GGCC of the two restriction endonu-
cleases *Hpa II* and *Hae III* respectively (Modrich, 1979). Both enzymes
make symmetrical double-stranded cuts. *Hpa II* cleaves CCGG between
the two cytosine residues on each strand while the cleavage of GGCC by
Hae III occurs between the central guanine and cytosine residues. The
CCGG recognition sites in the two structures have similar alternating
patterns of helix twist and roll angles (Figs. 7 and 8). The common features
of the other recognition sequence in the two structures, GGCC, are reflected in the helix twist angle whereas the roll patterns are different. For
both recognition sequences the absolute values of the twist angles for
corresponding steps in the two structures are significantly different. The
stacking patterns (Fig. 9) are also different. These differences may arise
from end effects and from the absence of connecting phosphate groups
in CCGG/CCGG.

The observation that the same recognition sites in different oligomers
have common conformational features suggests that these may be recognized by the enzyme. It is possible that the relative motion of the base
pairs, induced by the specific order of the bases, brings the functional
groups on the DNA into favorable geometry for binding to the protein.

The influence of the specific base sequence on the global features of
the double helix, such as groove width and helix bending, is also shown
by the crystal structures. Although GGCCGGCC and CCGG/CCGG helices are straight, the GGTATACC helix is bent symmetrically as a consequence of the interaction between adjacent guanine and thymine bases.
This bending results in a substantial widening of the major groove, which
makes the functional groups on the bases more accessible. Local bending
effects and changes in groove dimensions may also play an important role
in the recognition between DNA and proteins.

Wang et al. (1982a) have suggested that the stacking interaction between the terminal base pairs and the shallow minor grooves of neighboring duplexes observed in the A-DNA structures might serve as a model
for the interaction of nucleic acids with planar molecules such as carcin-
ogens (Jeffrey et al., 1976).

4. CONCLUSIONS

At the beginning of this chapter we asked the question: Do specific base sequences induce local variations in the DNA conformation and are such variations recognized by proteins in biological events? The comparative study of the A-type structures with this question in mind has led to the following conclusions:

1. There is strong evidence of local variability in the conformation of the double helix, which is related to base sequence.
2. Base stacking appears to be the primary driving force for these local variations. Energetically favorable stacking modes are achieved by concerted changes in the backbone torsion angles and in the orientation about the glycosidic bonds.
3. Local variations can lead to global changes in the double helix such as bending and groove widening. Both local and global variations may provide specificity for the recognition of DNA by proteins.

Although these findings arise from A-DNA structures they are also relevant to B-DNA and may be a general feature of other forms of DNA as well. Only very few single crystal analyses of DNA fragments have been published to date. Consequently, we do not yet have sufficient information to estimate the extent to which a specific base sequence rather than external conditions determine the type of helix to be adopted. We know that each type of helix is characterized and stabilized by a particular pattern of base stacking. Specific base sequences may favor certain stacking patterns, and it is possible that under the same conditions one base sequence is stable in one form and another sequence adopts a different form. This idea is supported by the systematic fiber diffraction studies of many synthetic DNA polymers of varied sequences, demonstrating the effect of the base sequence on the transitions between the several helical forms (Leslie et al., 1980).

The crystal structures of three regulatory proteins have been determined and are described in Chapters 5, 6 and 7. The structure analyses of *cro* and *lambda* repressors and the catabolic gene activator protein suggest binding to right-handed B-DNA (Anderson et al., 1981; Pabo and Lewis, 1982; McKay and Steitz, 1981). The structural diversity of DNA discovered in the various X-ray investigations promises much excitement during the coming years when new DNA sequences, DNA binding proteins, and their complexes will be analyzed at the molecular level.

Acknowledgments. We thank the United States/Israel Binational Science Foundation (BSF), Jerusalem, Israel, and the Medical Research Council of the United Kingdom for financial support. We thank our colleagues D. Rabinovich and F. L. Hirshfeld for useful comments, and Mrs. M. Ben-Ami for editing and typing the manucript.

REFERENCES

Anderson, W. F., Ohlendorf, D. H., Takeda, Y., and Matthews, B. W. (1981). *Nature (London)* **290**, 754–758.

Arnott, S. (1976). In *Organization and Expression of Chromosomes,* V. G. Allfrey, E. K. F. Bautz, B. J. McCarthy, R. T. Schimke, A. Tissieres, Eds. Dahlem Conference, Berlin, 1976, pp. 209–222.

Arnott, S., and Chandrasekaran, R. (1982). Personal communication.

Arnott, S., Smith, P. J. C., and Chandrasekaran, R. (1976). In *CRC Handbook of Biochemistry and Molecular Biology,* G. D. Fasman, Ed., 3rd ed., Vol. 2. Chemical Rubber Co., Cleveland, Ohio, pp. 100–110.

Calladine, C. R. (1982). *J. Mol. Biol.* **161**, 343–352.

Conner, B. N., Takano, T., Tanaka, J., Itakura, K., and Dickerson, R. E. (1982). *Nature (London)* **295**, 294–299.

Crawford, J. L., Kolpak, F. J., Wang, A. H.-J., Quigley. G. J., van Boom, J. H., van der Marel, G., and Rich, A. (1980). *Proc. Natl. Acad. Sci. USA.* **77**, 4016–4020.

Dickerson, R. E. (1983). *J. Mol. Biol.,* **166**, 419–441.

Dickerson, R. E. and Drew, H. R. (1981). *J. Mol. Biol.* **149**, 761–786.

Drew, H. R., and Dickerson, R. E. (1981). *J. Mol. Biol.* **151**, 535–556.

Drew, H. R., Takano, T., Tanaka, S., Itakura, K., and Dickerson, R. E. (1980). *Nature (London)* **286**, 567–573.

Drew, H. R., Wing, R. M., Takano, T., Broka, C., Tanaka, S., Itakura, K., and Dickerson, R. E. (1981). *Proc. Natl. Acad. Sci. USA.* **78**, 2179–2183.

Fratini, A. V., Kopka, M. L., Drew, H. R., and Dickerson, R. E. (1982). *J. Biol. Chem.* **257**, 14,686–14,707.

Fujii, S., Wang, A. H.-J., van Boom, J. H., and Rich, A. (1982). *Nucl. Acids Res. Symp. Ser.* **11**, 104–112.

Hendrickson, W. A., and Konnert, J. H. (1979). In *Biomolecular Structure, Conformation, Function and Evolution,* R. Srinivasan, Ed. Vol. 1. Pergamon, Oxford, pp. 43–57.

Jack, A., and Levitt, M. (1978). *Acta Crystallogr.* **A34**, 931–935.

Jeffrey, A. M., Jennette, K. W., Blobstein, S. H., Weinstein, I. B., Beland, F. A., Harvey, R. G., Kasai, H., Miura, I., and Nakanishi, K. (1976). *J. Am. Chem. Soc.* **98,** 5714–5715.

Klug, A., Jack, A., Viswamitra, M. A., Kennard, O., Shakked, Z., and Steitz, T. A. (1979). *J. Mol. Biol.* **131,** 669–680.

Konnert, J. H. (1976). *Acta Crystallogr.* **A32,** 614–617.

Langridge, R., Marvin, D. A., Seeds, W. E., Wilson, H. R., Hooper, C. W., Wilkins, M. H. F., and Hamilton, L. D. (1960). *J. Mol. Biol.* **2,** 38–62.

Leslie, A. G. W., Arnott, S., Chandrasekaran, R., and Ratliff, R. L. (1980). *J. Mol. Biol.* **143,** 49–72.

Levitt, M. (1978). *Proc. Natl. Acad. Sci. USA* **75,** 640–644.

Levitt, M., and Warshel, A. (1978). *J. Am. Chem. Soc.* **100,** 2607–2613.

McKay, D. B., and Steitz, T. A. (1981). *Nature (London)* **290,** 744–749.

Modrich, P. (1979). *Quart. Rev. Biophys.* **12,** 315–369.

Pabo, C. O., and Lewis, M. (1982). *Nature (London)* **298,** 443–447.

Pribnow, D. (1979). In *Biological Regulation and Development,* R. F. Goldberger, Ed. Plenum, New York, pp. 219–278.

Shakked, Z., Rabinovich, D., Cruse, W. B. T., Egert, E., Kennard, O., Sala, G., Salisbury, S. A., and Viswamitra, M. A. (1981). *Proc. R. Soc. Ser. B* **213,** 479–487.

Shakked, Z., Rabinovich, D., Kennard, O., Cruse, W. B. T., Salisbury, S. A., and Viswamitra, M. A. (1983). *J. Mol. Biol.,* **166,** 183–201.

Siebeblist, U., Simpson, R. B., and Gilbert, W. (1980). *Cell* **20,** 269–281.

Sundaralingam, M. (1969). *Biopolymers* **7,** 821–860.

Sussman, J. L., Holbrook, S. R., Church, G. M., and Kim, S.-H. (1977). *Acta Crystallogr.* **A33,** 800–804.

Viswamitra, M. A., Kennard, O., Jones, P. G., Sheldrick, G. M., Salisbury, S., Falvello, L., and Shakked, Z. (1978). *Nature (London)* **273,** 687–688.

Viswamitra, M. A., Shakked, Z., Jones, P. G., Sheldrick, G. M., Salisbury, S. A., and Kennard, O. (1982). *Biopolymers* **21,** 513–533.

Wang, A. H.-J., Fujii, S., van Boom, J. H., and Rich, A. (1982a). *Proc. Natl. Acad. Sci. USA* **79,** 3968–3972.

Wang, A. H.-J., Fujii, S., van Boom, H. J., van der Marel, G. A., van Boeckel, S. A. A., and Rich, A. (1982b). *Nature (London)* **299,** 601–604.

Wang, A. H.-J., Quigley, G. J., Kolpak, F. J., Crawford, J. L., van Boom, J. H., van der Marel, G., and Rich, A. (1979). *Nature (London)* **282,** 680–686.

Watson, J. D., and Crick, F. H. C. (1953). *Nature (London)* **171,** 737–740.

Wilkins, M. H. F. (1963). *Science* **140,** 941–950.

Wing, R. M., Drew, H. R., Takano, T., Broka, C., Tanaka, S., Itakura, K., and Dickerson, R. E. (1980). *Nature* **287,** 755–758.

2

Base Sequence, Helix Geometry, Hydration, and Helix Stability in B-DNA

RICHARD E. DICKERSON
MARY L. KOPKA
PHILIP PJURA
Molecular Biology Institute
University of California
Los Angeles, California

CONTENTS

1. BACKGROUND

Science thrives on the unexpected. Nearly 30 years of X-ray analysis of fiber diffraction patterns from stretched and oriented natural DNA had encouraged the belief that double-helical DNA would be found in one of two families: the high-humidity B form or the low-humidity A form. Yet after the methodology of DNA synthesis had progressed to the point where synthetic oligonucleotides could be prepared in quantity and purity suitable for crystallization, the first two double-helical structures to be solved, CGCGCG (Wang et al., 1979) and CGCG (Drew et al., 1980; Drew and Dickerson, 1981a), proved to belong to an entirely new family of left-handed Z-DNA. It must be admitted that such a result was not entirely unexpected; both research groups had independently chosen alternating polymers of C and G for study because of a well-documented literature indicating that poly(dC-dG) underwent some type of phase transition in solution in response to an increase in concentration of either salt or alcohol (Drew et al., 1978). Yet it was a surprise to find that the high-salt form of the alternating copolymer had a left-handed helix sense, and a zigzag backbone with two successive base pairs per helical repeat, as described in detail in Chapter 3.

The next double-helical oligonucleotide structure to be solved after Z-DNA was that of the DNA dodecamer CGCGAATTCGCG, and this is largely the subject of the present chapter. The molecule was synthesized by the liquid-phase triester method by Pyotr Dembek and Horace Drew under the guidance of Keiichi Itakura. The particular sequence was chosen for two reasons: (1) It combined Z-compatible CGCG ends containing alternating purines and pyrimidines with an AATT center that was incompatible with the Z structure because it did not alternate in base type. (2) It contained the GAATTC sequence that is the natural recognition and cleavage site for the Eco RI restriction enzyme. The investigators anticipated that the structure would be a broken or interrupted helix, with left-handed Z-helical CGCG ends and a disordered or looped-out center. But once again the unexpected occurred: The molecule proved to be a classical B helix, with overall average helix parameters similar to those predicted by fiber diffraction studies (Wing et al., 1980; Drew et al., 1981; Dickerson and Drew, 1981a). Other double-helical oligomer structures have since been found to adopt the A form (Chapter 1) or the Z form (Chapter 3), but to date CGCGAATTCGCG and its brominated variants are the only examples of a B helix, aside from the isolated central TA step of the daunomycin intercalation complex of CGTACG (Quigley et al., 1980). This situation undoubtedly will change in the near future, but at present one can only take a close look at one sequence, and the changes in structure that are produced by bromination (simulating the 5-methylcytosine known to be involved in control of gene expression), change of crystallizing medium, and temperature. These have already provided us with insights into the way base sequence can influence local helix geometry, the reason why B-DNA is stabilized by hydration, and the preferred mode of bending of the double helix. The analytical framework constructed for the study of this first sequence should be of general value in the study of other sequences as they become available.

2. INDEPENDENT REFINEMENT OF VARIANTS OF CGCGAATTCGCG

The double-helical B-DNA dodecamer structure has been refined independently in five variants, listed in Table 1 along with references to the original literature. Cycles of Jack–Levitt restrained least-squares refinement were alternated with examination of Fourier and difference Fourier maps to search for errors or to locate ordered solvent peaks. This process and the refinement parameters employed have been set out fully by Fratini et al. (1982) and will not be discussed further here.

One section through the full three-dimensional X-ray diffraction pattern from crystals of CGCGAATTCGCG is shown in Figure 1. This photograph contains features of intensity distribution resembling those of a fiber diffraction photo with the fiber axis vertical. Although the X-ray scattering occurs in the discrete spots characteristic of a crystal rather than the more diffuse maxima from a fiber, clusters of strong reflections define a cross pattern typical of a helix, and other clusters of strong intensities above and below the center suggest a vertical repeat in the molecule of around 3.3 Å, as would be expected from the stacked base pairs of B-DNA. Hence even at the crystal survey stage it was deduced, correctly, that the crystal would contain double-helical molecules, with their axes parallel to the vertical direction in this photograph.

Two avenues of approach were taken to obtain isomorphous heavy atom derivatives for phase analysis of the original CGCGAATTCGCG (designated as the Native structure): diffusion of cisplatin [*cis*-dichloro-diaminoplatinum(II)] into pregrown crystals, and *de novo* synthesis of the dodecamer with 5-bromocytosine substituted for cytosine at the first, third, or ninth position along each strand (termed the 1-Br, 3-Br, and 9-Br derivatives). The 1-Br crystals were of poor quality and were not studied further. The 3-Br crystals and those perfused with cisplatin provided two good isomorphous derivatives, and these were used in phase

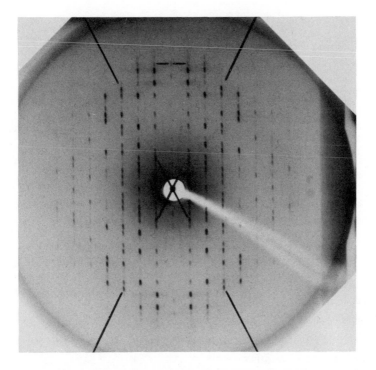

Figure 1. 20° precession camera survey photograph of the h0l zone of the diffraction pattern from crystals of CGCGAATTCGCG at room temperature. The crystals are orthorhombic, space group $P2_12_12_1$, with cell dimensions $a = 24.87$ Å, $b = 40.39$ Å, $c = 66.20$ Å. This photograph contains strong clues that the c axis, which is vertical, is also the axis of a B-type double helix. Particularly strong reflections occur in a cross pattern (marked by diagonal lines) with half-angle 27°, corresponding to the 27° pitch angle of the phosphate backbone in the double helix. Another group of strong reflections is found above and below the center near the 20th order along c^* and c, as marked by the two short horizontal lines near the top of the photo. This indicates the repetition of some prominent structural feature every 66.20 Å/20 = 3.3 Å along the helix axis, and in fact arises from the perpendicular stacking of base pairs. In an A-type helix this axial cluster of strong reflections would be less prominent because of the tilting of individual base planes. Compare this with a B-DNA fiber diffraction photograph such as plate I of Langridge et al. (1960).

Table 1. Comparison of Structure Refinement of B-DNA Dodecamers[a]

Dodecamer	Designation	Cell Dimensions (Å)			Nominal Resolution	No. of Reflections	
		a	b	c		>2δ	Total
Native at 20°C	Native	24.87	40.39	66.20	1.9Å	2725	5534
Native at 16 K	16K	23.44	39.31	65.26	2.7Å	1051	1824
Native plus cisplatin	Cisplatin	24.16	39.93	66.12	2.6Å	—	2012
9-Br in MPD at 20°C	MPD20	24.71	40.56	65.62	3.0Å	866	1579
9-Br in MPD at 7°C	MPD7	24.20	40.09	63.95	2.3Å	2031	3176

[a] Base sequences are either CGCGAATTCGCG (Native) or CGCGAATT[Br]CGCG (9-Br).
[b] Difference in inclination of base pairs C1·G24 and C11·G14, from Tables A1–A5.
[c] Diameter from packing in unit cell = $\frac{1}{2}(a^2 + b^2)^{\frac{1}{2}}$.

Final R. Factor		No. of Solvent Peaks	Axial Bend[b]	Mean Helix Rotation	Mean Base Pairs per Turn	Mean Rise per Base Pair	Crystal Cell Packing Diameter[c]	References
>2δ	All							
0.178	0.239	80	18°	38.75°	9.29	3.24Å	23.7Å	Wing et al., 1980; Drew et al., 1981; Dickerson and Drew, 1981a; Drew and Dickerson, 1981b; Dickerson and Drew, 1981b; Dickerson et al., 1981
0.151	0.210	83	22°	38.98°	9.24	3.21Å	22.9Å	Drew et al., 1982
—	0.164	121	17°	39.59°	9.09	3.24Å	23.3Å	Wing et al., 1984
0.130	0.174	44	14°	38.46°	9.36	3.27Å	23.7Å	Fratini et al., 1982
0.173	0.215	114	3°	37.10°	9.70	3.34Å	23.4Å	Dickerson et al., 1982a; Fratini et al., 1982; Kopka et al., 1983; Dickerson et al., 1983a; Dickerson et al., 1983b

analysis of the Native structure. Data could be collected out to 1.9 Å resolution before the pattern became too weak, and the structure was refined to a final crystallographic residual or *R* factor of 23.9% for all data, or 17.8% for reflections above two sigma.

Crystals of the Native dodecamer also were cooled to 16° K in a liquid-helium-cooled diffractometer (Drew et al., 1982). Data at this low temperature could be collected only to 2.7 Å resolution. This data set was refined independently, using the room temperature structure as a starting model, to an *R* factor of 21.0% for all data, or 15.1% for two sigma data. The cisplatin derivative itself was refined as an independent structure, and a 2.6 Å data set was brought to an *R* factor of 16.4% for all data.

The 3-Br derivative was so isomorphous with the Native molecule that further analysis was of little interest. But the 9-Br derivative, which had been synthesized too late for consideration in phase analysis, proved to have unexpected and ultimately informative special features. In all three of the structures described so far—Native, 16K, and cisplatin—the overall axis of the helix is bent by approximately 14°–22° over twelve base pairs. Two data sets were collected from the 9-Br compound: one at 7°C and another at 20°C after a mounted crystal had been allowed to come to room temperature on the diffractometer. Because these crystals were grown in 35% MPD (2-methyl-2,4-pentanediol) and transferred to 60% MPD after growth, the two data sets were named the MPD7 and MPD20 sets, respectively. A 2.3 Å data set could be collected for MPD7, but crystal degradation imposed a 3.0 Å limit on the MPD20 crystal.

In view of its lower resolution, the MPD20 analysis is of interest chiefly because it provided a bridge to the solution of the MPD7 structure. As described more fully by Fratini et al. (1982), the MPD7 structure could not be solved by using the Native double helix as a starting model. A difference map using MPD7-minus-Native amplitudes and Native phases failed even to reveal the bromine peak on each strand of the helix. The MPD20 structure, however, could be solved by this molecular replacement approach, and it was found to differ from the Native starting model mainly by a rotation and shift of the helix within the crystal unit cell. Application of multiples of this MPD20 rotation and shift ultimately proved to be the key to solving the MPD7 structure. The MPD7 helix, in addition to being moved within the cell, was seen to have lost its 19° bend in helix axis and to be nearly straight. Hence, although only one B-DNA double helix sequence is available for study to date, it is present in two

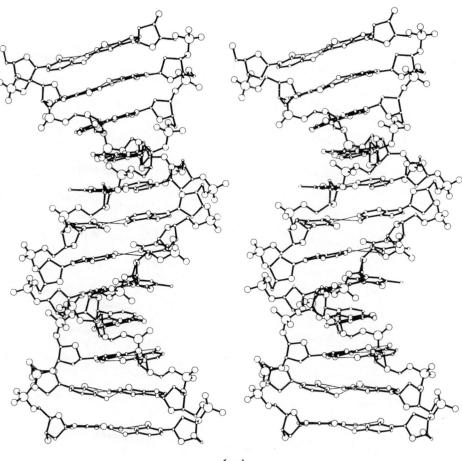

(a)

Figure 2. Ball and stick skeletal stereo pair drawings of DNA dodecamers, viewed into the minor groove at the center. (*a*) 16K structure with a 22° bend in overall helix axis between base pairs C1·G24 and C11·G14. *on following pages*: (*b*) MPD20 helix with 14° overall axial bend. (*c*) Nearly straight MPD7 helix with only a 3° bend. Bends are defined as in Table 1. Strand 1 of the double helix begins with base C1 at upper left and ends with G12 at lower left; strand 2 begins with base C13 at lower right and ends with G24 at upper right. Atoms in order of decreasing radius are Br (in *b* and *c* only), P, O, N, and C, with hydrogen atoms not shown. The two ends of each helix are related by an approximate twofold axis of symmetry perpendicular to the page through the center of the molecule. This symmetry axis, which relates corresponding features on the two chemically identical strands, is only approximate because of the bend in the double helix. It would presumably be exact for a molecule isolated in solution and free from perturbation by neighboring molecules.

45

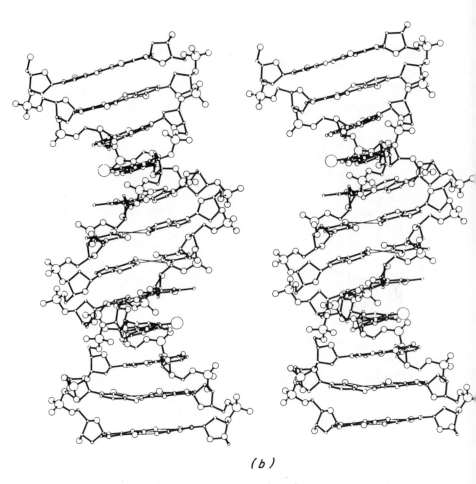

(b)

Figure 2.   (Continued)

structural forms: straight and curved. Comparison of the two forms enables us to understand how a double helix bends most readily.

Crystal and refinement parameters for the five refined B structures are compared in Table 1. It should be kept in mind that the quality of these five analyses differs. With similar efforts expended on the individual refinements, the credibility of the final structures varies approximately with the quantity of data employed, or with the nominal resolution. The low R factors tabulated for cisplatin and MPD20 are somewhat misleading because of the lower resolution of their data sets, and hence the smaller

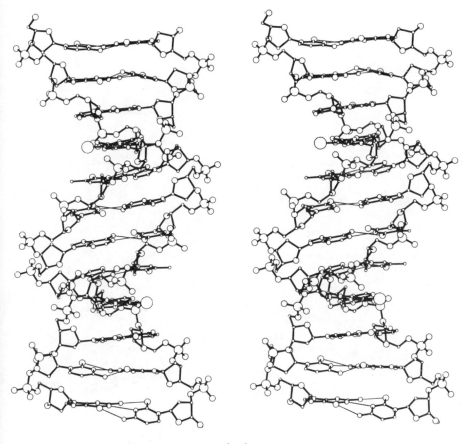

(c)

Figure 2.   (Continued)

number of data points that must be fitted successfully by the structural model. Root-mean-square errors in atomic position, as estimated by a Luzzati plot of R factor versus resolution, are 0.27 Å for the Native structure and 0.35 Å for the MPD7 structure at slightly lower resolution. These are the structures to which most credence can be given; the other three are best regarded as supplementary sources of information on specific points.

Three of the five refined dodecamer structures are compared in Figure 2: those with the greatest, intermediate, and least overall bend in helix

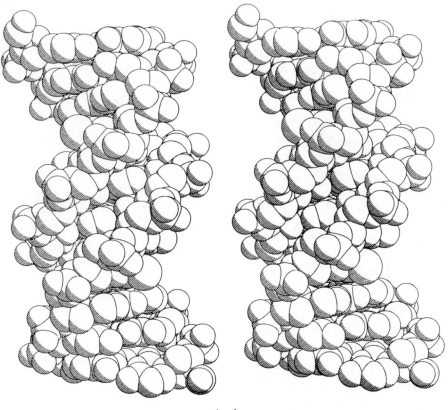

(a)

Figure 3. Space-filling stereo drawings of the MPD7 helix with unbent axis. Largest spheres are Br; others in order of decreasing size are P, C, N, and O. (*a*) View into the minor groove as in Figure 2. (*b*) ''Left side'' view with major groove above and minor groove below.

axis. The maximally bent 16K structure looks very much as though most of the bending occurs by compression of the major groove four base pairs down from the top; the exact nature of the bending will be discussed in Section 7. A better impression of the physical appearance of the helix, as if it were being viewed in some kind of X-ray microscope, is provided by the space-filling stereo drawings of Figure 3. These illustrations show how broad the major groove is, and how deep and narrow the minor groove is. Other skeletal and space-filling stereo drawings of the Native,

(b)

Figure 3.   (Continued)

16K, MPD20, and MPD7 helices may be found in the original literature, and Table 2 is intended as an index to aid in locating them.

At the crystal survey stage of the dodecamer structure analysis it was noticed that the c axis of 66.2 Å was almost exactly 24 times the expected rise per residue for an A helix, leading the investigators to think that two A-DNA dodecamers might be stacked atop one another along the c direction. As usual, reality was slightly more subtle: There are two double helices along the c direction, but they are B helices, and the greater intrinsic length of a B helix is accommodated by overlapping ends of the

Table 2. Index to Stereo Drawings of B-DNA Dodecamer

	Source References for Stereo Drawings			
	Native	16K	MPD20	MPD7
1. Ball and Stick Skeleton				
Minor groove	Dickerson and Drew, 1981a; Drew and Dickerson, 1981b; Dickerson et al., 1981; Fratini et al., 1982	Drew et al., 1982	Fratini et al., 1982	Dickerson et al., 1982a; Fratini et al., 1982
Right side[a]				Dickerson et al., 1983b
Major groove	Wing et al., 1980; Drew et al., 1981; Dickerson, 1981b; Dickerson et al., 1983b			Dickerson et al., 1983b
Left side[a]	Drew et al., 1981			
2. Space-Filling Helix				
Minor groove	Dickerson et al., 1981			Dickerson et al., 1982a; Fratini et al., 1982; Dickerson et al., 1983a

Right side		Dickerson et al., 1983b
Major groove	Wing et al., 1980	Dickerson et al., 1983b
Left side		Fratini et al., 1982
3. Space-Filling, Hydrated		
Minor groove	Drew and Dickerson, 1981b; Dickerson et al., 1981	Kopka et al., 1983; Dickerson et al., 1983a
Right side	Drew and Dickerson, 1981b	Kopka et al., 1983
Major groove	Drew and Dickerson, 1981b; Dickerson et al., 1981	
Left side	Drew and Dickerson, 1981b	

[a] "Right" and "left" sides are oriented with the minor groove face as the front and the major groove as the rear.

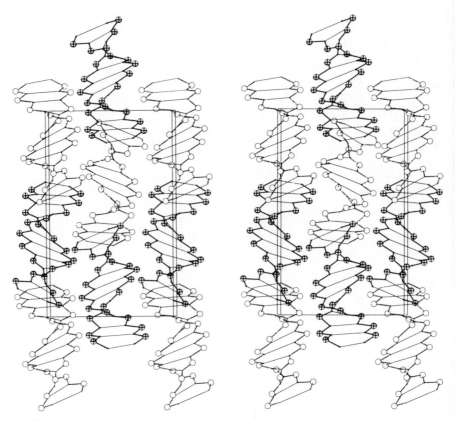

(a)

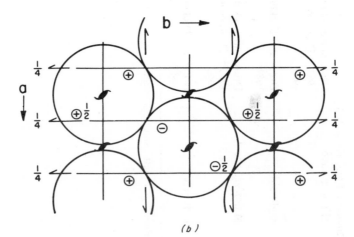

(b)

molecules in the manner shown in Figures 4 and 5. Guanine base edges in the minor grooves are paired between molecules, to form symmetrical N2 . . . N3 and N3 . . . N2 hydrogen bonds. The final base pair of one helix nests in van der Waals contact with one sugar-phosphate wall of the minor groove of its neighbor, and the final base pair of that molecule in turn is packed against the wall of the minor groove of the original double helix.* This efficient stacking undoubtedly contributes to the order and

---

* It is of interest that the molecular packing is identical in all three of the known A-DNA octamer structures—CCGG/CCGG, GGCCGGCC, and GGTATACC—and that it also involves nesting the outermost base pair of one double helix against one side of the minor groove of a neighbor (Shakked et al., 1981; Conner et al., 1982; Dickerson et al., 1982a; Wang et al., 1982a). This apparently is a very stable kind of stacking interaction, and it may mimic the nonintercalative binding of some antibiotics and antitumor drugs to the grooves of DNA.

Figure 4. (*Opposite*) Packing diagrams showing how individual dodecamer molecules are arranged within the unit cell. (*a*) "Side view" of the cell, with *b* horizontal and *c* vertical. Sugar-phosphate backbone chains are represented only by zigzag lines connecting phosphorus atoms (large open or crossed spheres) and sugar C1′ atoms, and base pairs are schematized by straight lines connecting C1′ atoms on opposite strands. Neighboring dodecamer molecules interlock along the *c* axis by meshing their minor grooves in the manner depicted in Figure 5, to form a continuous column. Successive molecules along one column, identified alternately by open and crossed phosphorus spheres, are related by a crystallographic twofold screw axis, involving rotation of 180° about the *c* axis and translation half a cell length along it. (*b*) "Top view" of the unit cell, showing the packing of columns of overlapping molecules. The *a* axis is vertical and the *b* axis is horizontal. Each column is represented by a circular cross section through a cylinder, and the two molecules related by a twofold screw axis along *c* are denoted symbolically by small circles. One small circle in each set has the indication "$\frac{1}{2}$" to show that its molecule has been shifted halfway up the *c* axis. Circles in one set of columns are labeled " + " and those in the other set are labeled " − " to indicate that the columns run in opposite directions along the *c* axis. Other twofold screw axes of symmetry are indicated by their standard international symbols. The entire 12-base-pair double helix is the crystallographically unique element of the structure. This same molecular packing scheme is employed in all five structures listed in Table 1, with minor alterations in cell dimensions, and with a shift in molecular position for MPD7 that accompanies the straightening of the helix. Figure 4*a* is of the Native structure; for an equivalent packing diagram of the MPD7 helix, see figure 11 of Dickerson et al. (1982a).

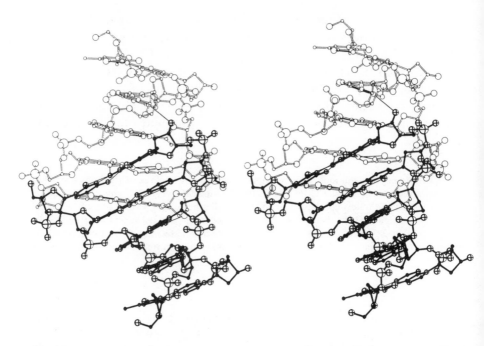

Figure 5.   Minor groove overlap of ends of Native dodecamer molecules along one vertical column of Figure 4*a*. The lower half of the open-bond molecule begins at the bottom with base pair G12·C13, and rises through T8·A15 before passing out of view at the top of the drawing. The upper half of the dark-bond molecule begins at its top with C1·G24, and continues downward through A5·T20 before disappearing off the bottom of the figure. Bases G12 on the light molecule and G2 on the dark are connected by pairs of N2 . . . N3 hydrogen bonds shown as thin lines, as are G14 on the light molecule and G24 on the dark. The 3'-OH of strand 1 of the light molecule is hydrogen bonded to the N2 amino group of G22 on the dark, and the 3'-OH of strand 2 of the dark molecule is similarly bonded to the ring N3 of G16 on the light (and not N2 of base G10 as strict twofold equivalence would suggest). For a view of the overlap 90° to the left, see figure 9 of Dickerson and Drew (1981a).

stability of the molecule, and it is tempting to suggest that the sequence CGCxxxxxGCG, where x represents any desired base, is the magic general formula for producing easily crystallizable and well-ordered B-DNA crystals. The overlapping of ends can have nothing to do with production of the bend in the helix axis, because essentially the same crystal packing is observed in all five structures, whether bent or straight.

3.   AVERAGE HELIX PARAMETERS AND FIBER COMPARISONS

A fiber diffraction pattern is inherently limited in the amount of information that it contains because the fiber itself has only limited order. The rotational orientation of each individual DNA filament relative to its neighbors is random, as is its position along the fiber. Hence the diffraction pattern itself is cylindrically randomized, and what would have been the sharp, discrete diffraction spots of an ordered structure are smeared out and overlapped into broader diffraction maxima. Accordingly the amount of information that can be reconstructed about the original scattering molecules is limited to an averaged structure. Overall helix parameters such as rotation and rise per base pair, diameter, and pitch angle can be deduced, and these averaged values can be observed to vary as the base composition and repeating sequence of a natural or synthetic DNA oligomer are altered. But individual variations in structure along the helix are invisible—effaced by the disorder inherent in the fiber. Structure analysis from fiber data must be an exercise in inspired model building. The ultimate goal is to demonstrate that the proposed model is compatible with the observed pattern and that it agrees with it better than does any other suggested model.

If the DNA can be induced to crystallize in an ordered and regular manner, then the quantity of information available rises by one or two orders of magnitude. Rotating a fiber about its long axis produces no new information; the diffraction pattern is unchanged because it already has been averaged by the cylindrical disorder in the fiber. But Figure 1 is only one slice through a three-dimensional collection of spots that make up the diffraction from a DNA crystal. Rotation of the crystal about the vertical axis produces a different zone, with different intensities and new information. The full diffraction pattern can contain thousands, or tens of thousands, of individual intensities, all of which must be matched correctly by a correct molecular structure model. Standard methods of single-crystal analysis can be used, such as have proven themselves over the years for proteins as well as smaller organic molecules. The limit of information available still depends on the degree of order among molecules within the crystal, and short-range molecular disorder is reflected in a fading of intensities at the outer edges of the pattern and a decrease in the nominal resolution of the data set. The scattering from crystals of

CGCGAATTCGCG in all its variants shows a marked decline in intensity beyond a resolution of 3.2 Å, visible in Figure 1, and the limit beyond which little or no useful information can be obtained lies around 2.3–1.9 Å. This would be insufficient to solve the structure of DNA if we had no idea of the nature of its component parts: deoxyribose, phosphates, and four well-characterized organic bases. But just as with proteins, this extra known information about the structure is used in building the original model from an electron density map and is incorporated into the refinement process in the form of restraints that keep bond lengths and angles within reasonable limits and maintain the flatness of aromatic rings of the bases. The result is a true molecular picture of the structure, in which local variations from the average helical structure are clearly visible. It is these local variations, their possible origin in base sequence, and their potential use in sequence readout that make single-crystal analyses of oligonucleotides particularly challenging. Table 1 lists the overall average helix parameters for the five refined variants of CGCGAATTCGCG as derived by a HELIX program written by John Rosenberg that searches for the best overall helix axis. The four bent helices show less than the 10 base pairs per turn expected from fiber studies, but straightening up the helix axis leads to a more canonical value of 9.7 base pairs per turn. The mean rise per base pair of 3.34 Å again is as expected, as is the packing diameter, defined as half the diagonal of the *ab* cell face in Figure 4*b*. But these mean values are not nearly so interesting as is the variation from the mean, discussed in Sections 5 and 6 below.

4. DNA HYDRATION AND HELIX STABILITY

Of the two types of DNA double helix known from fiber studies, A and B, the latter is stable under high humidity conditions, around 95% relative humidity (RH). In most cases, drying the fiber or oriented film to around 75% RH leads to a transition to the A form, but in other circumstances the B form stubbornly remains until excessive drying causes a breakup of helicity and order and a consequent destruction of the fiber pattern. Why is the B helix the high-humidity form, and why can some sequences be converted to the A form by drying while others cannot? Single-crystal X-ray analyses of the B-DNA dodecamer offer answers to these questions.

The hydration behavior of DNA in solution has been studied for more than 20 years. This work has been reviewed by Kopka et al. (1983) and is therefore only summarized here. Falk et al. (1962, 1963a, b) studied the hydration of DNA in solution using infrared spectroscopy and gravimetric methods, and suggested the following order of decreasing affinity of water for the double helix:

1. Free phosphate oxygen atoms.
2. Esterified O3' and O5' phosphate oxygens.
3. Sugar O4' ring atoms.
4. Base edge N: and C=O groups.
5. Base edge $-NH_2$ groups.

Wolf and Hanlon (1975) surveyed the preexisting literature and proposed that hydration be considered in terms of (1) a primary water layer with molecules immediately adjacent to the DNA and with infrared absorption suggesting an environment different from that of bulk water, and (2) a looser secondary shell with the characteristics of liquid water. They proposed 18–23 water molecules per nucleotide in the primary shell, of which 10–12 are bound directly to the DNA and a further 8–9 are less tightly bound.

The actual distribution of solvent peaks observed around the MPD7 molecule is depicted in Figure 6. The broad major groove is coated by a monolayer of water molecules that interact with the exposed C=O, N:, and $-NH_2$ of base edges, and with a less regular array of solvent molecules in a second hydration shell. The phosphate backbone also is extensively hydrated. The deep and narrow minor groove contains a particularly well-ordered chain of solvent peaks that has been called the spine of hydration, and which is believed to make a major contribution to the stability of the B form of DNA.

In order to assess the significance of the results, it is important to consider the way in which solvent peaks are found during refinement. After restrained least-squares refinement of the DNA itself, a search then can be made for those solvent molecules that are sufficiently ordered from one molecule to the next in the crystal to produce discrete peaks in the electron density map. "Shotgunning"—ascribing a water molecule to every visible feature in the solvent region of the map—will produce mis-

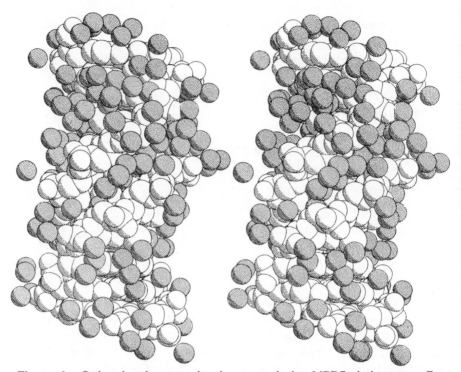

Figure 6. Ordered solvent molecules around the MPD7 dodecamer. Grey spheres are solvent peaks, and the DNA molecule is white. The view, into the minor groove, is identical to that of Figure 3a. Note the filling of the minor groove, the monolayer of water lining the major groove, and the frequent association of solvent peaks with phosphate oxygens. Most solvent peaks are assumed by their distance from DNA atoms to be water molecules, but a few peaks at distances of only 2 Å are probably Mg^{2+} ions. An example of this is the peak to the right of phosphate 22, near the upper right of the drawing.

leading results, but a cautious extension of the solvent shells during successive cycles of refinement can yield meaningful information about hydration. At any given stage of refinement of each of the dodecamer structures, a potential solvent peak was added to the analysis only if it appeared both in the current electron density map and the difference map and if it was within reasonable interaction distance of a preexisting DNA or solvent feature. This procedure led to a gradual buildup of the hydration structure from the surface of the DNA molecule outward, as is depicted in figures 2*a–d* of Drew and Dickerson (1981b). The process was halted when the remaining unexplained peaks were considered unacceptably diffuse.

The crystallographic *B* value or thermal factor is an exponential coefficient that measures the degree of smeared-outness of the peak image in the electron density map. Because this map is an average over all molecules in the crystal, if an atom does not occupy precisely the same position in every molecule, its collective image will be blurred and it will have a high *B* value. This is termed static disorder. Conversely, if all the atoms are vibrating about a mean position because of thermal energy, precisely the same blurred image will result, and the *B* value again will be high. At a given temperature there is no way of discriminating between thermal vibration and static disorder as the origin of a high *B* value for an atom, but these two components can be unscrambled if the temperature is lowered so as to damp down thermal motion.

In the CGCGAATTCGCG structure at room temperature, refined *B* values for atoms in base pairs are lower on the average than those for atoms in sugar rings, and these in turn are lower than those for the P or O of phosphates, as listed in Table 3. This suggests that the entire double helix is undergoing rigid-body libration about its axis, with the base pairs at the center of the molecule moving less than the phosphates on the periphery. Solvent atoms tend to have *B* values slightly larger than those of the atoms to which they are attached, reinforcing the impression of molecular libration. The search for meaningful solvent peaks was discontinued when the *B* values of the remaining peaks were one-and-a-half times the phosphate group average. Previously assumed solvent peaks whose *B* values rose beyond this level upon subsequent cycles of refinement also were deleted.

Table 3.   Mean Refined *B* Values in Dodecamer Structures

Refinement	Resolution (Å)	Mean B Value ($Å^2$)			Number of Solvent Peaks	Maximum Solvent B Value
		Bases	Sugars	Phosphates		
Native	1.9	28	42	51	80	74
MPD7	2.3	12	24	35	114	47
16K	2.7	-3^a	11	17	83	41

[a] The negative *B* for bases at 16K is zero within limits of accuracy of *B*'s.

Figure 6 shows the observed hydration in MPD7, which is the 9-Br derivative in 60% MPD at 7°C. Hydration in the Native dodecamer at 20°C was similar, with one puzzling exception: Almost no solvent peaks were found around the phosphate backbone, the very region that Falk and co-workers had predicted to be the most strongly hydrated. Drew and Dickerson (1981b) proposed that this signified, not the absence of solvent, but merely the absence of *ordered* solvent. With a mean *B* value of 51 for the phosphate backbone, the associated solvent molecules might be so disordered from one molecule to the next that their images would fail to register above the general background noise level of the electron density map. This was tested by cooling a dodecamer crystal slowly over a period of two days in a liquid helium diffractometer apparatus developed by Sten Samson, from an initial 23°C (296° K) down to the limit of the device, 16° K. As shown in Figure 7, the crystal cell volume shrank in a linear fashion until 200° K was reached, below which the crystal was es-

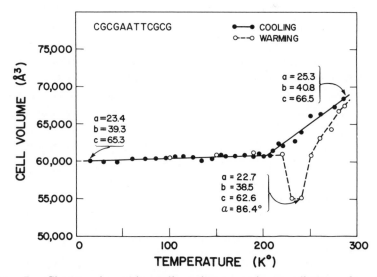

Figure 7. Change in unit cell volume and coordinates for Native CGCGAATTCGCG crystals during cooling from room temperature to 16° K (black dots) and rewarming to room temperature (open circles). The crystals have essentially solidified after cooling to 200° K; no further significant cell volume change occurs down to 16° K. Upon rewarming past 200° K the crystal changes briefly from orthorhombic $P2_12_12_1$ to monoclinic $P2_1$ with beta angle 86.4°, but reverts to the orthorhombic form at room temperature. From Drew et al. (1982).

Table 4. Solvent Temperature Factors in the MPD7 Structure and Local Solvent Environments[a]

Peak No.	B Value	Interactions	Peak No.	B Value	Interactions
35	12.4	mS	129	33.1	M
29	13.2	spermine	102	33.2	M2/M2
37	13.7	mS	99	33.3	P
63	14.4	spermine/(P)	103	33.4	(M2)/(M2)
42	15.0	mS	66	33.5	mS
30	16.1	spermine/P	92	33.9	P,3',m2
48	18.7	M	53	34.2	P/OH
26	19.1	M	56	34.2	M
25	22.2	P,3'/P/M2	67	34.6	mS
28	22.2	spermine/P	61	34.7	(3'),m2
47	22.6	P,(5'),M2	74	34.8	(P),M2
65	23.8	mS	91	35.1	(P),M2
31	24.2	mS/OH	76	35.2	m2
39	24.8	P,(5'),M2	98	35.2	M2
40	25.0	mS	125	35.3	(P)
32	25.6	m2	69	35.6	P, M2
70	25.7	(P),(M2)	77	35.6	P
41	26.0	(P),m2	95	35.6	P
38	26.6	P,(5'),M2	49	35.7	(M2)
80	26.6	M2	118	35.7	mS
36	28.0	M2	44	35.8	P,5'
43	28.1	M	100	35.9	M
107	29.2	P,5'/P,(3')/(M)	51	36.8	(P)/3',(O1')
62	29.4	spermine/P,(5')	119	36.8	(P)/P,(3'),m2
86	29.7	M2	114	37.2	spermine
105	30.6	m2,(3')	82	37.5	P
110	30.8	P,5', M2/(OH)	83	37.7	M2/M2
75	30.9	M2	54	37.9	M2/(OH)
60	31.3	M2	64	38.1	P,(3'),m2
90	31.3	m2	72	38.1	(3'),m2
104	31.9	M	52	38.4	P
88	32.1	M	68	38.5	M
59	32.4	M2	120	38.5	P,3',m2
45	32.5	mS	85	38.6	P, m2/OH
94	32.7	M	87	38.6	M2/(m2)
46	32.8	(mS)	71	38.8	(P)
93	32.8	M	79	38.8	P,M2/M2
108	33.0	P/P,m2	109	38.8	(P),(5'),M2/OH
34	33.1	M	73	39.0	P,(5'),(M)

(*Continued*)

61

Table 4.   (*Continued*)

Peak No.	*B* Value	Interactions	Peak No.	*B* Value	Interactions
89	39.1	M2	96	42.6	P,(5')/M2
116	39.1	(P),(3'),m2/(M2)	55	42.8	M2
124	39.1	(M)	81	42.8	(P),M2
97	39.4	spermine/(P)	78	42.9	P,(m2)/(m2)
115	39.5	P,3'/(OH)	101	42.9	P/(M2)
131	39.8	M	135	42.9	M2
126	39.9	(P)/M2/M2	130	43.1	M2
117	40.4	(P),m2	57	43.4	M
106	40.8	(P),(5')	121	43.4	M2
33	40.9	(P)/M2	58	43.9	(P),(3')/P
127	41.0	M2	111	44.3	(3')/P
27	41.1	(P)/M2	138	44.4	(P)/(M2)
113	41.1	mS/OH	139	45.2	3',(5'),mS
128	41.1	(P)/(P)	137	45.7	(M)
84	41.7	(P)/(P)	50	46.6	M2
122	42.2	M2	133	46.8	P,M2
132	42.3	(3')/M	112	46.9	spermine
136	42.4	(M2)	123	47.0	(M)

[a] Solvent molecules are listed in order of increasing *B* value. Environments (interactions) are abbreviated: P, phosphate; 3', O3' backbone atom; 5', O5' atom; M, first hydration shell of major groove; M2, upper hydration layers of major groove; mS, minor groove spine of hydration; m2, upper hydration layers of minor groove; OH, chain-terminal —OH group. Symbols in parentheses indicate distances between 3.5 and 4.1 Å; all others are 3.5 Å or less. Slashes separate distances from solvent peak to different dodecamer molecules.

sentially solid and additional cooling had little effect. A three-dimensional data set was collected at 16 K, but peak broadening limited it to a resolution of 2.7 Å, compared with 1.9 Å at room temperature. The structure was refined independently, using the Native coordinates as a starting point, and solvent peaks were added without reference to those of the Native structure (Drew et al., 1982). Refined mean *B* values for bases, sugars, and phosphates are all much lower than for the room temperature structure (Table 3), but the same pattern prevails of increasing *B* from bases to sugars to phosphates. This suggests either that very low frequency thermal modes of libration still are active at 16 K or that as the molecules come to rest during cooling, they do so in a random manner in many local minima that preserve on average some features of the pattern of vibration seen at higher temperatures. As expected, the hydration of the phosphate backbone now is clear (see figure 2*b* of Kopka et al., 1983), and resembles that for MPD7 in Figure 6.

Mean *B* values for bases, sugars, and phosphates in MPD7 are intermediate between those of the Native and 16 K structures but preserve the same trend. Individual *B* values for all solvent peaks in the MPD7 structure are listed in Table 4, along with the interactions that each peak makes with the DNA. With the list arranged in order of increasing *B*, a scan of the "interactions" column provides a quick impression of relative strengths of interaction. The most well-ordered solvent peaks, with low *B* values, are found along the minor groove spine of hydration (mS), in the image of the spermine molecule bridging the major groove, and in primary association with the phosphate backbone (P) and the major groove (M). At the end of the table, those least well-defined solvent molecules with high *B* values tend to be found in the second hydration layers of major and minor grooves (M2, m2). Symbols enclosed in parentheses, indicating long interaction distances between 3.5 Å and 4.1 Å, are also more prevalent at the high-B end of the table. All of these trends are particularly striking if the individual *B* values from Table 4 are added to the hydration diagrams of the phosphate backbone and major and minor grooves published by Kopka et al. (1983), from which the minor groove diagram is reproduced in Figure 8. A normal hydrogen bond length is around 2.8 Å, and allowing a variation of two standard deviations in the MPD7 analysis would permit distances up to 3.5 Å to be considered as possible hydrogen bonds. Distances beyond this out to 4.1 Å must be regarded as looser associations rather than true hydrogen bonding. The more distant and diffuse peaks with their high *B* values indicate a tendency for a solvent molecule to occupy a given region of space, but not at every DNA molecule in the crystal and, when present, not at precisely the same position.

Both the major groove and the phosphate backbone are extensively hydrated, but essentially in a monolayer, without ordered second shell hydration. A partial-occupancy spermine image spans the major groove from phosphate 2 to 22 (Drew and Dickerson, 1981b). Strings of reasonably ordered solvent molecules extend above the major groove, anchored to the phosphate backbone at either end, but with few connections to the monolayer at the bottom of the groove (see figure 6 of Kopka et al., 1983). These strings of molecules across the groove actually were anticipated from Monte Carlo simulations by Clementi and Corongiu (1981).

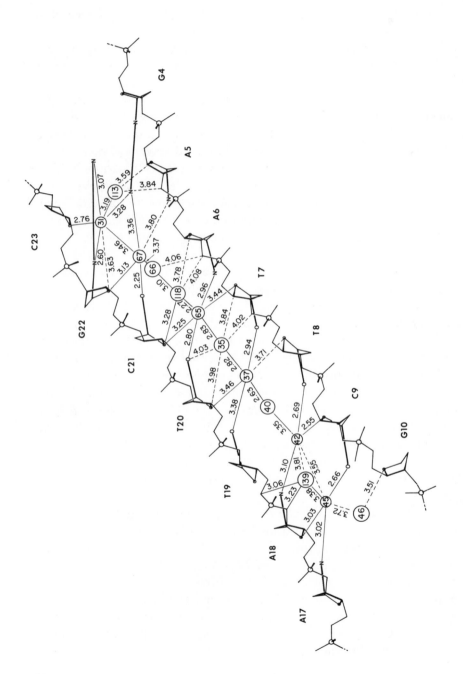

The minor groove exhibits a higher degree of solvent order. A zigzag chain of water molecules winds its way down the bottom of the groove in the AATT center of the dodecamer, as shown in stereo in Figure 9 and in unrolled cylindrical projection in Figure 8. Each adenine has an N3 atom as a potential hydrogen bond acceptor in the minor groove, and each thymine has an O2 (Figure 10). A water molecule forms a bridge between those N3 and O2 on opposite backbone strands at adjacent base pairs that are brought closer together by the rotation of the helix (Figure 9). These bridge waters in turn are connected by a second set to form a continuous zigzag chain that has been dubbed the "dragon's spine" or simply the spine of hydration. This spine begins to pull away from the bottom of the minor groove in the CGCG ends of the molecule, even before the groove is blocked by interdigitation from neighboring molecules.

The most likely cause of disruption of the minor groove spine of hydration in C·G regions of the molecule is the presence of the N2 amino group on guanines (Figure 10). This group is doubly intrusive: it physically gets in the way of a potentially bridging water molecule, and it provides a hydrogen bond donor where only acceptors formerly were found. The weakening of the spine as it leaves the AATT center at lower left and upper right of Figure 8 is borne out by the individual solvent B values. Peaks 35 and 37 in the center have low B's of 12 and 14, whereas the B's of peaks 139, 45, and 46 at the bottom are 45, 33, and 33, and that of peak 113 at the upper end is 41. In the center of the AATT region the water molecules of the spine are slotted into the bottom of the groove, enfolded on both sides by the sugar O4' atoms that line the walls, whereas at the two ends the spine pulls up and out of the groove altogether.

Drew and Dickerson (1981b) propose that this minor groove spine of hydration is the principal factor in stabilizing the B helix at high hydration and that disruption of the spine is a necessary first step in bringing about

---

Figure 8. (*Opposite*) Unrolled cylindrical projection of the central region of the minor groove, showing the spine of hydration as numbered spheres. All atoms in the sugar-phosphate backbone are depicted accurately in projection, but bases are schematized only by thick lines connecting their N3, N2, and O2 atoms to the sugar C1'. Phosphorus atoms are depicted by small circles along the backbone chain, sugar ring O4' atoms (sometimes designated as O1') are black dots, and other main-chain atoms are represented only by junctions of stick bonds. Solvent interactions of 3.5 Å or less are indicated by thin solid lines, and those between 3.5 Å and 4.1 Å are dashed.

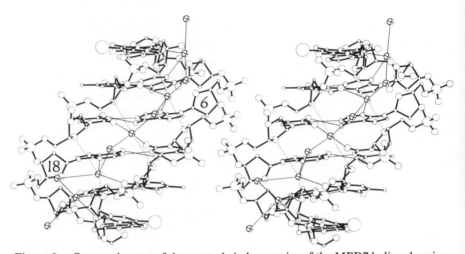

Figure 9.   Stereo closeup of the central six base pairs of the MPD7 helix, showing the spine of hydration (crossed spheres) down the bottom of the minor groove. The sugar rings of adenines 6 and 18 are labeled for reference. Water molecules at the very bottom of the groove are hydrogen bonded to the adenine N2 and thymine O2 atoms that are brought into closer proximity by helix rotation, on opposite strands of the helix at adjacent base pairs. These inner water molecules in turn are connected by an upper layer that gives them a local tetrahedral environment and forms a continuous zigzag spine of hydration down the groove. An identical spine structure has been found independently in each of the five dodecamer refinements listed in Table 1. For a corresponding view of the spine in the Native helix, see figure 5 of Drew and Dickerson (1981a).

the B-to-A helix transition. Supporting evidence for this proposal is provided by Leslie et al. (1980), who synthesized long polynucleotides of various repeating sequences, observed their B-type fiber diffraction patterns under high humidity conditions, and attempted to obtain the A pattern by drying. They found that, with one exception, every sequence that contained guanines with their N2 amine groups in the minor groove *could* be converted from B to A by drying (Table 5), and with one questionable and unrepeatable exception, every sequence lacking N2 amine groups in the minor groove could *not* be converted to the A form. The B pattern remained until excessive drying destroyed the order within the fiber and hence all pattern. Inosine, which simply is guanine without the N2 amine, behaved in its I·C base pairs like A·T and not like G·C.

The integrity of the minor groove spine of hydration is a cooperative affair. Each solvent link in the chain is stabilized by those to either side of it, and a breach in the spine at any point would weaken the bonds on both sides. Consequently one would predict not only that a high G·C content would destabilize the B helix relative to A but also that scattering G·C base pairs evenly along the chain would have a more disruptive effect than a clustering into G·C and A·T base pair regions, because in the latter regions the stabilizing spine could remain intact. One also would predict that in synthetic polymers with 2-aminoadenine or 2-aA, the 2-aA·T base pairs would behave like G·C in their disruption of the spine, and not like A·T.

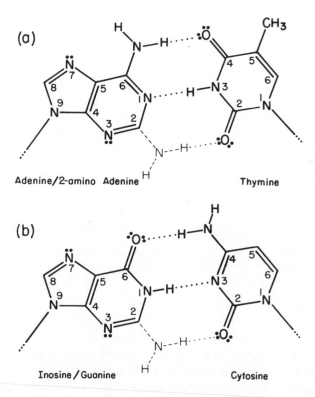

Figure 10. Base pairs and the presence or absence of N2 amine groups in the minor groove. (*a*) Adenine lacks the N2 amine, whereas 2-aminoadenine possesses it. (*b*) Inosine lacks the N2 amine, but guanine has it. Both A·T and I·C base pairs should favor the B conformation because they permit formation of an uninterrupted spine of hydration down the minor groove, whereas both 2-aA·T and G·C base pairs should interrupt the spine and destabilize the B structure.

Table 5.   Sequence Dependence of DNA Conformational Behavior[a]

DNA Polymer	Purine N2 Amino Group Present?	Observed Helix Conformations
(A)(T)	no	B
(I)(C)	no	B
(A-I)(C-T)	no	B
(A-T)(A-T)	no	B, (A)
(I-C)(I-C)	no	B
(I-T)(A-C)	no	B
(A-A-T)(A-T-T)	no	B
(A-I-T)(A-C-T)	no	B
(A-I-C)(I-C-T)	no	B
(I-I-T)(A-C-C)	no	B
(G)(C)	yes	B, A
(A-G)(C-T)	yes	B
(G-C)(G-C)	yes	B, A, Z
(G-T)(A-C)	yes	B, A, Z
(A-A-C)(G-T-T)	yes	B A
(A-G-T)(A-C-T)	yes	B A
(A-G-C)(G-C-T)	yes	B A
(G-A-T)(A-T-C)	yes	B A
(G-G-T)(A-C-C)	yes	B A

[a] Adapted from Leslie et al. (1980). The repeating unit in each polymer strand is given in conventional 5′ → 3′ order within parentheses. "B" includes all members of the B family of helices: B, B′, C, C′, C″, D, and E. "Z" includes the S helix. (A) indicates that the A form of poly(dA-dT) could not be obtained reproducibly; the A conformation of this sequence is stable only over a narrow range of relative humidity.

Kopka et al. (1983) have attempted to quantify the hydration of the MPD7 structure by making a census of potential hydration sites of various kinds and by determining how many sites were occupied by at least one solvent molecule at a distance of 3.5 Å or less, how many others had a closest approach of solvent between 3.5 and 4.1 Å, and how many potential sites had no solvent peak nearer than 4.1 Å (Table 6). Their order of decreasing relative strength of hydration is as follows:

1.   Minor groove spine (100% occupancy within 3.5 Å in A·T region).
2.   Free phosphate oxygen atoms (66% occupancy).

3. Sugar O4' atoms (50% occupancy, but of uncertain significance because this may be a secondary consequence of the presence of the spine of hydration).

4. Major groove base edge N and O atoms (42% occupancy).

5. Esterified O3' and O5' phosphate oxygens (23% occupancy).

This differs from the predictions of Falk et al. only in two respects: the dominance of the minor groove spine, which represents a level of structural detail that solution studies could not have revealed, and the relative lack of hydration at main chain oxygen ester sites. Examination of the dodecamer structure reveals no steric blocking of these sites, and their low level of hydration must be ascribed to an intrinsic unreactivity.

Crystals of CGCGAATTCGCG are grown under relatively low salt conditions: 4.8 mM magnesium acetate, 0.224 mM in spermine. Skuratovskii and colleagues (Skuratovskii et al., 1979; Bartenev et al., 1983) have examined fibers of B-form DNA under high salt conditions and find that Cs$^+$ ions now occupy the same positions down the minor groove that are occupied by first-layer water molecules of the spine, bridging purine N3 and pyrimidine O2 atoms (e.g., positions 45, 42, 37, 65, and 67 in Fig. 8). They calculate from optimum coordination distances that similar minor groove binding could be expected with all the alkali metal and alkaline earth cations except the smallest Li$^+$ and Mg^{2+}, and that this is why the B to A helix conversion takes place only under low salt conditions. The B helix is stabilized by a minor groove spine, whether it be made up of water molecules or cations; and dehydration will be ineffectual in bringing

Table 6. Percentage Occupancy of Hydration Sites in MPD7

DNA feature	Minimum Hydration Distance[a]				Percentage	
	<3.5 Å	3.5–4.1 Å	>4.1 Å	Total	<3.5 Å	<4.1 Å
Minor groove spine	10	0	0	10	100	—
Free phosphate O	29	10	5	44	66	89
Sugar O4'	7	6	1	14	50	93
Major groove bases	15	11	10	36	42	72
Phosphate O3'	5	7	10	22	23	55
Phosphate O5'	4	4	13	22	23	41

[a] Distances <3.5 Å represent good H bonds, distances of 3.5–4.1 Å represent more tenuous associations, and distances of >4.1 Å cannot properly be said to be hydration at all.

about conversion to the A form if cations are present to take the place of the water molecules.

5.   LOCAL VARIATION IN HELIX STRUCTURE

The number of helix parameters that can be characterized by fiber diffraction studies is inherently limited: mean helical rotation and rise per base pair, backbone pitch and radius, average backbone torsion angles and sugar conformation, and mean propeller twist angle. Single crystal studies, in contrast, can produce individual values at each step of the helix rather than mean values; they can also provide information about other structural aspects, such as roll and tilt of base planes and lateral displacement of base pairs out of an ideal helix stack. Even more significantly, explanations for these local variations in helix structure can be sought in base sequence, local crystal environment, or the presence of chemical substituents on base rings and binding by antibiotics and drugs.

Each of the five B-DNA dodecamer structures listed in Table 1 has been analyzed independently with a series of computer programs that first establishes a best overall helix axis and then examines the geometry of the molecule relative to that axis. The results are listed in the photoreproduced tables in the appendix to this volume. For comparison, results are also given for A and Z helices.

The initial choice of a best overall helix axis is not a trivial matter, particularly if the helix itself has a detectable curvature. The optimum procedure in such a case would be to find the best smoothly curved approximation to the local helix axis at each base pair. But with the dodecamer the difference between this procedure and that of simply fitting a best overall straight axis is small, in terms of the effects on local helix parameters, and the more complex procedure has not been followed. In the HELIX program written by John Rosenberg, the user identifies at least three marker atoms in one base pair, and the equivalent three atoms in a different helix repeat unit up the axis, usually but not necessarily the following base pair. The program then generates the three helix repeat vectors, brings them to a common origin, uses the ends of the vectors to define a plane, and chooses the helix axis to be the line through the origin of vectors, perpendicular to the vector plane. The rise per helix step is the distance from the origin to the vector plane, and the rotation can be found from a projection of the repeat vectors onto the plane. As many vectors as desired can be used, and the vector plane then is found by making the best least-squares fit of a plane through the ends of all the helix defining vectors. Local variation of exactly the type that is of interest here can make the overall helix axis sensitive to the particular choice of defining atoms. The best choices appear to be vectors from one C1′ atom to the equivalent atom in the following base pair, and similar vectors from the N1 or N9 atom at the other end of the glycosyl bond to its equivalent one step up the helix.

A second, entirely independent definition of the helix axis is provided by the instinctive feeling that it should pass through the center of the helix at all points, or that backbone atoms such as phosphorus should as nearly as possible be at a constant distance from the helix axis. This criterion cannot be applied to short oligomers such as dimers or tetramers, but it is a powerful one in a helix as long as the dodecamer. Uncritical use of the HELIX program with defining vectors from C1′ to C1′ and N1/9 to N1/9 at every step of the helix led to phosphorus radial coordinates that

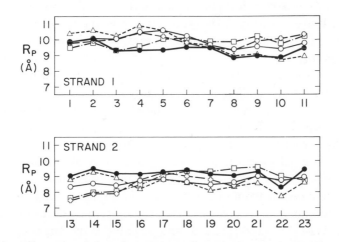

Figure 11. Distances of phosphorus atoms from the best overall helix axis in the five dodecamer structures. Phosphorus atoms here are identified by step number, 1–11 and 13–23. Circles and thin solid lines, Native helix; crosses and dashed lines, 16K; triangles and short dashed lines, cisplatin; squares and dot-dash lines, MPD20; closed circles and heavy solid lines, MPD7. Various choices of helix defining vectors were used with the HELIX program to make these curves as flat as possible. From Appendix Tables A1–A5.

drifted systematically from one end of the helix to the other, indicating that the helix axis was actually skewed relative to the helix considered as a cylinder. (Throughout this chapter, a step will be defined as the interval between one base pair and the next, and never used to identify an individual base pair. Helix step n is the interval between base pair n and base pair $n + 1$.) By trial and error, it was found that the best-centered overall helix axis was produced by the HELIX program in the MPD20 and MPD7 structures if steps 3, 4, 8, and 9 were omitted (as will be seen below, these are indeed atypical steps compared with the rest of the helix), and in the Native, 16K, and cisplatin structures if only the outermost steps, 1, 2, 10, and 11, were included for maximum leverage on the helix axis. With these optimal helix axes, the radial coordinates of the phosphorus atoms are as shown in Figure 11. Mean values for helical rotation and rise per base pair from the HELIX program are given in Table 1.

5.1.   Base Plane Orientation

One of the primary distinctions between A-DNA and B-DNA is the orientation of the base pairs; perpendicular to the helix axis in B, but tilted by 13°–19° in A. Figure 12 shows a convenient set of axes for describing base plane orientation. Helical *twist* is rotation of the base pair about the helix axis, *tilt* is rotation about the pseudo twofold axis that relates glycosyl C1′–N bonds at the two ends of the base pair, and *roll* is rotation about the third perpendicular axis, roughly the line from the C8 of purine to the C6 of pyrimidine. For single base pairs or individual bases, roll angle Φ_R is positive for counterclockwise rotation when viewed down the C8/C6 line from the strand 1 side of the base pair, and tilt angle Φ_T is positive for clockwise rotation when viewed into the minor groove. Roll and tilt angles θ_R and θ_T describe the *change* in orientation from one base pair to the next, or from one base plane to the next along an individual strand. Roll angle θ_R is positive if the angle between bases or base pairs opens toward the minor groove, and negative if the angle opens to the major groove; tilt angle θ_T is positive if the angle opens toward strand 1 of the double helix, and negative if toward strand 2. Appendix Tables A1–A11 list all four roll and tilt angles first for individual strands and then for the best least-squares plane through both bases of a pair, for the five dodecamer B helices, four A helices, and two Z helices.

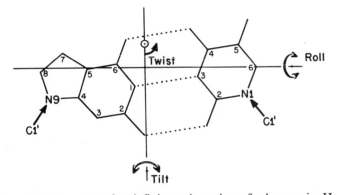

Figure 12. Reference axes for defining orientation of a base pair. Helical *twist* is a rotation about the helix axis, *tilt* is a rotation about the pseudo twofold axis in the plane of the base pair that relates the C1'-N glycosyl bond at one end of the pair to that at the other, and *roll* is a rotation about the third mutually perpendicular axis, which is closely approximated by the line from C8 of the purine (G or A) to C6 of a pyrimidine (C or T). Roll and tilt angles Φ_R and Φ_T define the orientation of a single base or base pair relative to the helix axis, and depend upon the choice of axis. In contrast, roll and tilt angles θ_R and θ_T define the *change* in orientation from one base or base pair to next along the helix, and are independent of the helix axis. Analytical definitions are to be found in Fratini et al. (1982).

Figure 13 shows the C6–C8 inclination angle at each base pair for seven helices. This is the angle that the C6–C8 line in one base pair makes with a plane normal to the helix axis, and is almost identical to $-\Phi_T$, as Appendix Tables A1–A11 indicate. The difference between them is mainly a matter of the choice of defining atoms: C6 and C8 for the inclination angle, and all atoms in the base pair for Φ_T. The negative inclination angles at the first four base pairs of all the B helices except MPD7 arise from the bend in helix axis visible in Figure 2. Most of the bending seems to occur in the top third of the molecule. Base pairs C1·G24 and C11·G14 are almost exactly one turn of helix apart, and their C6–C8 lines lie in the plane of bending. Hence it is convenient to measure the total bend angle over one turn of helix as the difference in C6–C8 inclination angles of these two base pairs. These are the numbers that are tabulated as "axial bend" in Table 1. In the four bent B helices the inclination angle begins negative, rises to a small positive value at the middle of the helix, and

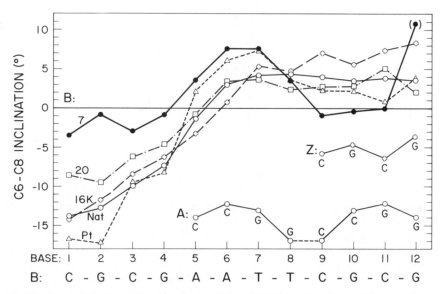

Figure 13. Variation along the helix of the C6–C8 inclination of a base pair, or the angle that the C6–C8 line makes with a plane normal to the helix axis. Symbol and line coding of the five dodecamer structures is as in Figure 11. As can be seen from Appendix Tables A5–A9, the C6–C8 inclination is nearly the negative of the base pair tilt angle, $-\Phi_T$. The negative inclination angles at the first four or five base pairs reflect the bend in helix axis that is present in all but the MPD7 structure. The large positive value for MPD7 at base pair 12 (in parentheses) is atypical and arises from an intermolecular hydrogen bond in the crystal involving base C13. The turned base plane of this base can be seen at the bottom of Figure 2c. The equivalent hydrogen bond is not present in the four bent helices. For comparison with the B helices, base pair inclination angles also are plotted for the A-helical CCGG and Z-helical CGCG.

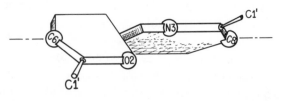

Figure 14. Propeller twist is a contrarotation of bases of a pair about their long C6–C8 axis. A positive angle is produced when the nearer base is rotated clockwise when viewed down the long axis, as here.

remains constant thereafter. The unbent MPD7 helix shows relatively little base inclination at its two CGCG ends but a quite definite positive inclination in the AATT center.

In crystals of A-helical CCGG (Conner, 1982; Conner et al., 1982, 1984), two double helical tetramer molecules are stacked together so as to continue the helix in a functional octamer, CCGG/CCGG, and the central G/C helix step without connecting phosphates is scarcely distinguish-

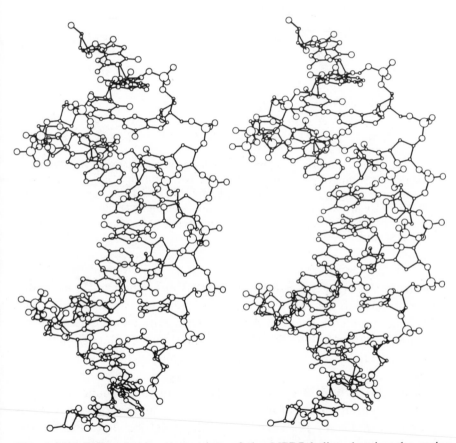

Figure 15. Inclined-axis stereo view of the MPD7 helix, showing the major groove in profile at the left. The bases of each strand of the double helix stack atop one another in an efficient manner, almost as if the other strand were not present. As a consequence, each individual base pair has a positive propeller twist. Base pair C1·G24 is at the top; G12·C13 is at the bottom.

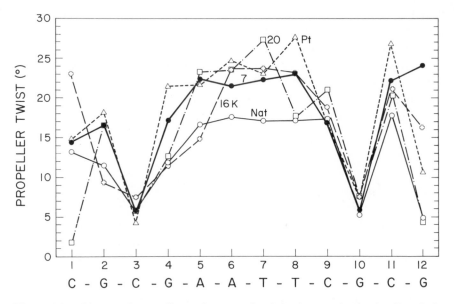

Figure 16. Observed propeller twist at each of the base pairs in the five dodecamer helices. Symbol and line coding in the five structures is as in Figure 11. Note the large propeller twist in the AATT center of each helix, the smaller mean propeller twist in the flanking CGCG regions, and the especially low propeller twist at the symmetrically arranged third and tenth base pairs. From Appendix Tables A5–A9.

able from the other six covalently linked steps. The C6–C8 inclination angles relative to the best overall octamer helix axis are shown in Figure 13 as a contrast with B-DNA. All of the base pairs are inclined uniformly between $-12°$ and $-17°$. Z-helical CGCG (Drew et al., 1978 and 1980; Drew and Dickerson, 1981a) also shows uniform inclination angles of around $-5°$.

Base pairs in both B-DNA and A-DNA show a pronounced propeller twist, or contrarotation, of individual bases about the long C6–C8 axis, as defined in Figure 14. Such a propeller twist stabilizes the helix by improving base stacking along each individual strand. The oblique view of the MPD7 double helix in Figure 15 makes it apparent that the bases of one strand stack atop one another and maximize their overlap, almost as though the other strand were not present. The rotation that improves this overlap leads to positive propeller twist. As we shall see in Section

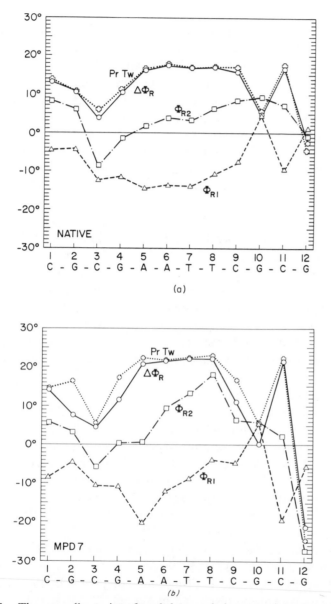

Figure 17. The propeller twist of each base pair is approximately equal to the difference in roll angles of its individual bases about the C6–C8 axis, or: Pr Tw $\simeq \Delta\Phi_R = (\Phi_R)_{\text{strand 2}} - (\Phi_R)_{\text{strand 1}}$. (*a*) Native dodecamer. (*b*) MPD7.

6, the presence of propeller twist is an important factor in translating base sequence into local variations of helix structure of a type that could potentially be read by repressors or restriction enzymes.

Observed propeller twist angles in the five B helices are tabulated in Appendix Tables A5–A9 and displayed in Figure 16. The same general pattern is seen in all cases: large and uniform propeller twist in the AATT center, from 17° to 23°, and a pronounced flattening of propeller twist down to 5°–8° at the third base pair in from each end of the helix. The propeller twist is very nearly equal to the difference in individual base plane roll angles on the two strands of the helix, or Pr Tw $\simeq \Delta\Phi_R = (\Phi_R)_{\text{strand 2}} - (\Phi_R)_{\text{strand 1}}$. Assuming this approximation to be valid, propeller twist has been factored into its two roll components for the Native and MPD7 helices in Figure 17. From these it can be seen that the flattening of propeller twist at the third and tenth base pairs is a result of a special orientation of the tenth base along each strand: G10 in strand 1 and G22 in strand 2. These guanines are turned back out of optimum stacking alignment with the cytosines above and below them. This probably arises because of steric hindrance with the two guanines paired with those cytosines on the other strand (see Figure 32, below); the effect will become clearer when we examine sequence/structure relationships in Section 6.

The behavior of roll angles from one base pair to the next in the five B-DNA helices is displayed in Figure 18. The most striking feature in the unbent MPD7 helix is a regular alternation of positive and negative roll angles; successive base pairs are tipped in opposite directions, making the angle between them open first toward the major groove and then toward the minor. The origin of this effect in terms of base sequence is discussed in Section 6. Superimposed on this is the effect of bending. The order of increasing overall bend as listed in Table 1 is: MPD7 (3°), MPD20 (14°), cisplatin (17°), Native (18°), and 16 K (22°). This also is precisely the order of increasing roll angle observed at step 8 of the double helix, between base pairs T8·A17 and C9·G16. Nearly the same order is preserved at steps 4, 3, 2, and 1. Elsewhere along the helix the changes are not as great. As will be discussed in Section 7, and as can be seen by examining the stereo drawings of Figure 2, bending the overall helix axis seems to involve compression of the wide major groove in the upper third of the molecule, with lesser change in the lower region.

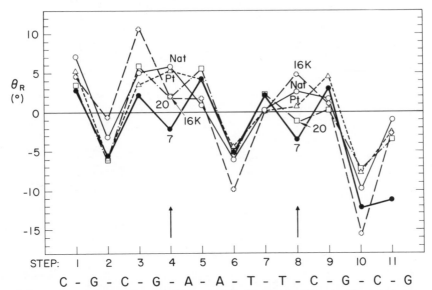

Figure 18. Roll angles between successive base pairs for the five dodecamer helices. The order of increasing overall axial bend is: MPD7–MPD20–cisplatin–Native–16K, which also is the order of increasingly positive roll angle, representing compression of the major groove, at step 8. The roll angle also becomes more positive in roughly this order at steps 1–4 at the upper end of the molecule, indicating a general compression of the major groove that is visible in Figure 2. From Appendix Tables A5–A9.

5.2. Cylindrical Helix Parameters

The natural coordinate framework within which to examine a helix is a cylindrical projection as in Figure 19. Every group then becomes visible from the same orientation, looking from the exterior toward the helix axis. Rotational and translational relationships of the helix backbone become especially clear.

Any atoms could be chosen as markers for studying the helix, but the sugar C1' atoms at the extremities of base pairs seem particularly appropriate when attention is focused on base stacking, and the phosphate P atoms are best when examining the backbone. Appendix Tables B1–B11 list the most useful cylindrical parameters based on both P and C1'. Rotation about the helix axis from one P to the next is called helix rotation, r, and that from one C1' to the next is helical twist, t. The distance up

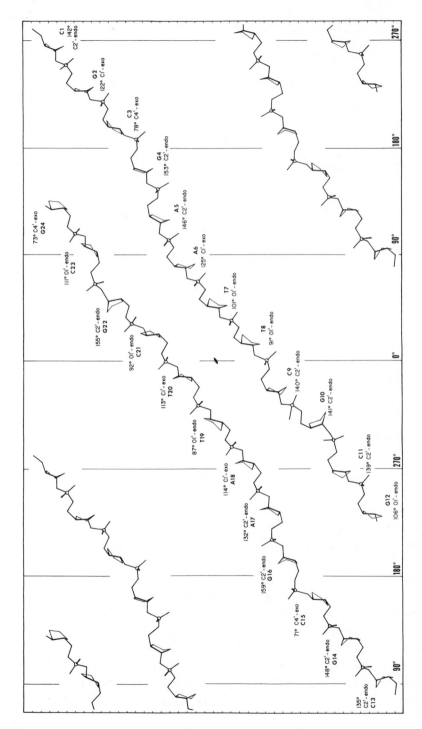

the helix axis from one P to the next is the helix rise, h, and that from one C1' to the next is $h(C1')$. As pointed out by Fratini et al. (1982), r and t can differ considerably at comparable positions along a strand. Angle r between phosphates measures the course of the backbone chain itself. The distance between successive P atoms, d in Tables B5–B9, is relatively constant. Hence in regions where the pitch angle of the backbone is shallow, r will be large and h will be small. Conversely, in local regions of steeper pitch, r will be small and h will be large. Local variation in pitch angle along each strand of the double helix is apparent in Figure 19, especially if one tilts the page and sights obliquely along one strand.

In contrast to rotation angle r, helical twist angle t between C1' atoms describes base stacking but reveals relatively little about backbone geometry. The C3'–C4' bond by which the deoxyribose ring is connected to the main chain is roughly vertical in Figure 19, and the sugar ring can swing around it like a gate around its hinges. Hence the position of the C1' atom (the latch on the gate, in this analogy) need bear little relationship to the steady progression of the backbone chain. For example, at position C9 in Figure 19 the sugar ring is swung to the left and at G10 the ring is swung to the right, leading to a small $t = 22°$. At the symmetrically related position on strand 2, C21 to G22, $t = 21°$ for the same reason. In contrast, at the steps just following these, the two C1' atoms are swung apart, so that $t = 49°$ for G10 to C11 in strand 1, and $t = 51°$ for G22 to C23 in strand 2. Helix rotation angle r measured between backbone phosphates

Figure 19. (*Opposite*) Unrolled projection of the backbone of the MPD7 helix on a cylinder of radius 8 Å. Because all backbone atoms lie within 2 Å of this 8 Å shell, distortions of lengths and angles in this projection are small. Gradations along the vertical axes are in Å, and this scale is approximately valid in the horizontal direction as well. The projection extends horizontally for more than a full revolution in order to display the two strands without interruption. The minor groove runs diagonally across the center of the projection from lower left to upper right. Bases are not shown because of distortions of the cylindrical projection, but are identified beside their sugar rings, along with the value of main chain torsion angle δ (C5'–C4'–C3'–O3') and the observed sugar ring conformation. Phosphorus atoms are indicated by small circles, and C3'–C4' bonds of sugar rings are darker for emphasis. Other bonds of the sugar ring, which sinks below the plane of the page, are lighter and tapered to suggest perspective. The most distant ring atom at the intersection of the lightest two lines is the C1' connection to the base.

show much less variation in these same regions, and this is a general observation; as judged by standard deviations from the mean, t is 1.6 times as variable as r.

The global twist angle, t_g, is defined as the angle between C1′–C1′ vectors in two successive base pairs viewed in projection down the helix axis. As Figure 20 illustrates, it is approximately equal to the average of the local t values for individual strands at the same base step. These two angles would be equal to one another, and to t_g, if the triangles defined by C1′ positions and the helix axis were congruent for the two base pairs. But if one base pair is shifted relative to the helix axis, along either its long direction or its short, then the three angles can differ. For example, in the MPD7 helix at the step between G2·C23 and C3·G22, for which t_g = 39.2° (see Appendix Table B9), t is 30.7° for strand 1 and 51.3° for strand 2. As will be seen in Section 6, translating a base pair along its

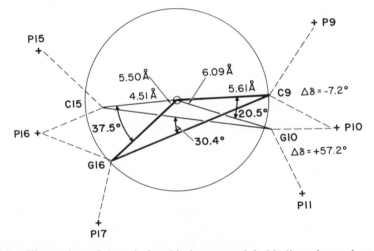

Figure 20. Illustration of the relationship between global helix twist angle t_g and the local twist angles on individual strands. In this view directly down the Native helix axis, circled at the center, phosphorus atoms are labeled P9, P10, P11, P15, P16, and P17, taking the numbering of the nucleoside that follows. Sugar C1′ atom positions are labeled C9, G10, C15, and G16. Local helix twist angles are t = 20.5° between C1′ atoms of bases C9 and G10 on strand 1, and t = 37.5° between bases C15 and G16 on strand 2. The global twist angle t_g between base pair C9·G16 and G10·C15, as measured by the change in orientation of C1′–C1′ vectors, is 30.4°. Global twist angle t_g is approximately equal to the average of the twist angles t from the individual strands. $\Delta\delta$ is the difference in main chain torsion angle δ between strands 1 and 2: $\Delta\delta = \delta_1 - \delta_2$.

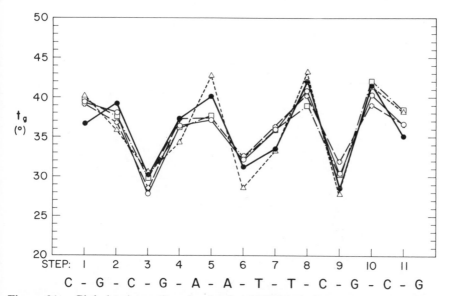

Figure 21. Global twist angle t_g for the five B-DNA dodecamers. Point and line conventions are as in Figure 11. The helix twist angle simulates a false periodicity every three base pairs; the true explanation in terms of base sequence is considered in Section 6. Helical twist is relatively unaffected by bending or straightening of the helix axis. Data from Appendix Tables B5–B9.

long axis tends to increase main chain torsion angle δ on the side being pulled out of the stack and to decrease δ by a similar amount on the other strand. In Figure 20, the difference in δ angles at the two ends of base pair C9·G16 is only 7.2°, but that for G10·C15, which has been translated along its long axis, is 57.2°.

Global helix twist angles t_g are compared for all five B dodecamers in Figure 21. There is very little difference from one structure to the other; twist angle apparently is relatively insensitive to bending or unbending of the helix axis. The t_g plots give an impression of periodicity every three base pairs, but this is misleading. A better explanation of the t_g behavior, derived from base sequence, is given in Section 6.

Figure 21 only shows the amount of local helical rotation, without specifying either the location and orientation of the rotation axis or the length of the translation down that axis. These quantities are displayed in the skeletal stereo drawings of Figure 22. Each local rotation vector is the result of using the HELIX program with two adjacent base pairs only. In general the local rotation axes are approximately parallel and

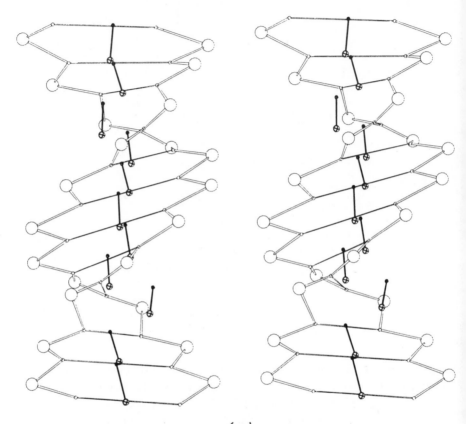

(a)

Figure 22. Skeletal stereo diagrams of the MPD7 helix, showing local helix twist axes. The sugar-phosphate backbone is represented only by spheres at phosphorus atoms and reentrant angles at C1' atoms, with open stick bonds connecting them. The two C1' atoms of each base pair are connected by a single thin line in lieu of the base pair. The local rotation axis from one base pair to the next is shown by a short vector with a crossed sphere below and a solid dot above. (a) View into the minor groove, as Figures 2c and 3a. (b) View to the "left" of the molecule, as Figure 3b. Note that the local rotation axes at steps 3 and 9 are displaced into the major groove (a) and inclined to the bases (b) in a manner more typical of the A helix than the B. The magnitude of the helix twist angle at these steps also is A-like (Figure 21). For similar diagrams of the Native helix, see figure 5 of Dickerson and Drew (1981a).

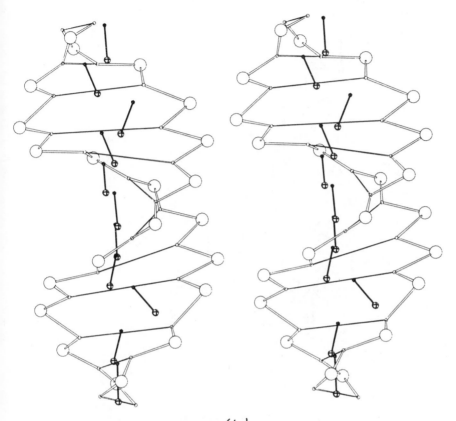

(b)

Figure 22. (Continued)

colinear, with two notable exceptions: steps 3 and 9, equidistant from top and bottom. In both cases the local rotation axis is displaced into the major groove and is tilted relative to the base pairs in a manner reminiscent of an A helix. The magnitude of rotation, 30.3° and 28.6°, also is A-like. Dickerson and Drew (1981a) originally suggested that this might be an example of variation in local helix type, from B to A, but it is more likely in the light of later data that this simply reflects the wide limits of variability of the B helix itself.

Although phosphate helix rotation angles *r* are less variable than C1′ twist angles *t*, they do exhibit one striking pattern of variation in cylindrical projection. As shown in Figure 23, the rotations are especially large

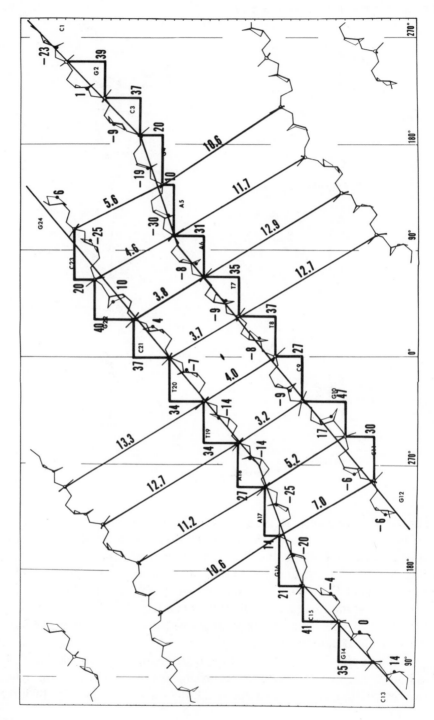

86

at G4 and A5 on one strand and at the symmetrically related G16 and A17 on the other. This is due to the slope of the phosphate backbone, which is markedly shallower here than along the rest of the helix and which in turn arises because of a narrowing of minor groove width in the AATT center of the molecule. The unusually large r values occur in the transition zone between this narrow AATT minor groove and a wider minor groove at the CGCG ends.

A complex relationship exists between propeller twist, minor groove width, and the minor groove spine of hydration. As Figure 24 shows, a positive propeller twist displaces the C1′ atoms at the ends of base pairs in a way that shifts the two backbone strands closer together, narrowing the minor groove. The twist also brings the N3 and O2 atoms of adenine and thymine to more favorable positions for binding the spine of hydration (Figure 8). The narrowing of the groove also helps stabilize the spine by bringing closer the sugar O4′ atoms that line the sides of the groove (Figure 9). It is not yet clear whether the lower average propeller twist in the CGCG ends is an intrinsic property of C·G base pairs, as was suggested by Dickerson and Drew (1981a), or whether it results simply from the alternation of purines and pyrimidines along the sequence, which maximizes steric clash in the manner described in Section 6. The particular sequence chosen, CGCGAATTCGCG, may be an extreme case in which the intrinsic effects of base pair type and purine/pyrimidine alternation reinforce one another. It would be interesting to examine the logical in-

---

Figure 23. (*Opposite*) Cylindrical projection of the MPD7 helix as in Figure 19, with numerical values of several parameters added. The minor groove is substantially narrower in the AATT center than in the two CGCG ends. This is apparent from the thin diagonal lines drawn from lower left to upper right along the mean phosphorus positions, approximating each phosphate backbone by three straight line segments. Diagonal lines from lower right to upper left connecting phosphates across the minor and major grooves are labeled with groove widths in Å (defined as the P–P separation less 5.8 Å for two phosphate group radii). In the heavy-line "stair step" along each phosphate backbone, the horizontal "treads" show helix rotation angles, r, and the vertical "risers" show rise per step, h. Positive numbers near intersections of these lines in the major grooves are pitch angles in degrees from one P to the next. Positive and negative numbers within the minor groove indicate ψ, the angle that the C4′–C3′ bond vector makes with the helix axis in cylindrical projection (vertical in this figure), with counterclockwise rotation being positive.

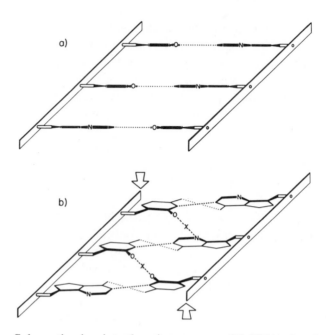

Figure 24.   Schematic view into the minor groove of B-DNA, showing the effect of propeller twist on groove width. Bases and C1′ atoms are depicted accurately; the remainder of the sugar ring and the phosphate backbone are only shown schematically. (*a*) Flat base pairs, viewed on edge. (*b*) Base pairs with positive propeller twist. Glycosyl bonds are displaced up and down at right and left, respectively, moving the backbone strands closer as indicated by the two arrows. Positive propeller twist also brings closer the pyrimidine O2 and purine N3 atoms that anchor the spine of hydration. First-level water molecules of the spine are located at positions *x*.

verse of the above sequence, CCGGATATCCGG, and this sequence is in fact now being synthesized for analysis.

5.3.   Torsion Angles and Sugar Conformation

The protein crystallographer needs only two torsion angles per amino acid residue to describe the path of the polypeptide chain completely. Polynucleotides are more complex, and six angles are required per base. These are identified as α–ζ in Figure 25, in the IUB/IUPAC recommended nomenclature. (In an older system, still used, angles α through ζ are called ω, ϕ, ψ, ψ', ϕ', and ω'.) This diagram also shows the classical B-DNA

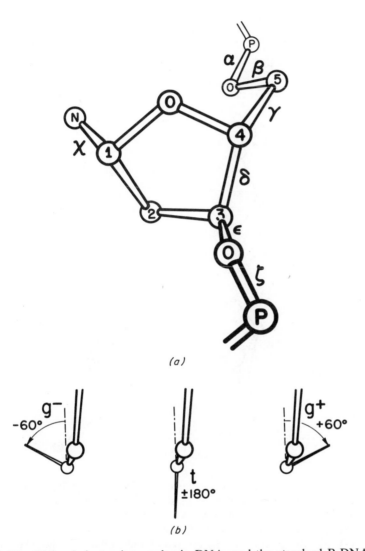

(a)

(b)

Figure 25. Main chain torsion angles in DNA, and the standard B-DNA backbone conformation. (*a*) In the IUB/IUPAC recommended nomenclature, torsion angles are labeled α through ζ from one P to the next in the indicated 5′-to-3′ direction. The angle is zero when the two bonds on either side of the bond under consideration are eclipsed, and it is positive for clockwise rotation of the more distant bond. (*b*) The three energetically favored staggered conformations about the central bond are given special names: *gauche⁻* or *g⁻* for angles near −60°, *trans* or *t* for 180°, and *gauche⁺* or *g⁺* for +60. Glycosyl bond orientation χ is defined by atoms O4′–C1′–N1–C2 for pyrimidines and O4′–C1′–N9–C4 for purines, with the same angle conventions.

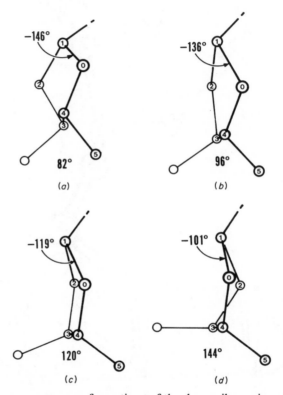

Figure 26. Four common conformations of the deoxyribose ring viewed directly down the helix axis, showing their relationship to main chain torsion angle δ. (*a*) C3′-*endo*, with δ = 82°. (*b*) O4′-*endo*, with δ = 96°. (*c*) C1′-*exo*, with δ = 120°. (*d*) C2′-*endo*, with δ = 144°. Atoms O4′ and C1′ through C5′ are labeled 0–5, and the O3′ atom is unlabeled. The glycosyl bond to the base (not shown) leads off to the upper right from C1′. The negative angle at upper left of each diagram is the ideal χ value for that ring conformation that would occur if base planes were perpendicular to the helix axis and if C3′–C4′ bonds were parallel to the axis in cylindrical projection.

torsion angle geometry expected from fiber studies, with bonds α through ζ in conformations: (g^-, t, g^+, t, t, g^-). Besides the six main-chain torsion angles, one further torsion angle is required to specify the conformation of a nucleoside: χ, the orientation about the glycosyl bond connecting base and sugar. Appendix Tables C1–C11 list all seven torsion angles for the eleven A, B and Z helices.

One of the most frequently discussed aspects of nucleic acid structure in the past has been deoxyribose ring conformation, or sugar puckering. Steric hindrance between carbon substituents in the ring requires that one of the five ring atoms lie out of the best plane defined by the other four. Sugar pucker can be described as either Y'-endo or Y'-exo, where Y' identifies the out-of-plane atom, and endo or exo describes whether it is out of plane on the same side of the ring as the C5' atom or on the opposite side. Four commonly encountered ring conformations are depicted in Figure 26 as they would appear to an observer looking straight down the helix axis, nearly along the C4'–C3' bond. This view is particularly informative because it reveals the close connection between sugar pucker and main chain torsion angle δ. If $\delta = 120°$ as in Figure 26c, then bonds C3'–C2' and C4'–O4' are parallel in projection, their four atoms lie in a common plane, and hence the C1' atom must be out of plane either in the C1'-exo direction shown here or in the C1'–endo direction (not shown). If torsion angle δ is opened up still further to 144°, as in Figure 26d, atom C2' is pushed to the right and the C2'-endo conformation results. If angle δ is closed down to 96° as in Figure 26b, atom O4' is displaced to the right and the O4'-endo pucker follows. Closing down δ all the way to 82° drives the C3' atom out of the plane, in the C3'-endo conformation. Hence sugar conformation and torsion angle δ are redundant, and we shall generally specify angle δ, with sugar conformation only implied in the manner shown here.

One of the first features of the dodecamer structure noted by Drew et al. (1981) was the strong correlation between glycosyl torsion angle χ and main chain torsion angle δ, as depicted in Figure 27a. Rather than clustering around the C2'-endo positon, as fiber diffraction model building studies had led one to expect, the points for the 24 bases are distributed over a broad range of conformations, from C2'-endo all the way through C1'-exo and O4'-endo (O1'-endo in an earlier ring atom labeling convention) to C3'-endo and even C4'-exo. The distribution is not random; purines have a systematic preference for high δ values in the neighborhood

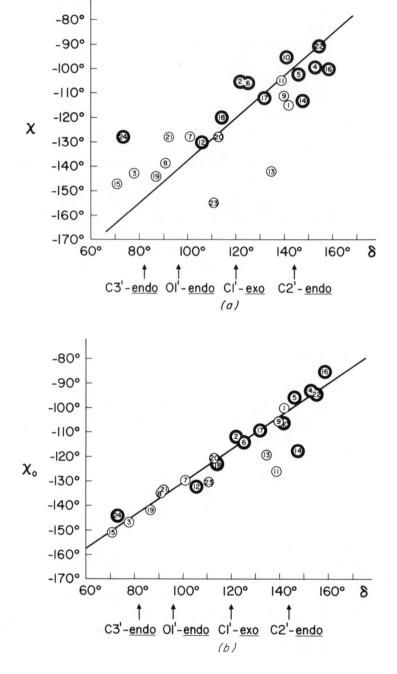

of the classical C2'-*endo* conformation, whereas pyrimidines generally are found at lower δ values. The data points for purine and pyrimidine of one base pair also tend to occur at comparable distances to the left and right of the center of the plot, whether far from the center (as 15 and 20, or 3 and 22), or close to the mean (as 12 and 13, or 23 and 2). This was noted by Drew et al. (1981) as the "principle of anticorrelation," but in fact it is only a natural consequence of shifts of base pairs along their long axes (Figure 28). As discussed in Section 6, base pairs frequently are translated along their long axes in the direction that pushes the purine out of the stack, because this helps decrease local purine-purine cross-chain steric hindrance. When this occurs, torsion angle δ on the purine end increases, and δ on the pyrimidine end decreases by an equivalent amount, providing a simple, one-step explanation both for the preference of purines for high δ values and for the principle of anticorrelation.

The extent of the base pair shift can be measured by Δ, the difference in torsion angle δ at the two ends of each base pair: $\Delta = \delta_1 - \delta_2$, as plotted in Figure 29. Except for atypical behavior at the two outermost base pairs, representing end effects, Δ tends to be positive at positions occupied by purines in strand 1 (whose sequence is given at the bottom of Figure 29) and negative at strand 1 pyrimidine positions, indicating that the motion is indeed a pushing of purines out of the base stack. The plot also is approximately antisymmetrical about its center, reflecting the approximate twofold symmetry between strands, or between the two ends of the double helix.

Most of the variation in main chain torsion angles occurs in the second half of each nucleotide. In the MPD7 helix, the first three torsion angles, α, β, and γ, exhibit standard deviations from the mean of only 11°, 17°, and 11°; they are essentially frozen in the g^-, t, and g^+ configuration. Variation is greater in the following three angles: 28° for δ, 31° for ϵ, and

---

Figure 27. (*Opposite*) Correlation between glycosyl angle χ and main chain torsion angle δ for the MPD7 helix. (*a*) Experimentally observed values, identified by base numbers 1–24. The least-squares linear regression line leads to a correlation coefficient of $R = 0.78$, which rises to 0.87 if end effects are minimized by using only the central 10 base pairs. (*b*) Improved correlation if the observed χ are modified to χ_0 by correcting for nonperpendicularity of base planes to the helix axis, and tilt of the sugar ring away from the vertical, as described by Fratini et al. (1982). The new regression line has a correlation coefficient of $R = 0.93$, and was used to find the ideal χ_0 values depicted in Figure 26.

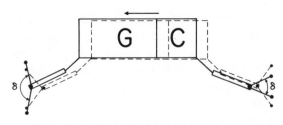

ANTICOMPLEMENTARITY OF TORSION ANGLES, δ

Figure 28.   If the phosphate backbone remains roughly constant, a shift of a base pair along its long axis so one end is pulled out of the stack will lead to an increase in main chain torsion angle δ at the end that is pulled out, and an equivalent decrease in δ at the other end. Hence the δ values at two ends of any base pair will tend to be equidistant on either side of a mean value representative of an unshifted base pair. This is what has been called the "principle of anticomplementarity." As Calladine has proposed (see Section 6), shifting a purine out of the stack usually decreases purine–purine steric clash. Hence purines generally have larger δ angles than pyrimidines, exactly as is observed in Figure 27. The base pair is represented by a flat plank in the center, and the two sugar rings by slabs to either side viewed on edge. The C3′–C4′ "hinge" of the slab is perpendicular to the plane of the page.

42° for ζ. Angle δ does indeed have a wide distribution, as Figure 27 has shown. But the distributions in ε and ζ, as we shall see, are not so much broad as they are bimodal, between two preferred states. Systematic analysis of relationships between main chain torsion angles (Fratini et al., 1982; Dickerson et al., 1982a) reveals only four pairings with correlation coefficients of magnitude 0.50 or greater: ζ with the preceding ε (-0.89) and δ (-0.71), and β with the preceding ζ ($+0.72$) and ε (-0.78). The β correlations are of doubtful significance because the actual variation in β is small. But those involving δ, ε, and ζ tell us something about B-DNA backbone conformations.

A correlation plot of ε versus ζ (Figure 30) reveals two preferred conformations. Most of the points cluster at the lower right in the region where ε is *trans* and ζ is *gauche⁻*. Because this is the dominant conformation, it has been designated B_I. A different geometry is found at only two loci along the helix, G10 on strand 1 and the symmetric G22 on strand 2. In this minority conformation, designated B_{II}, ε is *gauche⁻* and ζ is *trans*. Figure 31 shows the structural differences between the two conformations: In B_I the C3′-O3′-P "elbow" lies in the surface of a cylinder enclosing the helix and points in the direction of the minor groove, whereas in B_{II} the elbow is turned so that it points inward toward the helix axis. The difference also can be seen in Figure 32, which offers a stereo closeup of backbone strand 1 in the vicinity of G10 and C11. B_I has long been the expected conformation from classical fiber diffraction work (e.g., Arnott and Hukins, 1972), but Gupta et al. (1980) have more recently proposed B_{II} as the standard conformation in B-DNA. Our results, while demonstrating that the B_{II} conformation can exist under special circumstances, come down decidedly on the side of Arnott and co-workers. The B_I geometry is seen experimentally to be compatible with a wide range of values of sugar conformations and hence of torsion angle δ, whereas the two observed examples of B_{II} conformation have δ around

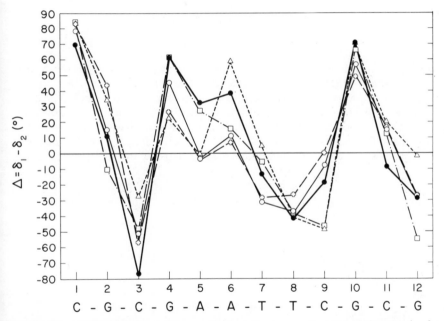

Figure 29. Difference in torsion angle δ at the two ends of each base pair, for the five B-DNA dodecamer structures. Point and line coding are as in Figure 11. Note the approximate antisymmetry through the center of the diagram: a large negative peak at base pair C3·G22 at left is matched by a large positive peak at the symmetrically disposed G10·C15 at right. In each case the value of Δ indicates that the base pair has been slid along its long axis so that the purine (G) is partially removed from the stack, as in Figure 28.

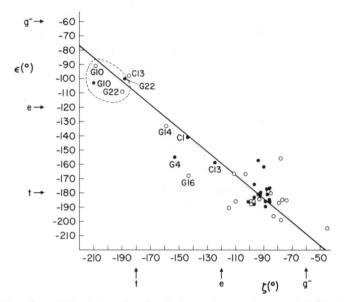

Figure 30. Correlation plot of main chain torsion angles ϵ and ζ, for the Native (black dots) and MPD7 (open circles) helices. The correlation coefficient for the line as drawn is $R = -0.92$. Of the 44 data points for the two helices, 34 cluster in the $(\epsilon, \zeta) = (t, g^-)$ region, five others including G10 and its symmetrically related G22 on the other strand cluster in a region with reversed (g^-, t) configuration, and five others are scattered along the intermediate region. The majority (t, g^-) and minority (g^-, t) conformations are denoted by B_I and B_{II} respectively.

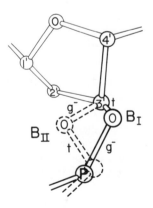

Figure 31. The majority or B_I conformation (solid lines) and the minority or B_{II} conformation (broken lines) at backbone torsion angles ϵ and ζ. The shift from B_I to B_{II} is essentially a rotation of the C3'–O3'–P "elbow" from alignment along the surface of the helix so that it points into the interior instead.

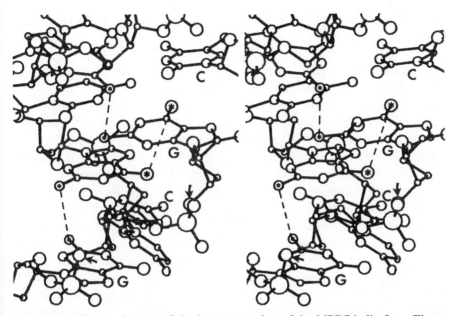

Figure 32. Stereo closeup of the bottom portion of the MPD7 helix from Figure 15, showing examples of B_I and B_{II} conformations, and illustrating the steric clash between purines that is brought about by positive propeller twist. This is a view into the minor groove, and the labeled bases from top to bottom at right are C9, G10, C11, and G12. O3' atoms illustrating the B_I and B_{II} backbone conformations are indicated by arrows. The lower one at base C11 is type B_I, with the O3' bond elbow pointing along the surface of the helix toward the minor groove. The one just above it at G10 (partially obscured by an unbridged phosphate oxygen) is type B_{II}, with the O3' elbow pointing inward toward the helix axis. Dashed lines connecting starred atoms indicate close purine-purine contacts. The upper and lower two occur in the minor groove at CpG steps, and the middle one lies in the major groove at a G-C step. The flattening of propeller twist in base pair G10·C15, noted earlier in connection with Figure 16, is itself a consequence of the two upper dashed steric repulsions from neighboring guanines.

145° and C2'-*endo* puckering. Figure 31 suggests that this may be an example of general behavior: pushing the C3'-O3'-P elbow back toward the interior of the helix turns atom C3' in such a way that the C3'-C2' bond is rotated to the *endo* side of the sugar ring. A pure B_{II}-DNA helix would be limited to uniform C2'-*endo* sugar puckering. Although this is what had been expected from fiber studies, it has been shown by the single-crystal analyses to be too simple.

6. SEQUENCE DEPENDENCE OF LOCAL HELIX GEOMETRY

One of the most interesting questions in DNA structure is the manner in which base sequence is recognized by restriction enzymes, repressors, and other control proteins that recognize and bind to specific regions of DNA. One obvious possibility is that the double helical DNA plays a purely passive role, allowing proteins to interact with it and search out a particular pattern of hydrogen bonding in the major and minor grooves. In the major groove, as Figure 10 reveals, an A·T base pair has a hydrogen bond acceptor (N:) on one side, a donor ($-NH_2$) in the center, and another acceptor (:O:) on the other side. In the inverted T·A base pair the order still is acceptor-donor-acceptor, but with slightly different geometry. A G·C pair has a different order: acceptor-acceptor-donor, and reversal of bases in C·G reverses the hydrogen bonding pattern: donor-acceptor-acceptor. Hence the four-symbol (A-G-C-T) code of bases is translated into a two-symbol (acceptor-donor) code, but one with the added dimensions of spatial arrangement down the bottom of the major groove. The minor groove is much less rich in information, the main difference being the presence of a donor $-NH_2$ in G·C but not in A·T. For this reason, and because the minor groove apparently has another important role in stabilizing the B helix via its spine of hydration, it seems likely that the major groove is most important in the sequence read-out process by control proteins. Because this process involves intrinsic physical and chemical properties of both DNA and protein, it can be termed intrinsic readout, in contrast with the extrinsic readout used in translating codon information into amino acid sequence, which requires a considerable battery of intermediate RNA molecules and enzymes for its operation (Dickerson, 1983a,b).

But is this the end of the story? Is the DNA present only in a passive role, or does the base sequence also affect the local helix structure in a manner that contributes to the intrinsic read-out process? Single-crystal X-ray analysis of B-DNA and A-DNA oligomers suggests the latter alternative. Much of the helix structure variation presented in Section 5 can be related directly to base sequence in a way that would make it easily recognized by other macromolecules.

All of the B- and A-helical oligomers whose molecular structures have been solved by single-crystal methods have proved to have an appreciable propeller twist, positive in the sense defined in Figure 14. As mentioned earlier, this is so because a positive propeller twist in a right-handed double helix increases the overlap of bases down each strand of the helix individually, and consequently helps to stabilize the helix. However, as Calladine (1982) has pointed out, an alternating purine-pyrimidine sequence also leads to steric clash between purines on opposite helix strands at adjacent base pairs. A double-ring purine extends past the center of the base pair. If a purine occurs on the opposite strand at the next base pair, then the fact that base planes on opposite strands are propeller twisted in opposite directions means that the purines will be brought into unacceptably close contact. In Figure 32, the N2 amine of base G16 in base pair C9·G16 clashes in the minor groove with the N2 of base G10 in pair G10·C15 (upper dashed line). At the next step, the O6 atom of base G10 clashes in the major groove with the corresponding atom of G14 in pair C11·G14. In the final step, the N2 amines of bases G14 and G12 clash in the minor groove. In general, steric hindrance will occur in the minor groove at Y-R steps (Y = any pyrimidine; R = any purine), and in the major groove at R-Y steps. Calladine used principles of elastic beam mechanics to analyze this clash and ways of relieving it, and proposed that the strain could be ameliorated by four strategies:

1. Flatten the propeller twist in one or both base pairs.
2. Open up the roll angle between base pairs on the side where clash is found.
3. Translate one or both base pairs along its long axis in a direction that pushes the pyrimidine out of the base stack, as in Figure 28.
4. Decrease the local helix twist angle at the step at which clash occurs.

The efficacy of strategy 1 is obvious. Strategy 2 implies that roll angle θ_R should be made more positive at Y-R steps, and more negative at R-Y steps. Figure 28 shows that strategy 3 will lead to an increase in δ at the purine and a decrease at the pyrimidine. The way in which a decrease in helix twist (strategy 4) lessens purine-purine cross-chain clash, no matter whether in major or minor groove, is best seen from base pair projections such as Figure 3 of Dickerson (1983a) or Figure 11 of Dickerson et al. (1983b). From his analysis of the behavior of δ, Calladine deduced that minor groove clashes at Y-R steps were twice as severe as major groove

	R	-	R	-	Y	-	Y	-	Y	-	Y	-	R	-	R
Σ_1 (twist)		+1		−2		+1				+2		−4		+2	
Σ_2 (roll)		+1		−2		+1				−2		+4		−2	
Σ_3 (δ)			+1		−1					−2		+2			
Σ_4 (pr. tw.)			−1		−1					−2		−2			

Figure 33. Algorithms for generating sum functions Σ_1–Σ_4, which use base sequence to derive expected local variation in helical twist, base plane roll, difference in torsion angle δ, and propeller twist, respectively. For implementation, see text.

clashes at R-Y steps, and this permitted an almost quantitative calculation, from base sequence alone, of the behavior of helix twist angle t_g (Calladine, 1982).

Dickerson (1983a,b) has quantified and extended this analysis by defining four simply constructed sum functions, $\Sigma_1 - \Sigma_4$, by means of which base sequence can be used to calculate the expected local variation in helix twist (Σ_1), base plane roll (Σ_2), torsion angle difference Δ at the two ends of a base pair (Σ_3), and flattening of propeller twist (Σ_4). The algorithms for constructing these sum functions are tabulated in Figure 33, and the process for twist function Σ_1 is as follows:

1. At every R-Y step, assume that the local helix twist angle is decreased by two arbitrary units because of rotations of the two base pairs, and that the helix twist at the two flanking steps is increased by one unit each as a consequence. Hence the perturbation at these three base steps is: $+1, -2, +1$, as depicted in Figure 33.

2. At every Y-R step, assume a similar contribution to the sum function, but doubled in magnitude because of the greater seriousness of minor groove clashes. Hence the sum function contributions at the Y-R step and its two flanking steps will be: $+2, -4, +2$.

3. Ignore homopolymer R-R and Y-Y steps because they involve no steric hindrance.

4. Repeat this process at every step of the sequence, and add the contributions to obtain the total sum function, Σ_1.

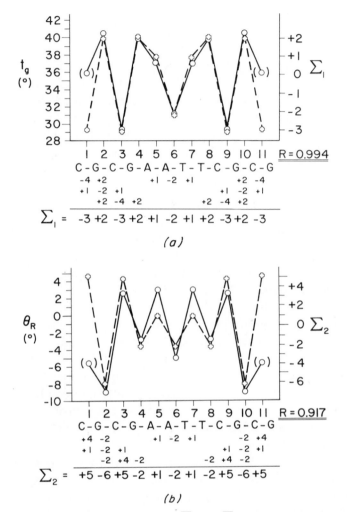

Figure 34. Application of sum functions $\sum_1$ and $\sum_2$ to the B-helical MPD7 structure. (*a*) Sum function $\sum_1$ and global twist angle, t_g. (*b*) Sum function $\sum_2$ and base roll angle, θ_R. In Figures 34–36, solid lines represent observed parameter values after end-for-end averaging, with scale at left, and broken lines display the sum function, generated as shown at the bottom of each graph, with scale at right. R at lower right of each plot is the linear correlation coefficient between observed parameter values and the calculated sum function. Outermost base pairs (data points in parentheses) in Figures 34 and 35 are omitted from the correlation, since they are perturbed by end effects and are in atypical conformations.

Figure 34a shows at the bottom how Σ_1 is generated for the sequence CGCGAATT^{Br}CGCG, and the plot above compares Σ_1 with the observed global twist angles, t_g. Because the two ends of the helix, which are nonequivalent in the crystal, would be entirely equivalent in isolated dodecamer molecules in solution, the observed values of t_g (and of other variables later) have been averaged end-for-end before being compared. Observed and calculated values agree quite well, with a linear correlation coefficient of $R = 0.994$. (The two outermost values, plotted within parentheses, were not used in the correlation because they represent end effects arising from the overlap of molecules and are in atypical geometries.)

The derivation of roll sum function Σ_2 is similar to Σ_1, except for a reversal of signs for the Y-R contribution, reflecting the fact that the roll angle is decreased at R-Y steps but increased at Y-R steps. The predictions of sum function Σ_2 are compared with experimental roll angles for CGCGAATT^{Br}CGCG in Figure 34b. Again the agreement is excellent, with a correlation coefficient of R = 0.917. (Outermost base pairs again omitted from correlation.)

The schemes for generating base shift or Δ sum function Σ_3 and propeller twist function Σ_4 are depicted in Figures 33 and 35, and justified in Dickerson (1983a). These are not as successful as Σ_1 and Σ_2, but the correlation between prediction and observation still is highly significant. The fit between propeller twist and Σ_4 is greatly improved, and the correlation coefficient rises from 0.68 to 0.88, if the contributions from the first base step at each end of the helix are omitted from the sum function. Σ_4 then has the terms 0 -1 -3 -2 0 -1 -1 0 -2 -3 -1 0 and matches the observed deep minima in propeller twist at base pairs 3 and 10. As can be seen from Figure 32, these minima arise experimentally because the guanines of base pairs 3 and 10 are subjected to flattening constraints on both sides.

The success of one straightforward physical assumption regarding steric clash in explaining four different effects in B-DNA is gratifying. But what is more pleasing, and not a little surprising, is to find that two of these sum functions, Σ_1 and Σ_2, also hold for three totally unrelated A-DNA structures: CCGG (Conner et al., 1982, 1984), GGTATACC (Shakked et al., 1981, 1983), and GGCCGGCC (Wang et al., 1982a; Fujii et al., 1982), with partial success for Σ_1 and complete success for Σ_2 in the case of an RNA/DNA hybrid helix, r(GCG)d(TATACGC) (Wang et

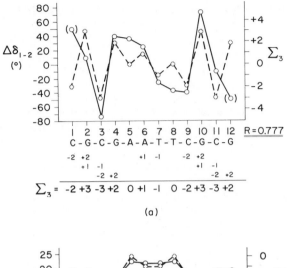

(a)

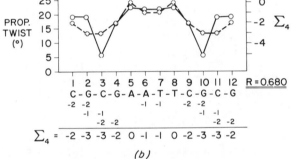

(b)

Figure 35. Application of sum functions $\sum_3$ and $\sum_4$ to the MPD7 helix. (a) Sum function $\sum_3$ versus Δ, the difference in main chain torsion angles δ at the two ends of the base pair. (b) Sum function $\sum_4$ versus propeller twist. Graph conventions as in Figure 34.

al., 1982b; Fujii et al., 1982). These analyses are presented in detail in Dickerson (1983a), but they can be summarized by Table 7 and Figure 36 for the case of CCGG.

CCGG crystallizes as double helical tetramers, but pairs of tetramers in the crystal are stacked end-for-end in a manner that continues a functional CCGG/CCGG octamer without interruption, even though the central G/C helix step has no covalent phosphate connections. This was ap-

Table 7.   Correlation Between Observed and Predicted Helix Variations[a]

x	y	Molecule	R	S	T	Significant?
Helix twist, t_g	$\sum_1$	C-G-C-G-A-A-T-T-^{Br}C-G-C-G	0.994	35.6	2.1	yes
		^{I}C-C-G-G	0.977	34.1	0.7	yes
		G-G-C-C-G-G-C-C	0.991	33.2	3.7	yes
		G-G-T-A-T-A-C-C	0.915	32.1	0.9	yes
		G-G-^{Br}U-A-^{Br}U-A-C-C	0.808	32.2	1.0	yes
		RNA/DNA hybrid	0.180	33.4	0.1	no
		Hybrid, DNA steps only	0.980	31.2	0.7	yes
Base roll, θ_R	$\sum_2$	C-G-C-G-A-A-T-T-^{Br}C-G-C-G	0.917	−1.1	1.1	yes
		^{I}C-C-G-G	0.995	6.3	1.6	yes
		G-G-C-C-G-G-C-C	0.896	6.3	2.1	yes
		G-G-T-A-T-A-C-C	0.984	6.6	1.3	yes
		G-G-^{Br}U-A-^{Br}U-A-C-C	0.996	6.6	1.1	yes
		RNA/DNA hybrid	0.835	9.8	0.8	yes
Base pair shift, $\Delta\delta$	$\sum_3$	C-G-C-G-A-A-T-T-^{Br}C-G-C-G	0.777	0.0	15.6	yes
		^{I}C-C-G-G	0.062	0.0	1.0	no
		G-G-C-C-G-G-C-C	0.442	0.0	15.2	no
		G-G-T-A-T-A-C-C	−0.611	0.0	−0.4	no
		G-G-^{Br}U-A-^{Br}U-A-C-C	0.184	0.0	0.2	no
		RNA/DNA Hybrid	0.365	0.0	0.9	no
Propeller twist	$\sum_4$	C-G-C-G-A-A-T-T-^{Br}C-G-C-G	0.680	24.3	3.6	yes
		^{I}C-C-G-G	−0.134	15.5	−0.5	no
		G-G-C-C-G-G-C-C	−0.383	6.6	−2.0	no
		G-G-T-A-T-A-C-C	−0.135	12.4	−0.3	no
		G-G-^{Br}U-A-^{Br}U-A-C-C	−0.359	10.0	−0.8	no
		RNA/DNA Hybrid	−0.943	2.6	−3.6	no

[a] Linear correlation: x = S + Ty; R correlation coefficient.

Raw helix parameters from the crystal structure analyses have been averaged about the twofold symmetry axis relating the two ends of the helix prior to correlation with $\sum_j$, since a free helix in solution would possess such twofold symmetry.

Data from Dickerson (1983a), Fujii et al. (1982), Shakked et al. (1983), and Appendix.

parent from inspection of crystal packing, but it was a surprise to find that the helical twist angle and base plane roll angle across the unbonded G/C step were exactly what would be expected had the tetramers been covalently connected into a continuous octamer. From the correlation plots of t_g (Figure 36a) and θ_R (Figure 36b) with their respective sum

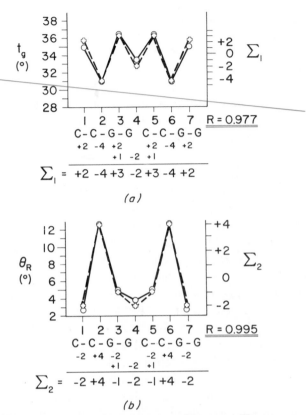

Figure 36. Application of sum functions Σ_1 and Σ_2 to the A-helical CCGG/CCGG. (*a*) Sum function Σ_1 and global twist angle, t_g. (*b*) Sum function Σ_2 and base role, θ_R. Two double helical CCGG tetramers stack atop one another in the crystal in a way that continues the helix in a functional octamer. The central G/C step, not connected by phosphate groups, is seen from these plots to be as typical a helix step as those with phosphate connections. Base stacking seems to be more important than covalent backbone connectivity.

functions, one could not have discerned that the central steps were unconnected. From the standpoint of local helix geometry, base stacking seems to be more important than is the presence or absence of connecting phosphate backbone.

Base shift or torsion angle sum function Σ_3 fails for A-DNA because torsion angles δ in the A-helical structures do not exhibit the broad range of variation seen in B-DNA; they all generally cluster in the vicinity of

$\delta = 82°$ and the C3'-*endo* sugar puckering (Appendix Tables C1–C4). Conversely, propeller twists in A-DNA do not correlate well because they are more variable than sum function Σ_4, for reasons that are not yet clear. Simple purine–purine repulsion seems not to dominate propeller twist when base pairs are steeply inclined to the helix axis. In the RNA/DNA hybrid, an A helix, helical rotation sum function Σ_1 holds only in the pure DNA center of the molecule, but roll angle sum function Σ_2 remains valid in both RNA and DNA regions.

In summary, four different local variations in helix structure can be predicted successfully for B-DNA: helical rotation angle, base plane roll, behavior of torsion angle δ, and propeller twist. Two of these, helix rotation and base roll, are equally sequence-determined in A-DNA, and base roll remains predictable from sequence even in an RNA/DNA hybrid. Local structure variations are firmly rooted in base sequence. But are these variations used as part of the intrinsic process of information readout? A definitive answer will have to await the availability of crystal structure analyses of control proteins such as repressors complexed with their own operator DNA sequences. But actual observed helix twist angles in oligomers are observed to differ by as much as $\pm 15°$ from their ideal fiber-derived values, and base pair roll angles of $\pm 8°$ from standard values are common. A recognition protein, as it binds to such a helix, must either make use of this variation or eliminate it by bringing the parameters back to standard values, and the former alternative seems much more likely. What matters in conferring selectivity and efficiency in the recognition of DNA by protein is not merely the pattern of hydrogen bonding donors and acceptors on the DNA, but their orientation in space as perceived by the protein. In the concluding words of Dickerson (1983a): "This paper is a first step in establishing a sequence/structure vocabulary. The next key question is: Do repressors and operators speak this language?"

7. MAJOR GROOVE SUBSTITUENTS AND THE GEOMETRY OF BENDING

Both of the heavy atoms Br and Pt that were introduced into the Native CGCGAATTCGCG structure as an aid in crystallographic phase analysis are found within the major groove. The bromine atom of the 3-Br derivative is covalently bonded to the C5 atom of cytosine, as is also the case with the 9-Br derivative that was studied and refined as the MPD7 structure. Bromine is the largest atom in Figure 3*b*, visible in the major groove above and to the right of center. An equivalent bromine on the other strand is out of sight around the back of the helix, and both bromines can be seen just around the edge of the phosphate backbone in Figure 3*a*. The presence of these bulky bromine atoms is believed to assist in opening up the major groove and eliminating the axial bend observed in the Native helix.

7.1. Cisplatin Binding

Cisplatin [*cis*-dichlorodiaminoplatinum(II)] is of interest not only because of its usefulness in phase analysis but because it currently is the most widely used inorganic antitumor drug in cancer chemotherapy (Pres-

Table 8. Binding of Cisplatin to B-DNA Dodecamer as a Function of Concentration[a]

Crystal Form	Native	Pt1	Pt2	Pt3
Cell dimensions (Å)				
a	24.87	24.36	24.33	24.16
b	40.39	40.05	40.08	39.93
c	66.20	66.13	66.26	66.12
Resolution (*d* in Å):	1.9	2.5	2.2	2.6
Current residual error, R^a	23.9%	20.0%	27.0%	16.4%
Percentage occupancy at site				
G16	—	20%	38%	49%
G4	—	10%	17%	21%
G10	—	—	13%	19%
G22	—	—	—	7%
Pt–N7 bond length at site G16 (Å)	—	2.51	2.43	2.26
Occupancy ratios[b]				
G16/G4	—	2.0	2.24	2.33
G10/G22	—	—	—	2.71
G16/G10	—	—	2.92	2.58
G4/G22	—	—	—	3.0

[a] R factor for all data, including unobserved reflections.
[b] The average value of ratio for sequence-equivalent positions on the two chains (G16/G4 and G10/G22) is 2.3, and that between the fourth and the third base pairs in from each end of the helix (G16/G10 and G4/G22) is 2.8.

tayako et al., 1980). From X-ray structural studies of complexes of cisplatin with guanine or inosine (Goodgame et al., 1975; Gellert and Bau, 1975; Cramer et al., 1980), one would expect the Pt to coordinate to the N7 position of guanine at a distance of 2.0 Å by displacing one of the four original square planar ligands. These studies also suggest that the Pt atom need not lie in the plane of the base but can be displaced out of plane by as much as 0.1–0.6 Å (see, Wing et al., 1984 for details).

Cisplatin derivatives of CGCGAATTCGCG were prepared by soaking pregrown DNA crystals in cisplatin solutions for varying lengths of time and varying concentrations. The maximum amount of drug was severely limited because prolonged soaking or high cisplatin concentrations led to fading of the diffraction pattern and increased sensitivity to radiation damage by X-rays. This again is completely in accordance with earlier solution studies indicating that platinum binding to DNA tends to destabilize and

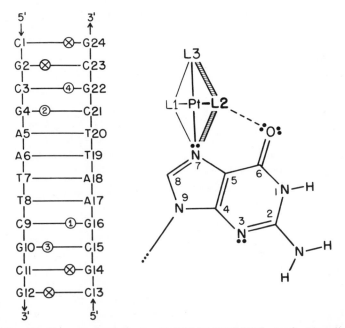

Figure 37. Binding of cisplatin to CGCGAATTCGCG. *Left*: Unrolled ladder view into the major groove of the helix, with guanine N7 positions represented as circles. Numbers 1–4 are in order of decreasing extent of substitution by cisplatin, and × indicates that no binding occurs. *Right*: Schematic view of possible ligand bridging between guanine N7 and O6. L2 probably is a water molecule.

even cleave the helix at ratios of platinum to DNA base above 0.05 (Horacek and Drobnik, 1971; Munchausen and Rahn, 1975; Mansy et al., 1978;
Ushay et al., 1981). Three separate cisplatin data sets were collected as
listed in Table 8 and refined independently to find the extent of substitution at each site. Cisplatin distributed itself unevenly at the eight guanine
N7 positions within the major groove (Figure 37), with strongest occu-

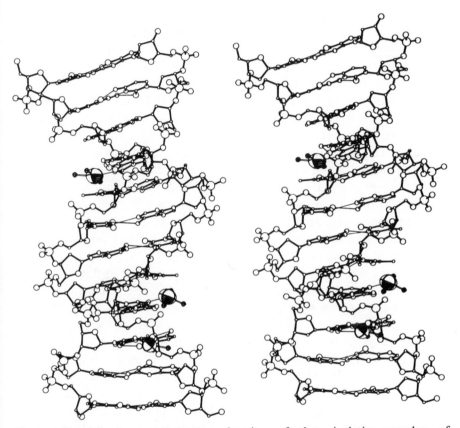

Figure 38. Minor groove stereo drawing of the cisplatin complex of
CGCGAATTCGCG, in the same style as Figure 2. Platinum complexes are shown
bound at guanines 4, 10, and 16. Cisplatin probably exchanges Cl^- for H_2O to
become the diaquo complex in solution, then binding to DNA by replacing one
H_2O by the N7 of guanine. Platinum atoms (large crossed spheres) are located
accurately, and are not necessarily in the plane of the base. The three Pt ligands
have only been added provisionally from density distributions in the electron
density map.

pancy always at G16 on strand 2, secondary occupancy at the symmetrically related G4 on strand 1, lesser occupancy at G10, and a very weak peak that could be an Mg^{2+} ion on G22. The four sites on the outermost two base pairs at each end of the helix (marked by $\times$ in Figure 38) were never occupied by cisplatin.

The most highly substituted derivative, Pt3, had a total Pt:DNA molar ratio of only 1:1 summing over all four sites, and this was achieved at the price of destroying useful diffraction information beyond a resolution of 2.6 Å. Higher resolution could be obtained only by shorter soaking times or lower ciplatin concentrations, which also meant lighter substitution. The most highly substituted data set, Pt3, has been given the most careful refinement, and is the set that is meant when "cisplatin" is mentioned elsewhere in this chapter. The refined Pt3 structure is shown in Figure 38. The overall axial bend of 17° is intermediate between those of the Native helix and MPD20 structures.

The platinum positions are probably the most accurately known features of the structure, but the combination of low substitution and low resolution means that the ligands of the platinum atoms are less well defined. Three ligands were placed around each Pt atom in a roughly square planar array, as suggested by the electron density maps, and then were allowed to refine without bond constraints. They remained in approximately square planar geometry but with erratic bond lengths that simply represent the limits of the power of refinement at this resolution. Platinum positions are secure; ligand positions are intended only to be suggestive. What does result from this is the information that Pt is too far away from the guanine O6 atom for a direct metal–oxygen ligand bond, as has occasionally been suggested for cisplatin binding to guanine in DNA. Conversely, the Pt–O distance is too short to allow a bridging ligand if all atoms lie in the plane of the base. The answer, suggested by single-crystal analyses of guanine and inosine complexes as well as this dodecamer study, seems to be that the ligand plane is rotated relative to that of the base, creating an out-of-plane bridge between Pt and O as shown in Figure 37. If the diaquo complex, with chlorines replaced by water molecules, is the reactive species in solution, then ligands L1 and L3 in Figure 37 probably are the original NH_3, and L2 probably is H_2O.

The pattern of occupancy of potential Pt sites in Table 8 exhibits two trends: decreasing affinity for Pt as one moves from the center of the double helix toward the two ends, and a superimposed preference for the lower half of the helix over the upper. Occupancy ratios between sites on the fourth and third base pairs in from each end of the helix are roughly constant in Pt3: G16/G10 = 2.6 and G4/G22 = 3.0. Similarly, ratios between corresponding sites on the bottom and top halves of the helix also are approximately constant: G16/G4 = 2.3 and G10/G22 = 2.7. (The G22 peak, not shown on the stereo in Figure 38, is so weak that it could be an Mg^{2+} ion rather than Pt.)

A clue to the preference of cisplatin for binding to sites nearer the center of the double helix may be provided by the way in which the apparent Pt–N7 bond length at the most highly occupied G16 site varies with percent occupancy (Figure 39). The Pt–N bond in Pt1 is 0.5 Å too long, but it decreases as occupancy increases in Pt2 and Pt3, in a manner that would extrapolate to the expected 2.0 Å at 100% occupancy. The image observed in an X-ray electron density map is an average over all the molecules in the crystal. Each individual molecule, of course, either has Pt bound to it at G16 or it does not. But the composite image will represent a weighted average of the two extremes, with the weighting factor being the percentage of occupancy. Hence if binding of Pt requires that the guanine be pulled slightly out of the base stack into the major groove, then the Pt position in fully substituted molecules will be more than 2.0 Å away from the guanine N7 position in the image of the original, unsubstituted Native structure. The greater the degree of substitution in the crystal, the more the averaged image of the guanine ring will be perceived as shifted from its Native position toward that which it would occupy if substitution could be made complete. Hence the expected behavior of the Pt–N7 bond distance will be as shown in Figure 39.

Direct evidence for the above explanation is provided by a comparison of the Pt3 cisplatin structure and the Native structure after optimally superimposing one molecule on the other by least-squares (Fig. 40). The two base pairs to which cisplatin has complexed have both moved slightly into the major groove, whereas the uncomplexed base pair nearest the viewer has moved in a different direction, rotating slightly and translating along its long axis perpendicular to the major groove. It appears that ligation of cisplatin is accompanied by a shift of the guanine ring out of the base pair stack toward the major groove, perhaps to make room for the bulk of the square planar complex.

Why should this requirement for a shift in guanine position favor ligation to base pairs nearer the middle of the double helix? As Figures 4

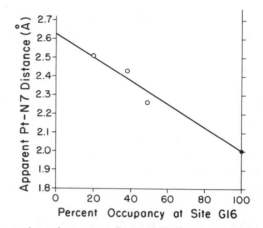

Figure 39. Shortening of apparent Pt–to–N7 distance at the percentage of occupancy of the G16 site increases. At an extrapolated 100% binding, the bond length shrinks to its expected 2.0 Å. Longer distances for partial occupancy occur because the electron density map is an average over an unshifted guanine image for no substitution and a shifted image for those molecules at which Pt binding occurs.

and 5 indicate, the first two base pairs in from each end of the helix are anchored because of hydrogen bonded overlap of their minor grooves with neighboring molecules in the crystal. Shifting these guanines into the major groove means pulling against the restraining hydrogen bonds. This may be just enough to decrease the affinity of cisplatin for these loci, resulting in preferential binding at less restricted sites: the third and, even more, fourth base pairs in from each end of the helix.

Any explanation for the preferential binding of cisplatin in the lower half of the molecule must consider why the two ends of the helix should differ at all in the crystal. They presumably are identical when the molecules are isolated in solution. Why should crystallization induce asymmetry? The most apparent asymmetry in all the dodecamer helices save MPD7 is the bend in helix axis. From Figures 2 and 38 it would appear that most of the bending takes place in the upper half of the molecule in the neighborhood of base pairs 3, 4, and 5. The one obvious distinction between this region and that of base pairs 7, 8, and 9 at the other end of the molecule is the presence of a spermine molecule, bound across the major groove from P2 to P22. All dodecamer crystals were grown in the presence of spermine, and at the conclusion of solvent refinement in the

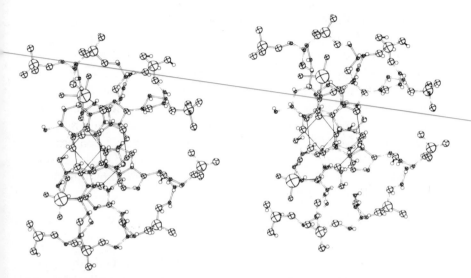

Figure 40. Stereo view down the top of base pairs G8·C17, C9·G16, and G10·C15, in descending order into the page, showing cisplatin binding to G16 and G10 and the shifts that are observed as a consequence of binding. Crossed spheres: Pt3 cisplatin complex, with shifted bases. Small open spheres: original Native dodecamer structure.

Native structure, a close-knit string of solvent peaks bridging the major groove in the upper half of the molecule was interpreted as an imperfectly ordered image of a spermine molecule (Drew and Dickerson, 1981b). Similar strings of peaks have been observed in all of the other dodecamer structures, with the notable exception of cisplatin. Fratini et al. (1982) propose that the binding of a single spermine molecule across one end of the major groove of molecules in solution might close down the major groove somewhat to produce a slightly bent molecule that by chance was particularly amenable to ordered, regular packing within a crystal. If, in contrast, binding of a second spermine across the major groove at the other end led to an even more bent molecule that now would not pack well into a lattice, then the following equilibrium would be set up:

$$\text{Helix} \underset{-\text{Sp}}{\overset{+\text{Sp}}{\rightleftharpoons}} \text{Helix·Sp} \underset{-\text{Sp}}{\overset{+\text{Sp}}{\rightleftharpoons}} \text{Helix·2Sp}$$
$$\text{Crystals}$$

All molecules in the crystal then would have a single bound spermine, at the same end of the helix. In support of this hypothesis, it was observed that a spermine : dodecamer ratio of a little less than 1 : 1 favored crystal formation, whereas an excess of spermine impeded it. The lack of a spermine image in the cisplatin structure probably represents extraction of spermine by the Pt atom, with which the spermine molecule is a strong chelator. The cisplatin crystals originally were grown in the presence of spermine. Once formed, they could be frozen in their bent conformation, and the compression of the major groove in the top half of the molecule would lower the affinity of this region for cisplatin coordination by comparison with the undistorted lower half.

7.2. Helix Bending in the Dodecamer

The cisplatin discussion suggests that CGCGAATTCGCG double helices crystallize in the bent form, with bending possibly being induced by 1 : 1 binding of spermine, and that this bent conformation, once formed, persists in the crystal even after displacement of spermine by cisplatin. What induces the axis to straighten out when the 9-Br derivative is introduced into 60% MPD at 7°C (MPD7)? The presence of the large bromine atom in the major groove is not sufficient by itself; the 9-Br molecules in 35% MPD remain bent, and they do not unbend until transferred to 60% MPD. Cisplatin in the major groove is insufficient to unbend the molecules even in 60% MPD. All of the cisplatin derivatives were prepared in the cold room by transferring crystals from the 35% MPD solution in which they were grown into cisplatin solutions that had been made 60% in MPD to ensure that the crystals would not redissolve. X-ray data were collected at room temperature because those crystals photographed in the cold room seemed to be extraordinarily sensitive to X-radiation. There may be a functional parallel between crystals with cisplatin in 60% MPD at room temperature and the MPD20 set, which is the 9-Br derivative in 60% MPD at room temperature. Both helices have axial bends, 17° and 14° respectively, that are slightly lower than that of the Native structure at room temperature (18°), and substantially less than that of the same molecule at 16° K (22°). Had it been possible to collect a cisplatin data set at 60% in the cold, it might have displayed an unbent axis by analogy with MPD7. The three elements that appear to be required simultaneously for a straightening of the helix axis are

1. Bulky additions within the major groove.
2. High MPD concentration, possibly as a means of lowering water activity.
3. Cooling below room temperature.

The unbent form, because it is produced within the crystal lattice after crystallization, may represent a metastable state induced by crowding within the major groove; this is reversed by thermal energy when the crystals are warmed to room temperature.

Although we cannot say with certainty why unbending occurs, a comparison of the five refined structures reveals quite clearly how that bending takes place. The stereo skeletal drawings in Figures 2 and 38 give the strong impression that bending occurs by compression of the major groove near base pairs 3–5 in the upper half of the molecule. Figure 18, showing the change in roll angle from one helix to the next, reinforces the idea that the top half sees the greater change, but also indicates that compression of the major groove occurs at step 8 as well. One disadvantage of roll angles is that they are measured relative to a coordinate frame that rotates with the helix. A more objective display of base plane orientation relative to an external, fixed reference frame is provided by Figure 41. These diagrams are obtained by plotting the normal vectors to the best least-squares planes through all base pairs of a helix from a common origin, extending these normal vectors until they intersect a sphere about the origin, and examining the intersection points (numbered 1–12) in projection down the helix axis. The distance from the origin to a point, gives the sine of the angle between the normal vector and the helix axis. Distances between normal vector points similarly are an approximate measure of the angles between the two normal vectors. The horizontal in these diagrams is approximately parallel to the long axes of base pairs C1·G24 and C11·G14, one turn of helix apart.

The most striking feature of these base normal plots is the steady drift of points from left to right in the 16K, Native, and cisplatin helices, the diminished drift in MPD20, and the clustering of MPD7 points in a tightly restricted area (with the exception of base pairs 11 and 12, whose abnormal orientations are clearly visible in Figure 2c). This is a direct measure of the bending in overall helix axis: large in the first three structures, smaller in MPD20, and essentially zero in MPD7. The distance between points 1 and 11 on each diagram is an indication of the axial bending that

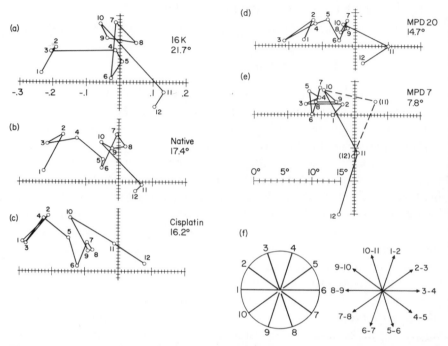

Figure 41. Orientation of base pair normal vectors, viewed directly down the helix axis, *z*. (*a*) 16K, (*b*) Native, (*c*) cisplatin, (*d*) MPD20, (*e*) MPD7, (*f*) helix base pair locator diagram. For each base pair in (*a*) through (*e*), the *x* and *y* direction cosines of the normal to the best mean plane through all atoms of the pair are plotted along *x* and *y* axes. Each point, numbered 1–12, therefore represents the tip of one base pair normal vector, with all 12 vectors assumed to be brought to a common origin. Distances between points in this projection down the helix axis can be converted approximately into angles between normal vectors by means of the scale at the bottom. The steady progression of points from left to right in (*a*) through (*c*) and to a lesser extent in (*d*) reflects the overall bend in helix axis. Angles at right are total angles between normal vectors of base pairs 1 and 11, one turn of helix apart. Solid lines leading to points 11 and 12 in (*e*) MPD7 are for the normals to both bases of the pair; dashed lines are for G14 in base pair 11 and G12 in base pair 12, eliminating the abnormal orientations of their hydrogen bonded partners that are visible in Figure 2*c*. The helix base pair locator diagrams in (*f*) show, at left, the orientation of the long axes of base pairs relative to parts (*a*) through (*e*), and at right, the direction of vectors representing positive roll angles θ_R between base pairs, opening into the minor groove.

is present in one full turn of helix. These distances are given at the right of each diagram.

At first glance the steps between points 1 and 11 on each diagram have the appearance of a random walk, but comparison with the two helix diagrams in Figure 41f shows that these steps are anything but random. The left diagram gives approximate orientations of long axes of base pairs relative to the normal vector diagrams in Figure 41a–e. The right diagram shows the direction of roll into the minor groove for each base step. For example, the vector marked 1–2 is perpendicular to a line that bisects the base pair axis lines for base pairs 1 and 2 in the drawing at left. This would be the orientation of the line between points 1 and 2 if the change of orientation of base pairs C1·G24 and G2·C23 involved only an opening of roll angle θ_R toward the minor groove, with no tilt component involved. This opening of θ_R also means that the major groove is compressed between base pairs 1 and 2.

If one now examines all 50 steps between base pairs 1 to 11 in the five structures, one finds that 41 of these steps involve a nearly pure rolling motion, with little tilt component, and that another six steps are borderline cases (Table 9). Only in three instances, all involving step 7-8 (base pair 7 to 8), does the change in orientation involve tilt rather than roll, and these

Table 9. Behavior of Base Pair Normal Vectors at Successive Helix Steps[a]

Step	Bases	16K	Native	Cisplatin	MPD20	MPD7
1	1–2	+	+	+	+	+
2	2–3	−	−	−	−	−
3	3–4	+	+	+/0	+	+
4	4–5	+	+	+	+	−
5	5–6	+/0	+/0	+	+	+
6	6–7	−	−	−	−	−
7	7–8	0	0	0	+	+/0
8	8–9	+	+	+	−	−
9	9–10	+/0	+	+/0	+	+
10	10–11	−	−	−	−	−

[a] Data are from Figure 41. + = change in orientation from one base pair to the next, that can be described as base roll with the angle opening into the minor groove, or a compression of the major groove. − = base roll opening into the major groove, and compression of the minor groove. 0 = change in base plane orientation that is best described as tilt rather than roll.

are less than 5°. Everywhere else in all five structures, one principle seems to be ironclad: *Bending in the double helix is produced by local roll of base pairs about their long axes.* Large roll angles occur at positions where they would contribute to the overall bend (as at step 3-4 in the 16K structure) and small roll angles occur two or three steps later along the helix where they would produce sideways bending (as at step 5-6 in the 16 K and Native structures). This combination of large steps in the direction of bend and small steps in the perpendicular direction results in a net overall bending. The process visible in Figure 41 can be described not as a random walk but as a random roll, with both perpendicular and retrograde steps but with a net direction of motion.

One further, less stringent tendency should be noted. The broad, open major groove should be inherently easier to compress than the narrow minor groove, filled as it is with the spine of hydration. Indeed, 23 of the 41 roll angle changes involve compression of the major groove ("+" in Table 9), and only 18 compress the minor groove ("−" in Table 9). The three steps that contribute most to the overall helix bend in the 16 K, Native, and cisplatin structures are steps 3-4 and 4-5 in the top half of the molecule, compressing the major groove, and step 10-11 at the bottom of the molecule, compressing the minor groove. The latter is always preceded by two steps, 8-9 and 9-10, which, although involving the easy process of compressing the major groove, are retrograde steps as far as overall helix axis bending is concerned.

The process of straightening out the helix, from Figures 41a through 41e, is one of confining these "random roll" steps to a smaller and smaller perimeter but preserving their individual character: a forward step at 1-2, a retrograde step at 2-3 that is dictated by cross-chain purine clash (see Section 6), forward steps at 3-4 and 4-5, a perpendicular "noise" step at 6-7, a small forward step at 7-8, two retrograde steps at 8-9 and 9-10, and one final forward step at 10-11. The final result in MPD7 is a set of base pair normal vectors that is confined to lie within a circle of radius 3° except for the last two base pairs with their anomalously turned cytosines.

Because all of these motions involve roll angle almost exclusively, with little or no base tilt component, the roll angle plot in Figure 18 actually contains nearly the whole story. Figures 18 and 41 are complementary; the former shows particularly clearly whether compression is occurring in the major (positive θ_R) or minor (negative θ_R) groove, and the latter is better in revealing whether the local orientation change is in a forward or retrograde direction with respect to the overall helix bend. Figure 18 is especially clear in showing how the introduction of bromines at the 5 position of cytosines 9 and 21 assists the straightening of the helix. From Figures 2c and 3b one can see that the bromines in each case are in close contact with the next base pair in toward the middle of the helix, or that steric hindrance will occur at steps 4 and 8. These are the positions that show a decrease in θ_R, or an opening up of the major groove, in the MPD7 structure.

In summary, bending in double helical B-DNA apparently occurs only by rolling adjacent base pairs over one another along their long axes; the lifting apart of ends that would be required by tilt angle contributions is too costly in free energy and does not occur. Roll angles at base steps can be positive (compression of major groove) or negative (compression of minor groove), with the former somewhat preferable. Individual steps may advance or oppose the overall direction of bend or may make lateral excursions, but the result of this "random roll" series of steps is the production of a net bending in the helix axis. The "annealed kinking" model proposed by Fratini et al. (1982) was suggested from the observation that a major bend at a natural roll point is flanked by decreasing roll angles at the steps to either side. The random walk model suggested by Figure 41 would describe this as a decreased roll angle as the helix step rotates toward a direction perpendicular to the overall bend.

8. EXTENDED HELICES: A, B, AND Z

Crystal structure analyses of short double helical DNA oligomers over the past five years have greatly enriched our understanding of DNA structure by showing us the way in which local helix structure deviates from the average structure and how this deviation can be attributed to base sequence. It is difficult, however, to obtain from an isolated oligomer an impression of what an infinite helix based on the same structural principles would look like. Hence we have attempted to simulate infinite helices while retaining the individuality of the oligomer molecules, by extending the longest DNA oligomer from each family in a systematic manner. The results are displayed in Figures 42–44. To produce these drawings, refined atomic positions in each helix were converted to cylindrical coordinates, the final base pair at each end of the helix was deleted to eliminate possible

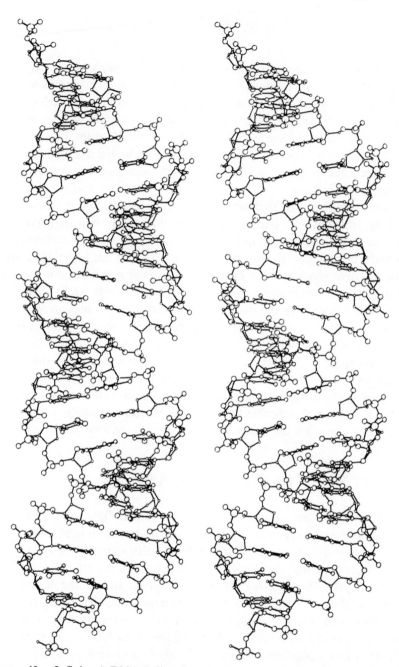

Figure 42. Infinite A-DNA helix of sequence . . . (GTATAC)$_n$ . . . , generated by repetition of the central six base pairs of the octamer GGTATACC. Coordinates courtesy of Z. Shakked and O. Kennard.

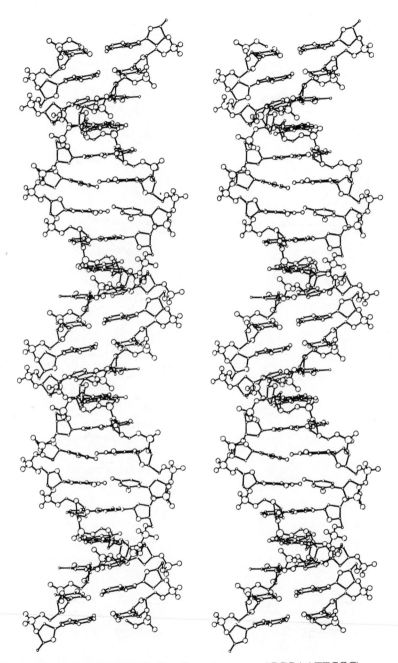

Figure 43. Infinite B-DNA helix of sequence . . . (GCGAATTCGC)$_n$ . . . , generated by repetition of the central 10 base pairs of the dodecamer CGCGAATT^{Br}CGCG.

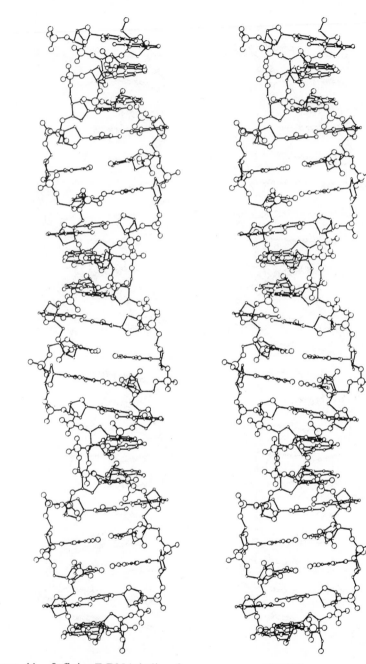

Figure 44. Infinite Z-DNA helix of sequence . . . (GCGC)$_n$ . . . , generated by repetition of the central four base pairs of the hexamer CGCGCG. Coordinates courtesy of A.H.-J. Wang and A. Rich.

end effect distortions, and the truncated helix then was rotated and translated up its axis until atoms at the bottom of the shifted image overlapped with equivalent atoms at the top of the original image. Positions of these overlappping atoms were averaged, and extraneous atoms above and below were deleted. This process was continued for as many repetitions as required to produce a continuous helix of 20–24 base pairs.

These stereo drawings combine the best features of fiber and single crystal analysis: infinite helices, but with true local structure variation. The overall features of each helix type are readily seen: the broad but short (for a constant number of base pairs) A helix, with its deep major groove lined with phosphates and its very shallow minor groove winding around the outside, the slimmer but taller B helix with a wide major groove and narrow minor groove of similar depth, and the thin and elongated Z-helix, with its deep minor groove lined by phosphates and its very shallow major groove. Z-DNA also is distinguished by its left-handed helix sense and by the fact that the helical repeat unit in the idealized case is two successive C and G steps rather than a single base pair as in the idealized A and B helices.

The A helix shows the tilting of bases ("inclination" in our nomenclature) that is characteristic of A-DNA, whereas base pairs in B and Z are nearly perpendicular to the helix axis. Individual base pairs in the two right-handed helices, A and B, exhibit a positive propeller twist that improves base stacking along each individual strand of the helix, whereas the propeller twist in Z-DNA is small. Base stacking in this latter helical form is discontinuous: the two cytosines on opposite strands at a CpG step are stacked atop one another (see Figure 44, and figure 5 of Drew and Dickerson, 1981a), these are sandwiched between two guanines from adjacent base pairs, and the outer two guanines are stacked against the O4′ atoms of deoxyribose rings. Hence a continuous Z helix as in Figure 44 is built up from short parallel stacks of sequence: O4′-G-C-C-G-O4′. Altering base pair propeller twist would do nothing to improve these contacts. The B helix also exhibits another sequence-based structural feature, which is particularly striking in Figure 43 because the 10 base pair infinite repeat, AATTCGCGCG, means that the helix always keeps the same sequence toward the viewer; the minor groove is much narrower in AT regions than in GC regions. The reason for this, as discussed in Section 5.2, is also visible in Figure 43: AT regions have noticeably greater propeller twist than GC regions.

There now appear to be three distinct families of double helical DNA: A, B, and Z. After the surprise of the Z helix, it would be a rash investigator indeed who asserts flatly that these three families are the end of the story. Yet it seems likely that the most important task for DNA structure analysis in the immediate future will be to learn more about sequence–structure relationships, and especially the way in which particular DNA sequences are recognized and bound by protein molecules such as repressors. The components have been studied separately; now one needs to examine their complexes.

Acknowledgments. We thank Chun Yoon for preparing the extended helix stereo drawings and Kathy Campbell of Sunrise Graphics for preparing the line drawings. This work was supported by National Science Foundation Grant No. PCM82-02775, National Institutes of Health Grant No. GM-30543, and a grant from The Upjohn Company.

REFERENCES

Arnott, S., and Hukins, D. W. (1972). *Biochem. Biophys. Res. Commun.* **47,** 1504–1509.

Bartenev, V. N., Golovanov, E. I., Kapitonova, K. A., Mokulskii, M. A., Volkova, L. I., and Skuratovskii, I. Yy. (1983). *J. Mol. Biol.* **169,** 217–234.

Calladine, C. R. (1982). *J. Mol. Biol.* **161,** 343–352.

Clementi, E., and Corongiu, G. (1981). In *Biomolecular Stereodynamics,* R. H. Sarma, Ed., Vol. 1, Adenine Press, New York, pp. 209–259.

Conner, B. N. (1982). "Iodo-CCGG: A Single Crystal Structure of A-DNA," Ph.D. Thesis, California Institute of Technology, University Microfilms, Ann Arbor, Michigan.

Conner, B. N., Takano, T., Tanaka, S., Itakura, I., and Dickerson, R. E. (1982). *Nature* **295,** 294–299.

Conner, B. N., Yoon, C., Dickerson, J. L. and Dickerson, R. E. (1984). *J. Mol. Biol.* **174,** 663–695.

Cramer, R. E., Dahlstrom, P. L., Seu, M. J. T., Norton, T., and Kashiwagi, M. (1980). *Inorg. Chem.* **19,** 148–154.

Dickerson, R. E. (1983a). *J. Mol. Biol.* **166,** 419–441.

Dickerson, R. E. (1983b). *Sci. Amer.* (December), 94–111.

Dickerson, R. E., Conner, B. N., Kopka, M. L., and Drew, H. R. (1982). In *Nucleic Acid Research: Future Developments,* I. Watanabe, Ed. Academic Press, Tokyo.

Dickerson, R. E., and Drew, H. R. (1981a). *J. Mol. Biol.* **149,** 761–786.

Dickerson, R. E., and Drew, H. R. (1981b). *Proc. Natl. Acad. Sci. USA* **78,** 7318–7322.

Dickerson, R. E., Drew, H. R., and Conner, B. N. (1981). In *Biomolecular Stereodynamics,* R. H. Sarma, Ed., Vol. 1. Adenine, New York, pp. 1–34.

Dickerson, R. E., Drew, H. R., Conner, B. N., Kopka, M. L., and Pjura, P. E. (1983). *Cold Spring Harbor Symp. Quant. Biol.* **47,** 23–34.

Dickerson, R. E., Kopka, M. L., and Drew, H. R. (1983). In *Structure and Dynamics: Nucleic Acids and Proteins,* E. Clementi and R. H. Sarma, Eds. Adenine, New York.

Drew, H. R., and Dickerson, R. E. (1981a). *J. Mol. Biol.* **152,** 723–736.

Drew, H. R., and Dickerson, R. E. (1981b). *J. Mol. Biol.* **151,** 535–556.

Drew, H. R., Dickerson, R. E., and Itakura, K. (1978). *J. Mol. Biol.* **125,** 535–543.

Drew, H. R., Takano, T., Tanaka, S., Itakura, K., and Dickerson, R. E. (1980). *Nature (London)* **286,** 567–573.

Drew, H. R., Samson, S., and Dickerson, R. E. (1982). *Proc. Natl. Acad. Sci. USA* **79,** 4040–4044.

Drew, H. R., Wing, R. M., Takano, T., Broka, C., Tanaka, S., Itakura, K., and Dickerson, R. E. (1981). *Proc. Natl. Acad. Sci. USA* **78,** 2179–2183.

Falk, M., Hartman, K. A., and Lord, R. C. (1962). *J. Am. Chem. Soc.* **84,** 3843–3846.

Falk, M., Hartman, K. A., and Lord, R. C. (1963a). *J. Am. Chem. Soc.* **85,** 387–391.

Falk, M., Hartman, K. A., and Lord, R. C. (1963b). *J. Am. Chem. Soc.* **85,** 391–394.

Fratini, A. V., Kopka, M. L., Drew, H. R., and Dickerson, R. E. (1982). *J. Biol. Chem.* **257,** 14,686–14,707.

Fujii, S., Wang, A. H.-J., van Boom, J. H., and Rich, A. (1982). *Nucl. Acids Res. Symp. Ser.* **11,** 104–112.

Gellert, R. W., and Bau, R. (1975). *J. Am. Chem. Soc.* **97,** 7379–7380.

Goodgame, D. M. L., Jeeves, I., Phillips, F. L., and Skapski, A. C. (1975). *Biochim. Biophys. Acta* **378,** 153–157.

Gupta, G., Bansal, M., and Sasisekharan, V. (1980). *Proc. Natl. Acad. Sci. USA* **77,** 6468–6490.

Horacek, P., and Drobnik, J. (1971). *Biochim. Biophys. Acta* **254,** 341–347.

Kopka, M. O., Fratini, A. V., Drew, H. R., and Dickerson, R. E. (1983). *J. Mol. Biol.* **163,** 129–146.

Langridge, R., Wilson, H. R., Hooper, C. W., Wilkins, M. H. F., and Hamilton, L. D. (1960). *J. Mol. Biol.* **2,** 19–37.

Leslie, A. G. W., Arnott, S., Chandrasekaran, R., and Ratliff, R. L. (1980). *J. Mol. Biol.* **143,** 49–72.

Mansy, S., Chu, G. Y. H., Duncan, R. E., and Tobias, R. S. (1978). *J. Am. Chem. Soc.* **100,** 607–616.

Munchausen, L. L., and Rahn, R. O. (1975). *Biochem. Biophys. Acta* **414,** 242–255.

Prestayako, A. W., Crooke, S. T., and Carter, S. K., Eds. (1980). *Cisplatin: Current Status and New Developments.* Academic, New York.

Quigley, G. J., Wang, A. H.-J., Ughetto, G., van der Marel, G., van Boom, J. H., and Rich, A. (1980). *Proc. Natl. Acad. Sci. USA* **77,** 7204–7208.

Shakked, Z., Rabinovich, D., Cruse, W. B. T., Egert, E., Kennard, O., Sala, G., Salisbury, S. A., and Viswamitra, M. A. (1981). *Proc. R. Soc. Ser. B* **213,** 479–487.

Shakked, Z., Rabinovich, D., Kennard, O., Cruse, W. B. T., Salisbury, S. A., and Viswamitra, M. A. (1983). *J. Mol. Biol.,* **166,** 183–201.

Skuratovskii, I. Yy., Volkova, L. I., Kapitonova, K. A., and Bartenev, V. N. (1979). *J. Mol. Biol.* **134,** 369–374.

Ushay, H. M., Tullius, T. D., and Lippard, S. J. (1981). *Biochemistry* **20,** 3744–3748.

Wang, A. H.-J., Fujii, S., van Boom, J. H., and Rich, A. (1982a). *Proc. Natl. Acad. Sci. USA* **79,** 3968–3972.

Wang, A. H.-J., Fujii, S., van Boom, J. H., van der Marel, G. A., van Boeckle, S. A. A., and Rich, A. (1982b). *Nature* **299,** 601–604.

Wang, A. H.-J., Quigley, G. J., Kolpak, F. J., Crawford, J. L., van Boom, J. H., van der Marel, G., and Rich, A. (1979). *Nature* **282,** 680–686.

Wing, R. M., Drew, H. R., Takano, T., Broka, C., Tanaka, S., Itakura, K., and Dickerson, R. E. (1980). *Nature* **287,** 755–758.

Wing, R. M., Pjura, P., Drew, H. R., and Dickerson, R. E. (1984) EMBO J. Vol. 3, pp. 1201–1206.

Wolf, B., and Hanlon, S. (1975). *Biochemistry* **14,** 1661–1670.

3

The Structure of the Z
Form of DNA

ANDREW H.-J. WANG
ALEXANDER RICH
Massachusetts Institute of Technology
Cambridge, Massachusetts

CONTENTS

1. INTRODUCTION

A helix has handedness; it can twist. The DNA double helix, a chiral assembly, can exist in both right-handed and left-handed conformations. This idea was considered shortly after the formulation of the double helix by Watson and Crick in 1953. In this early period attention was focused on the question whether the X-ray diffraction patterns of DNA fibers were consistent with the initial proposal. During the decade following, investigators produced ample experimental evidence that, with some modification of details, the Watson–Crick right-handed helical model was essentially consistent with the experimental X-ray data (Arnott, 1970). Although DNA fiber analysis provided physical data in agreement with the Watson–Crick structure, it was understood by workers in the field that it did not unambiguously prove the structure. The limited resolution of the fiber X-ray diffraction patterns simply required too many assumptions to interpret the data. One constraint in particular was that the structure was perfectly regular with an asymmetric unit consisting of one nucleotide. Agreement between model and fiber diffraction data was reasonable, however, and this eventually led to the general adoption of the right-handed B-DNA conformation as the predominant form in biological systems.

Substantial interest was inspired, however, by observations that under certain conditions, alterations in conformation appeared. The most important of these was the A-DNA pattern, originally derived from air-dried fibers (Franklin and Gosling, 1951). The greater wealth of diffraction data contained in the A pattern carried with it more extensive structural information. The relative ease with which the A form converted to the B pattern reinforced the idea that these were two conformational variants, and it was shown that both structures could still be accommodated by a right-handed double helix.

Although the possibility of left-handed helices was considered, particularly in the period immediately following the description of the double helix in 1953, few systematic studies of left-handed helices were conducted during the next two decades. It might have been anticipated that left-handed double helical DNA regions would facilitate fork movement and chain separation during DNA replication, but flexibility as a mechanism for generating structural recognition signals along the DNA molecule or for regulating superhelical topology was not discussed seriously. It is only rather recently that there has been increased emphasis on alternative conformational models for DNA (Hopkins, 1981; Sasiskeharan, 1982). Part of the stimulus arose from the postulate that DNA might not exist as a continuously coiled double helix but that it might adopt a side-by-side conformation (Rodley, et al., 1976; Gupta, et al., 1980). Confronted with these new challenges, it became apparent that fiber diffraction, with its limited information content, could no longer resolve questions regarding the detailed conformations that DNA is capable of assuming.

2. OLIGONUCLEOTIDES AND SINGLE CRYSTALS

Chemical methods developed over the past few years for synthesizing DNA make it possible to produce oligonucleotides in sufficient quantity for crystallization experiments. Unlike fibers, single crystals can diffract to high resolution, and they provide large numbers of reflections or experimental measurements. This makes it possible to solve structures and observe fine details of DNA conformation in which very little or no interpretation is required. In an atomic resolution (<1.0 Å) electron density map, every atom is seen directly and no assumptions are required to obtain bond angles, distances, ring pucker, and other features. Studies of single crystals bear the promise of resolving issues regarding the fine details of DNA conformation.

Single crystal X-ray studies of nucleic acid bases, nucleosides, and monomeric nucleotides have been carried out for many years. Single crystal diffraction studies have been conducted in many laboratories that show the purine-pyrimidine base pairs to have many types of hydrogen bonds (Voet and Rich, 1970). The first visualization of the double helix at atomic resolution, however, occurred only 10 years ago with the solution of the structures of rApU (Rosenberg et al., 1973; Seeman et al., 1976) and rGpC (Day et al., 1973; Rosenberg et al., 1976), which were found to form right-handed double helical fragments of the type that had been anticipated for RNA molecules. This work on ribonucleotide fragments laid the groundwork for the subsequent and more extended analysis of deoxyoligonucleotides.

In our research, we have collaborated with J. H. van Boom and his colleagues at Leiden University who developed many of the methods for synthesizing oligonucleotides. Several years ago, van Boom and his colleagues synthesized all four of the self-complementary deoxynucleotide tetramers containing only guanosine and cytosine residues. It seemed likely that these sequences might produce stable oligonucleotide double-helical fragments that could then be crystallized. Three of the four were crystallized and one of them, d(CpGpCpG), crystallized with apparent ease. This suggested that it might be profitable to explore the hexamer d(CpGpCpGpCpG) seen in Figure 1. It was discovered that crystals could be formed readily using magnesium and spermine as cations (Wang et al., 1979). These crystals diffracted to 0.9 Å resolution. It was clear that solution of this crystal structure would not only make it possible to visualize the details of the double helix but would also yield information about hydration and ions.

The hexamer crystal structure was solved by the conventional multiple isomorphous replacement method. Various crystalline heavy metal cation derivatives were obtained to solve the structure. The asymmetric unit of

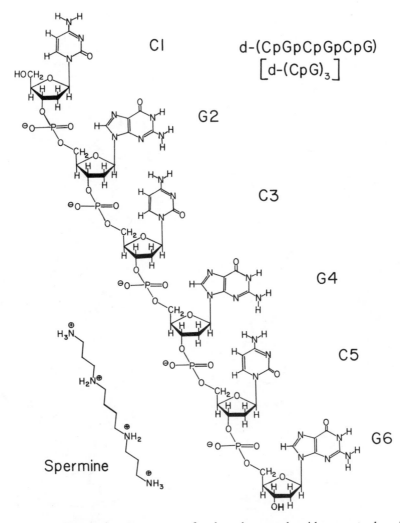

CI

G2

C3

G4

C5

G6

d-(CpGpCpGpCpG)
[d-(CpG)₃]

Spermine

Figure 1. Chemical structure of the hexanucleoside pentaphosphate d(CpGpCpGpCpG) and spermine. Note that only the amino group hydrogen atoms of spermine are shown.

the crystal had one double-helical fragment of DNA containing six base pairs, two spermine molecules, a hydrated magnesium ion, and 62 water molecules. The asymmetric unit had a molecular weight of almost 5,000 daltons and thus was as large as some of the smaller proteins.

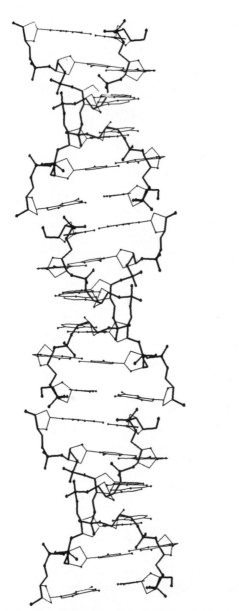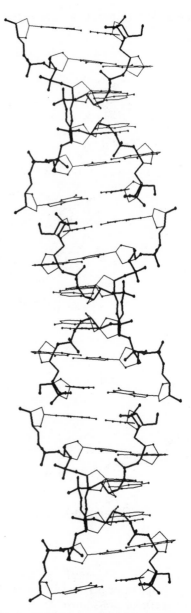

Figure 2. Stereo diagram of three molecules in the Z-DNA crystal lattice as they are stacked along the *c* axis in what looks like a continuous double helix. However, a phosphate group is absent every sixth residue along the chain. Successive base pairs along the helix have the deoxyribose rings oriented so that O1' is alternately pointing up or down and the same orientation is found in both sugar residues across the base pair. This indicates that the helix asymmetric unit is two nucleotides rather than one as in B-DNA.

3.   THE DOUBLE HELIX TURNS LEFT-HANDED

The structure was quite unusual (Figure 2). It was a left-handed double helix. However, it was not a simple left-handed helix but one in which the alternating purine-pyrimidine sequence of the oligonucleotide was used in a unique way. A van der Waals drawing is shown in Figure 3 alongside a drawing of right-handed B-DNA. A comparison is listed in

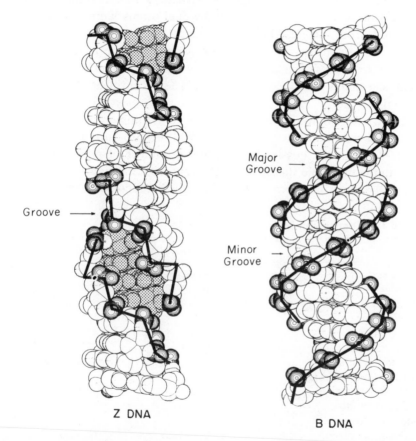

Figure 3.   Van der Waals side views of Z-DNA and B-DNA. The irregularity of the Z-DNA backbone is illustrated by the heavy lines, which go from phosphate to phosphate residues along the chain. This includes positions where the phosphate residues are missing in the crystal structure but would be occupied in a continuous double helix. The shaded groove in Z-DNA is quite deep. In contrast, B-DNA has a smooth line connecting the phosphate groups and it has two grooves.

Table 1. The left-handed helix has the phosphates arranged in a zigzag array; hence the name Z-DNA. The Z-DNA helix has 12 base pairs per turn of the helix, which adopts a length of 44.6 Å. This is in contrast to the 10 base pairs per turn with a pitch of 34 Å in right-handed B-DNA. The same Z structure has been found in crystals of tetramers d(CpGpCpG) (Drew et al., 1980; Crawford et al., 1980) as well as in other hexamer crystals (Wang et al., 1981) and in fibers (Arnott, et al., 1980).

We could be confident that the structure was correct because it was solved at atomic resolution. Individual atoms were seen distinctly separated from neighboring atoms and their assignment into structure was straightforward. This is not necessarily the case for medium or lower resolution single crystal analyses. For example, at 3 Å resolution one might have been considerably less confident about the form of the molecule.

The structure of left-handed Z-DNA differs in numerous and significant ways from right-handed B-DNA. Why had the structure not been anticipated? The answer lies in the fact that a number of profound changes occur in the molecule that make it different from the right-handed double helix. One does not form Z-DNA simply by reconstructing the right-

Table 1. Comparison of B-DNA and Z-DNA

	B-DNA	Z-DNA[a]
Helix sense	Right-handed	Left-handed
Residues/turn	10	12 (6 dimers)
Diameter	~20 Å	~18 Å
Rise/residue	3.4 Å	3.7 Å
Helix pitch	34 Å	45 Å
Base pair tilt	6°	7°
Rotation/residue	36°	−60° (per dimer)
Glycosidic torsion angle		
Deoxyguanosine	*anti*	*syn*
Deoxycytidine	*anti*	*anti*
Sugar pucker		
Deoxyguanosine	C2'-*endo*	C3'-*endo*
Deoxycytidine	C2'-*endo*	C2'-*endo*
Distance of P from axis		
dGpC	9.0 Å	8.0 Å
dCpG	9.0 Å	6.9 Å

[a] Average values are given and end effects are excluded.

handed double helix in a left-handed fashion. A number of other conformational alterations are also required.

In the Z-DNA crystal, the backbone has guanosine residues in an unusual conformation. This is illustrated in Figure 4, which shows the disposition of deoxyguanosine as seen in Z-DNA and B-DNA. In Z-DNA the guanine bases adopt the *syn* conformation. The *anti* conformation is found in all the bases in B-DNA and in the cytosine residues in Z-DNA. Nuclear magnetic resonance (NMR) studies in solution have shown that purine residues can adopt the *syn* and *anti* conformations with equal ease (Son et al., 1972). An early analysis of the rotational barriers about the glycosyl bonds suggested that purines can rotate more easily than pyrimidines into the *syn* conformation (Haschemeyer and Rich, 1967).

Further differences are found in the pucker of the sugar ring. The guanosine in Z-DNA (Fig. 4) has the C3'-*endo* conformation (which is the preferred conformation for ribonucleotides), whereas in B-DNA the C2'-*endo* conformation is adopted for all nucleotides. However, the cytidine residues in Z-DNA have the C2'-*endo* conformation of the sugar. The asymmetric unit in Z-DNA is thus a dinucleotide consisting of a cytidine residue in one conformation and a guanosine residue in a different conformation. This is seen diagrammatically in Figure 5, which shows the sugar conformations in B-DNA and Z-DNA molecules. Another feature distinguishing B-DNA from Z-DNA is the orientation of the base pairs relative to the sugar-phosphate backbone. In order to convert from B-DNA to Z-DNA it is necessary for the base pairs to turn over or flip by 180° so that the relationship of the sides of the base pairs is inverted. This is shown diagrammatically in Figure 6, in which four base pairs in the center of B-DNA form Z-DNA. This occurs schematically by an inversion of the base pairs, as shown in the diagram. This "flipping over" of the base pairs is brought about by rotating the purine residues about their glycosyl bond going from *anti* to *syn*. In the case of the pyrimidine residues to which they are paired, both the sugar and the base are turned upside down. It is this inversion of the sugar which introduces a zigzag into the sugar-phosphate backbone in Z-DNA.

These changes in nucleotide conformation result in significant modifications in the stacking relation between successive base pairs along the helix. Figure 7 shows the stacking of the CpG and GpC sequences for Z-DNA and B-DNA. The guanine residues are in the *syn* position, and this results in the imidazole ring of guanine occupying a position near the

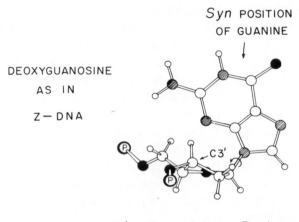

Syn POSITION
OF GUANINE

DEOXYGUANOSINE
AS IN
Z — DNA

C3' endo Sugar Pucker

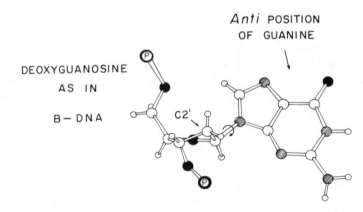

Anti POSITION
OF GUANINE

DEOXYGUANOSINE
AS IN
B — DNA

C2' endo Sugar Pucker

Figure 4. Conformation of deoxyguanosine in B-DNA and in Z-DNA. The sugar is oriented so that the plane defined by Cl′–O1′–C4′ is horizontal. Atoms lying above this plane are in the *endo* conformation. The C3′ is *endo* in Z-DNA, whereas in B-DNA the C2′ is *endo*. These two different ring puckers are associated with significant changes in the distance between the phosphorus atoms. In addition, Z-DNA has guanine in the *syn* position, in contrast to the *anti* position in B-DNA. A curved arrow around the glycosyl bond indicates the site of rotation.

136

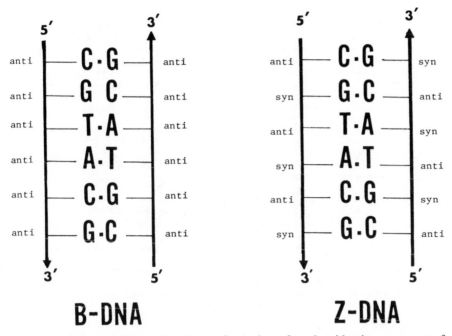

Figure 5. Diagram illustrating the conformation of nucleosides in a segment of B-DNA and Z-DNA. The purines occupying alternative positions along the chain of Z-DNA are in the *syn* conformation.

outside of the helix with the atom C8 near the periphery. This is in contrast to its position in B-DNA, where it is in van der Waals contact with the sugar-phosphate backbone. The stacking of CpG differs considerably from GpC in Z-DNA even though they are fairly similar in B-DNA. The CpG sequence is sheared so that the cytosine residues stack upon each other, but the guanine residues do not stack upon bases at all. Instead, they stack on the sugar of the next residue. This results in a stacking interaction between O1′ of the sugar and the pyrimidine-purine ring. The GpC residues, on the other hand, have the bases on each strand stacking upon the bases of the base pair below in a manner similar to that found in right-handed B-DNA.

The groove in Z-DNA is fairly deep compared with that of B-DNA. Figure 8 shows a van der Waals diagram containing three base pairs of Z-DNA. The axis of the molecule is indicated by the solid dot and the

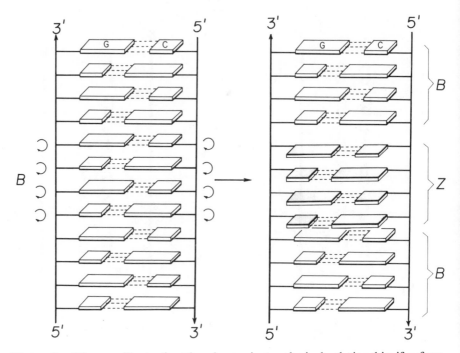

Figure 6. Diagram illustrating the change in topological relationship if a four-base-pair segment of B-DNA were converted to Z-DNA. This conversion could be accomplished by rotation of the bases relative to those in B-DNA. This rotation is shown diagrammatically by shading one surface of the bases. All of the dark shaded areas are at the bottom in B-DNA. In the segment of Z-DNA, however, four of them are turned upward. The turning is indicated by the curved arrows. Rotation of the guanine residues about the glycosyl bond produces deoxyguanosine in the *syn* conformation while for dC residues, both cytosine and deoxyribose are rotated. The altered position of the Z-DNA segment is drawn to indicate that these bases may not be stacking directly on the base pairs in the B-DNA segment.

depth of the groove can be seen quite clearly. Unlike B-DNA, the bases are displaced away from the center of the axis. The helix twists in a clockwise direction moving toward the reader. Thus, the three phosphate groups are visible on the left side while only one is seen on the right. The axis is found very close to the O2 atom of the cytosine residue. In this diagram, the imidazole ring of guanine is shown projecting on the outside of the molecule where it is readily accessible to other reagents, ions, or ligands. Figure 9 shows a stereo diagram of Z-DNA tilted by about 30°

so that the groove can be visualized as an indentation in the molecule. It can be seen to extend almost to the center.

End views of Z-DNA and B-DNA are compared in Figure 10. Here the purine residues are shaded and the backbone is drawn in slightly heavier lines. In B-DNA, the purine residues are clustered near the center of the axis while the sugar-phosphate backbone is organized on the outer perimeter of the molecule. In Z-DNA, on the other hand, the purine residues are located near the perimeter of the molecule with the imidazole groups pointing to the outside. This diagram also illustrates the fact that

Z-DNA

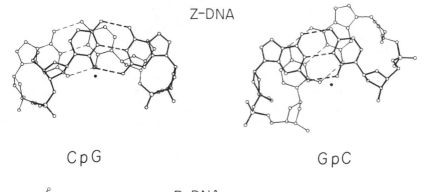

CpG   GpC

B-DNA

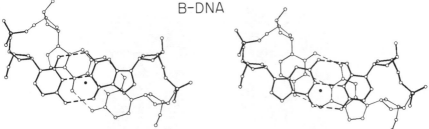

Figure 7. Stacking diagram illustrating the overlap of successive bases along Z-DNA and B-DNA chains. The base pair drawn with heavier lines is stacked above the pair drawn with lighter lines. The left-hand column represents d(CpG) sequences of both Z-DNA and B-DNA, while d(GpC) sequences are on the right. The direction of the deoxyribose–phosphate chains is the same in all these diagrams. Note that the minor groove in B-DNA is found at the top of the B-DNA diagrams while the analogous side of the base pairs is found at the bottom of the Z-DNA diagrams. The solid black dot indicates the helical axis.

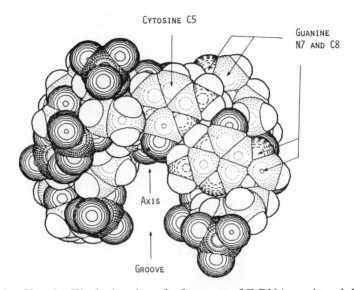

Figure 8. Van der Waals drawing of a fragment of Z-DNA as viewed down the axis of the helix. Three base pairs are shown, and the deep groove is seen to extend almost to the axis of the molecule. In these three base pairs the groove rotates clockwise, moving toward the reader. For that reason, three phosphates are visible on the left and only one on the right. The N7 and C8 atoms of guanine are near the outer wall of the molecule. The solid black dot indicates the axis of the molecule.

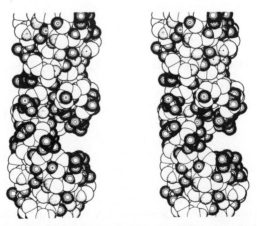

Figure 9. Van der Waals stereo drawing of Z-DNA in which the axis of the molecule is tipped away from the reader in such a way that the depth of the helical groove can be visualized. The prominent ridges of the sugar-phosphate backbone are readily visible.

140

Z-DNA has a slightly smaller cross-sectional diameter than B-DNA. The van der Waals diameter of B-DNA is close to 20 Å whereas Z-DNA has a diameter of only 18 Å.

The different forms of DNA are distinctive with regard to the position of the molecular axis around which the base pairs turn. This is shown in Figure 11, which illustrates the axes of A-DNA, B-DNA, and Z-DNA superimposed on a GC base pair. In B-DNA the axis is near the center of the base pair, and this accounts for the clustering of the purine residues, shaded in the center of the diagram shown in Figure 10. In Z-DNA, on the other hand, the axis is close to the O2 atom of the cytosine residue and it is no longer on the perpendicular bisector of the C1′–C1′ line in Figure 11. In both A-DNA and B-DNA, where the asymmetric unit is one base pair, the axis must be on this perpendicular bisectrix. In Z-DNA,

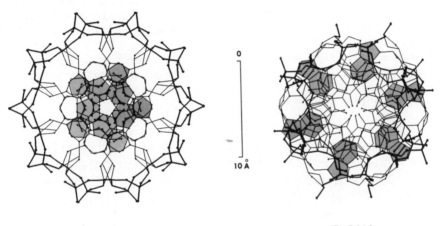

B DNA   Z DNA

Figure 10. End views of B-DNA and Z-DNA are illustrated in which the guanine residues of one strand have been shaded. The Z-DNA figure represents a view down the complete c axis of the crystal structure encompassing two molecules. The shaded guanine residues illustrate the approximate sixfold symmetry. The imidazole part of the guanine residue forms a segment of the outer cylindrical wall of the molecule together with the phosphate residues. The B-DNA figure represents one full helix turn. In contrast to Z-DNA, the guanine residues in B-DNA are located closer to the center of the molecule and the phosphates are on the outside. The B-DNA diagram is drawn from idealized regular coordinates obtained from fiber diffraction studies. The actual molecule is likely to have irregularities similar to those seen in Z-DNA.

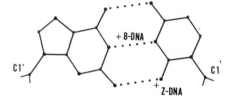

Figure 11. Diagram illustrating the positions of the axes in three major DNA conformations. The axes of A-DNA and B-DNA are found on a line perpendicular to C1'–C1' and halfway between them. This is necessary because of the pseudodyad found in both of these structures in the plane of the base pair. The axis of Z-DNA is not on that line because the asymmetric unit consists of two bases and there is no longer a pseudodyad in the plane of the pairs. The base pairs are far removed from the A-DNA axis, so that the major groove is deep and the minor groove on the external surface in the molecule is shallow. The Z-DNA axis is also somewhat removed from the center, so that its groove is deep and the external surface of Z-DNA is what corresponds to the major groove side of the base pair. In B-DNA the axis goes through the middle of the base pairs, giving rise to two grooves of slightly unequal size.

where the asymmetric unit contains two base pairs, the axis can be anywhere—it is not confined to the perpendicular bisectrix of the C1'–C1' vector. In A-DNA, the axis is well above the base pair, reflecting the fact that the double helix has a very deep major groove in A-DNA and a very shallow minor groove.

In a regular B-DNA helix there are 10 residues per turn of the helix, which spans 34 Å. The twist angle is the angle between adjacent base pairs going along the helix. For a 10-fold helix this would be 36°. The twist angle is constant in going from base pair to base pair along the helix in idealized B-DNA. The situation with Z-DNA is quite different. Figure 12 shows that, in contrast to the constant relationship between the base pairs, the twist angle between adjacent base pairs is the same for CpG as it is for GpC, that is, +36° (Wang et al., 1982). The sequence CpG has very little rotation between adjacent base pairs while the sequence GpC has a strong negative rotation. This is evident from looking at the stacking diagrams in Figure 7 and is shown quantitatively in Figure 12, which plots

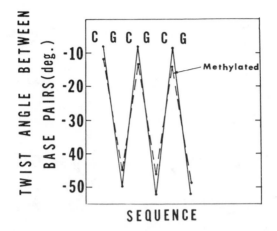

Figure 12. Twist angle between adjacent base pairs is plotted for methylated and nonmethylated Z-DNA crystal structures. The points are positioned between the bases listed in the sequence, and each point represents the twist angle between the two base pairs on either side of the point. The twist angle is the angle between adjacent lines connecting C1′ atoms of nucleoside base pairs.

the twist angle of adjacent sequences along the helix axis. The solid line shows that the twist angle is close to $-8°$ for the CpG sequences and close to $-52°$ for the GpC sequences.

4.   CONFORMATIONAL CHANGES IN Z-DNA

There is a pseudo twofold, or pseudo dyad, axis between the base pairs which relates one chain to another. These dyads are not used in the symmetry of the crystal lattice but are a consequence of interactions within the double helix. The extent to which these are true dyads can be seen in Figure 13, which shows the conformation of one nucleotide strand (C1 to G6) compared with the other strand (C7 to G12) (Wang et al., 1981). In creating Figure 13, the two strands, which are antiparallel to each other in the crystal lattice, were rotated to show the same view of both. The two strands have a remarkably similar conformation, reflecting the internal regularity of the molecule. A significant exception is the conformation of the phosphate group between G4 and C5 on one strand and G10 and C11 on the other. The conformation of G10pC11 is similar to the

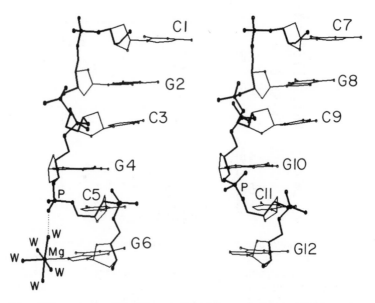

Figure 13. Diagram showing the conformation of the two independent hexa-nucleotide molecules in the spermine-magnesium Z-DNA crystal. In the crystal the two strands are antiparallel, but in this diagram they have been arranged in a parallel alignment to show the similarity of the two chains. The two independent chains are very similar except for the linkage G4pC5. That phosphate group on the left has rotated in such a manner that its oxygen is forming a hydrogen bond with a water molecule (W) in the octahedral coordination shell surrounding the magnesium ion, which is complexed to N7 of guanine 6. This conformation is called Z_{II}, while the conformation for the rest of the phosphate groups is called Z_I.

conformation of the other two GpC phosphates in the molecule. However, the phosphate group in G4pC5 has rotated and the phosphorus atom is approximately 1 Å away from the position it occupies in the other three GpC linkages. Further, the C3′-O3′-P-O5′ dihedral angle has rotated about 125°. In the G4pC5 step, the phosphate group is in a position where it forms a hydrogen bond to a hydrated magnesium ion, which is in turn complexed to the imidazole N7 of guanine 6, as shown in Figure 13. The magnesium ion has octahedral coordination in which one ligand is guanine N7 and the other five are water molecules. We designate the majority conformation found in linkages G2pC3, G8pC9, and G10pC11 as Z_I and the conformation found in G4pC5 as Z_{II}. Conformation Z_I is *ga-*

uche(−)*trans* for the phosphodiester conformation, while Z_{II} is *ga-uche*(+)*trans* (Sundaralingam, 1969).

Analysis of several crystals suggests that left-handed Z-DNA molecules may exist in a mixture of conformations ranging from pure Z_I through mixtures of Z_I and Z_{II} to pure Z_{II}. The exact proportions of these two may depend on the ionic composition and hydration structure of the lattice.

One of the principal differences between Z-DNA and B-DNA is the distance between phosphate groups on opposite strands of the molecule. The distance of the closest approach in B-DNA is across the minor groove, where phosphate groups are 11.7 Å apart. In marked contrast, the closest phosphate groups are only 7.7 and 8.6 Å apart in Z_I-DNA and Z_{II}-DNA. These are distances between the phosphates of CpG sequences. The closest distances across the chains between phosphates of CpG and GpC sequences are 11.2 and 13.5 Å for Z_I and Z_{II}, respectively. The longer distance for Z_{II} reflects the fact that its GpC phosphates are rotated away from the groove.

The differences between Z_I and Z_{II} are shown in Figure 14 which has end views of both conformations together with B-DNA (Wang et al., 1981). As these are all idealized conformations, Z_I and Z_{II} have perfect sixfold symmetry and B-DNA has tenfold symmetry. The large number of differences among A-DNA, B-DNA, and Z-DNA are summarized in Figure 15. In B-DNA, dyad axes are located in the plane of the base pairs

Figure 14. End views of the regular idealized helical forms of Z_I, Z_{II}, and B-DNA. Heavier lines are used for the phosphate-ribose backbone. A guanine-cytosine base pair is shown by shading. The difference in the positions of the base pairs is quite striking; they are near the center of B-DNA but at the periphery of Z-DNA.

as well as between the base pairs. In Z-DNA the dyad axis is preserved between the base pairs but no longer is there a twofold symmetry element in the plane of the base pairs.

Figure 15 further demonstrates that there are appreciable differences in the organization of the polynucleotide chains in Z-DNA compared with B-DNA. In a sense, these differences are so numerous that it is easy to understand why the Z form was not suggested by model building. In retrospect, a number of clues in the literature might now be interpreted as anticipating the Z-DNA structure. For example, the manner by which the sugar pucker could change from C3'-*endo* to C2'-*endo* and the observation of *syn* conformations in purines but not in pyrimidines were well known. Further, the possibility of forming left-handed structures did receive due consideration. Evidence for conformational changes from solution studies of polymers with alternating guanine and cytosine residues was a strong indication. Nevertheless, it would have been difficult to predict from a study of B-DNA all of the necessary changes required to produce the Z structure.

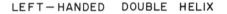

LEFT—HANDED   DOUBLE   HELIX

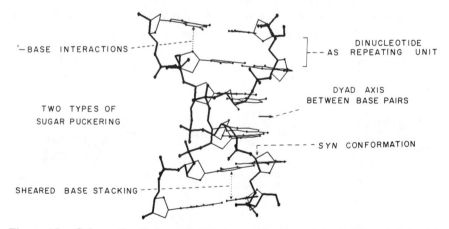

Figure 15. Schematic diagram showing a projection of the left-handed double helix of the spermine-magnesium d(CpGpCpGpCpG) hexamer. Seven structural features found in this left-handed DNA conformation are different from those found in right-handed B-DNA, as indicated in the diagram and discussed in the text.

Table 2. Crystal Data of the Z-DNA Family

Compound	Cell Constants			Space Group	Resolution (Å)	Current R value	Reference
	a(Å)	b(Å)	c(Å)				
Orthorhombic Lattice							
d(CGCGCG) [Mg^{2+} spermine]	17.88	31.55	44.58	$P2_12_12_1$	0.9	0.13	Wang et al., 1981
d(CGCGCG) [Mg^{2+} spermidine]	17.96	31.19	44.73	$P2_12_12_1$	1.3	0.17	Wang et al., 1981
d(CGCGCG) [Ba^{2+} spermine]	17.94	31.54	44.68	$P2_12_12_1$	1.1	0.18	Wang et al., 1981
d(CGCGCG) [Mg^{2+} (and/or Na^+)]	17.98	30.94	44.81	$P2_12_12_1$	1.8	0.13	Wang et al., 1981
d(m5CGm5CGm5CG) [Mg^{2+} spermine]	17.76	30.57	45.42	$P2_12_12_1$	1.3	0.16	Fujii et al., 1982
d(brCGTAbrCG) [Mg^{2+} / Na^+]	17.85	30.74	44.78	$P2_12_12_1$	1.5	0.17	Wang et al., 1984
d(m5CGTAm5CG) [Mg^{2+} / Na^+]	17.91	30.43	44.96	$P2_12_12_1$	1.2	0.16	Wang et al., 1984
d(CGCG) [Mg^{2+} (and/or Na^+)]	19.50	31.27	64.67	$C222_1$	1.5	0.20	Drew et al., 1980
Hexagonal Lattice							
d(CGCGCGCG) [Mg^{2+} (and/or Na^+)]	31.27	31.27	43.56	$P6_5$	1.6	0.19	Fujii et al., 1984
d(CGCATGCG) [Mg^{2+} (and/or Na^+)] (MPD-isopropanol form)	30.90	30.90	43.14	$P6_5$	2.5	0.16	Fujii et al., 1984
d(CGCATGCG) [Mg^{2+} (and/or Na^+)] (MPD-$MgCl_2$ form)	31.05	31.05	43.24	$P6_5$	1.9	0.24	Fujii et al., 1984
d(CGCG) [Mg^{2+} spermine]	31.25	31.25	44.06	$P6_5$	1.5	0.19	Crawford et al., 1980
d(CGCG) [Mg^{2+} (and/or Na^+)]	30.92	30.92	43.29	$P6_5$	1.5	0.21	Crawford et al., 1980

5. LATTICES

Z-DNA structures have been found in crystalline tetramers, hexamers, and octamers. In all of these crystals (Table 2) two types of lattices are found, orthorhombic and hexagonal (Crawford et al., 1980). The most common type of orthorhombic lattice and the single hexagonal lattice are shown in Figure 16. The structure of a tetramer $(dC-dG)_2$ has been solved in the hexagonal lattice, and the hexamer $(dC-dG)_3$ in the orthorhombic lattice. There are two types of molecules in the hexagonal unit cell: those at the corners of the diamond-shaped figures, and those that are internal. The molecules at the corners are symmetry-related by sixfold screw axes and the molecules inside, by threefold screw axes. The c axis of 44.6 Å is such that it will just accommodate a helix with 12 base pairs per turn, as observed in the hexamer crystal (Wang et al., 1979). Thus, the symmetry used by the internal molecules (threefold screw axes) allowed precisely what one would expect with three segments of tetramer forming a left-handed helix with 12 base pairs per turn of the helix.

6. STUDIES OF Z-DNA IN SOLUTION

The earliest evidence suggesting the existence of Z-DNA was the work of Pohl and Jovin (1972). They showed that increasing the NaCl concentration of a poly(dG-dC) solution to 4 M produced a near inversion of the circular dichroism spectrum. In Z-DNA, the phosphate groups on op-

---

Figure 16. (*Opposite*) Diagram of molecular packing of Z-DNA molecules as viewed down the helix axis. (Left) Packing of the hexamer d(CG)$_3$ is shown in the outlined orthorhombic unit cell. (Right) Closely related packing of the tetramer d(CG)$_2$ is shown in a hexagonal unit cell. The horizontal axes in both the orthorhombic and hexagonal unit cells are close to 31.5 Å. The hexagonal unit cell is somewhat larger than the orthorhombic and it can be seen to contain two different kinds of molecules: those found at the corners of the diamond-shaped unit cell and the two that are internal. Those molecules found at the corners of the unit cell are related by sixfold screw axis symmetry whereas the internal molecules by threefold screw axis symmetry. The molecules in the orthorhombic hexamer crystal are related by twofold screw symmetry. Only the oligonucleotides are plotted while spermine ions and solvent molecules are omitted. It is interesting that all of the molecules in both lattices are completely sheathed in water, with no direct contacts between the oligonucleotides. The apparent contact between the hexamer nucleotides does not actually exist because the two parts of the molecule that appear to be in contact are actually separated by half the unit cell.

Orthorhombic Crystal
of   d-(C G)₃

Hexagonal Crystal
of   d-(C G)₂

149

posite chains are closer together than they are across the minor groove of B-DNA. This implies that the structure is likely to be stabilized by increasing the concentration of cations. Thus it was reasonable to believe that the high salt form of poly(dG-dC) might be Z-DNA. Conclusive proof was obtained from the observation that the Raman spectra of the high salt and the low salt forms of poly(dG-dC) differ from each other in a distinctive manner (Pohl et al., 1973). Examination of the Raman spectrum of the Z-DNA hexamer crystals revealed that they had a spectrum almost identical to that of the high salt form of poly(dG-dC) and quite different from that of the low salt form (Thamann et al., 1981). Thus the high salt form of poly(dG-dC) was established to be Z-DNA.

In solution, there is an equilibrium between B-DNA and Z-DNA for poly(dG-dC), where the equilibrium point is determined by the environment. The midpoint for the conversion is 2.5 M Na$^+$ or 0.7 M Mg^{2+} (Pohl and Jovin, 1972). Evidence for the existence of an equilibrium is provided by the fact that the original solution of the hexamer which crystallized as Z-DNA was a low salt solution (Wang et al., 1979) and there is no evidence for Z-DNA formation under those conditions, as judged by the circular dichroism. The crystals were probably nucleated by a minor component of Z-DNA in the low salt solution, and as the crystals grew by mass action all of the material was converted into the Z form. Recognition of this equilibrium is a useful step toward developing an understanding of the effect of various modifications of the polymer on the ability to form Z-DNA. It is fair to presume that there is a similar equilibrium in native DNA, which is strongly influenced by base sequence as well as methylations, as discussed below.

7. TORSIONAL ANGLES IN Z-DNA

We have summarized the distribution of torsion angles in Z-DNA in Figure 17. It contains plots of the various torsion angles listed against each other

Figure 17. (*Opposite*) Torsion angles are shown for the various linkages in the Z-DNA crystal, d(CG)$_3$. The numbers refer to the nucleotides, which are 1–6 on one chain and 7–12 on the other. These are shown explicitly in Figure 13. The torsion angles are plotted against each other in order. The conformational change that gives rise to Z$_{II}$ (Figure 13) produces the torsion angles where residues 4 and 5 stand alone. For comparative purposes, A-RNA (or A-DNA) and B-DNA are included. The shading is added to illustrate the clustering of the data.

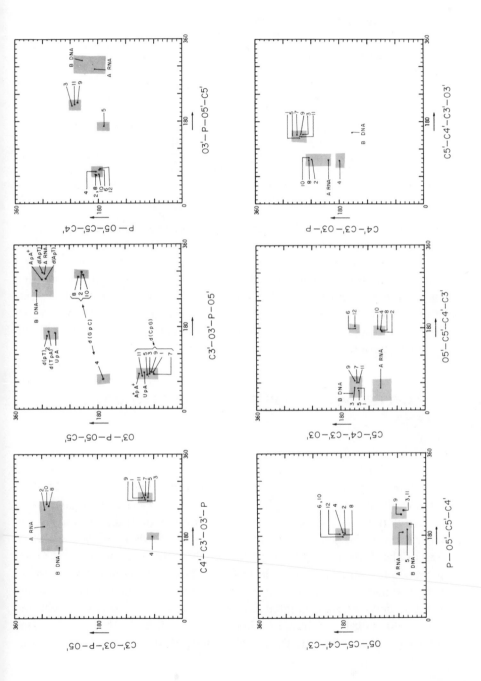

Table 3. Dihedral Angles of DNA

	α	β	γ	δ	ε	ξ	χ
Z_I-DNA							
C	− 137	− 139	56	138	− 94	80	− 159
G	47	179	− 169	99	− 104	− 69	68
Z_{II}-DNA							
C	146	164	66	147	− 100	74	− 148
G	92	− 167	157	94	− 179	55	62
A-DNA	− 90	− 149	47	83	− 175	− 45	− 154
B-DNA	− 41	136	38	139	− 133	− 157	− 102

for the most highly refined (dC-dG)$_3$ structure. The pronounced regularity of the structure is due to the clustering together of angles in groups. Notable exceptions associated with the Z_I and Z_{II} conformation are seen in torsion angles of nucleotides 4 and 5 (see Figure 13 for the numbering system). For comparative purposes, torsion angles are listed for right-handed B-DNA and A-RNA (which is very similar to A-DNA) (Table 3). The differences are apparent and they represent basic alternatives in the way the two molecules are folded.

8. METHYLATION OF CYTOSINE STABILIZES Z-DNA

In higher eukaryotes, one of the most important modifications of DNA is methylation of cytosine on the 5 position where the cytosine is followed by a guanine. Many studies have shown that methylation of DNA is often but not always associated with cessation of transcription. Conversely, demethylation is associated with the initiation of transcription or the expression of the gene (Razin and Riggs, 1980; Ehrlich and Wang, 1981). One of the features apparent in the Z-DNA structure was that CG sequences play an important role in fixing its conformation. Accordingly, it is reasonable to ask to what extent methylation might modify the distribution between right-handed B-DNA and left-handed Z-DNA. Behe and Felsenfeld (1981) addressed this problem directly by synthesizing poly(dG-m^{5}dC), which is fully methylated. Methylation of the cytosine residue on the 5 position has a profound effect in altering the equilibrium between B-DNA and Z-DNA in solution. If one has poly(dG-dC) and poly(dG-m^{5}dC) in 50 mM NaCl, both molecules are in the right-handed B-DNA conformation. In order to convert poly(dG-dC) to left-handed Z-DNA, the magnesium concentration must be raised to 760 mM (Pohl and Jovin, 1972). However, for the fully methylated polymer, 0.6 mM is adequate to convert it to Z-DNA. There is a decrease by three orders of magnitude in the amount of magnesium ions needed to stabilize Z-DNA if the polymer is methylated. Z-DNA is the form of poly(dG-m^{5}dC) in a physiological salt solution.

Figure 18 shows a van der Waals diagram of Z-DNA and the crystal structure of (m^{5}dC-dG)$_3$ (Fujii et al., 1982). The molecules have the same orientation. The arrow on Z-DNA points to a depression on the surface of the molecule, while on the right the methyl group of m^{5}C can be seen to fill the depression. Stereo views are presented in Figure 19 of the unmethylated molecule, in comparison to the methylated molecule in Figure 20. Table 4 compares features of the methylated and nonmethylated hexamer crystals.

The general form of the methylated Z-DNA molecule is similar to that seen in the unmethylated molecule. This is consistent with the observation that antibodies raised against nonmethylated Z-DNA can also recognize Z-DNA formed by the methylated polymer (Lafer et al., 1981). However, there have been some alterations in the geometry of the molecule, principally reflecting the methyl group proximity to the carbon atoms C1' and C2' of the adjacent guanosine residue, and this has produced minor alterations in the helix. There are slight changes as well in the relative positions of adjacent base pairs in the two structures and also some differences in the ring pucker. The major difference is in the relative flatness of the C3'-*endo* conformation of the deoxyribose ring of deoxyguanosine in the methylated polymer as compared with the nonmethylated polymer. The reason for the observed change in the ring is its interaction with the methyl group. There is also a slight change in the helical twist angle, as shown in Figure 12. The net effect of changing the twist angle is to bring the two methyl groups together. In Figure 20 it can be seen that the methyl groups on opposite strands are rather close to one another. The carbon atoms are 4.6 Å apart, almost in van der Waals contact. If one placed a methyl group on the C5 position of cytosine in the unmethylated structure,

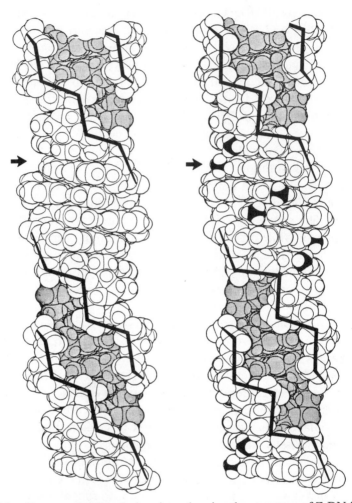

Figure 18. Two van der Waals drawings showing the structure of Z-DNA in both its unmethylated and methylated forms as determined in single-crystal structures of (dC-dG)$_3$ and (m^{5}dC-dG)$_3$, respectively. The groove in the molecule is shown by the shading. The black zigzag lines go from phosphate group to phosphate group to show the arrangement of the sugar-phosphate backbone. In the methylated polymer, the methyl groups on the C5 position of cytosine are drawn in black. The arrow shows that a depression found in the unmethylated polymer is filled by methyl groups. The methyl group indicated by the arrow is in close contact with the imidazole ring of guanosine above it and the C1′ and C2′ atoms of the sugar ring.

the two methyl groups would be a distance of 5.2 Å apart. Thus a shortening of almost 0.6 Å in the distance between methyl groups on opposite strands is a consequence of the change in twist angle described above.

Figures 18 and 20 show that the methyl group occupies a somewhat protected position, recessed slightly on the surface of the molecule so that it is under the imidazole group of guanine with which it is in van der Waals contact (3.4 Å). The methyl group is 3.6–3.8 Å away from C2' of guanosine and 4.2 Å from C1'. There is thus a close contact between the methyl

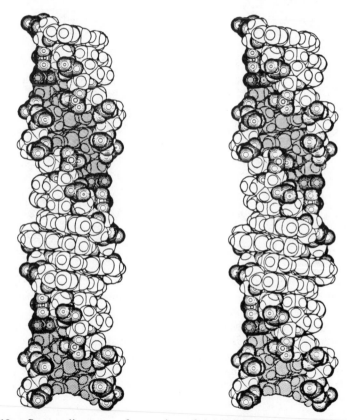

Figure 19.   Stereo diagrams of a portion of the Z-DNA helix found in unmethylated hexamer crystals. Van der Waals models are drawn in which the oxygens in the backbone are indicated by circles and the phosphate groups by circles with crossed lines. A slight depression on the convex outer surface of the molecule occurs because the guanine imidazole rings project farther away from the axis than the cytosine rings.

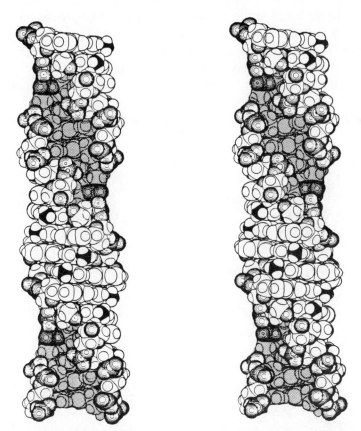

Figure 20. Stereo diagram of the methylated hexamer. The methyl groups are shaded solid black. Three hexamer segments are shown in each diagram just as they appear in the crystal. Every sixth phosphate group is missing because the molecule in the crystal is a hexanucleoside pentaphosphate. The methyl groups are close together and they fill part of the depression on the surface caused by the overhanging imidazole rings of guanine, which protrudes from the center of the molecule. This can be seen easily at the side of the molecule where the methyl groups effectively fill a depression which is visible in the nonmethylated polymer of Figure 18.

group and the sugar residue of the adjacent guanosine (5′ side) as well as its imidazole ring. The reason for the slight conformational change of the methylated hexamer compared with the nonmethylated molecule can be seen if one places a methyl group on the 5 position of cytosine on the unmethylated polymer. When such a structure is made, the distance be-

Table 4. Comparison of $(m^5dC-dG)_3$ and $(dC-dG)_3$

	$(m^5dC-dG)_3$	$(dC-dG)_3$
Cell constants		
a (Å)	17.76	17.88
b (Å)	30.57	31.55
c (Å)	45.42	44.58
Space group	$P2_12_12_1$	$P2_12_12_1$
Helical twist angle ($\pm \sigma$)		
CpG	$-13°$ (1)	$-8°$ (1)
GpC	$-46°$ (1)	$-51°$ (2)
Sugar conformation[a,b]		
deoxycytidine	C2'-endo	C2'-endo
	$\delta = 141°$	$\delta = 146°$
	($P = 149°, \tau_m = 41°$)	($P = 153°, \tau_m = 35°$)
deoxyguanosine	C3'-endo	C3'-endo
(excluding terminal)	$\delta = 94°$	$\delta = 97°$
	($P = 30°, \tau_m = 19°$)	($P = 27°, \tau_m = 31°$)
Glycosyl orientation[a]		
cytosine	anti	anti
	$\chi = -157°$	$\chi = -151°$
guanine	syn	syn
	$\chi = 69°$	$\chi = 67°$

[a] Torsional angles are defined as $O3'--P\overset{\alpha}{-}O5'\overset{\beta}{-}C5'\overset{\gamma}{-}C4'\overset{\delta}{-}C3'\overset{\epsilon}{-}O3'\overset{\xi}{-}P--O5'$, and χ is the glycosyl torsion angle.
[b] P and τ_m are, respectively, the phase angle of pseudo rotation and the degree of pucker.

tween this methyl group and guanosine C2' is 3.2 Å, which is too short a distance for van der Waals contact. That distance is relaxed in the actual structure of the methylated polymer to a distance of 3.6 to 3.8 Å in the various residues. In order to relieve an unacceptable van der Waals contact, the molecule has readjusted itself slightly to produce a change in the helix twist angle and flatten the guanosine sugar ring somewhat.

The disposition of the methyl group of 5-methylcytosine in Z-DNA, compared with the position occupied by that methyl group in right-handed B-DNA, is illustrated in the stereo diagrams of Figures 20 and 21. The black methyl groups on opposite strands are almost touching in Z-DNA (Fig. 20) and are far apart in B-DNA (Fig. 21). The change in the molecule is associated with only small alterations in the lattice. A projection down the c axis is shown in Figure 22. There are small changes, but generally the packing interactions are similar.

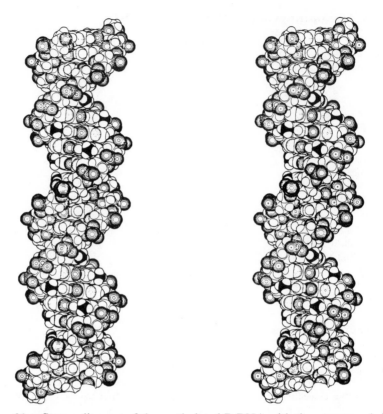

Figure 21. Stereo diagram of the methylated B-DNA with the sequence (m⁵dC-dG)ₙ. Note that the black methyl groups on cytosine residues of opposite strands are far away from each other, in contrast to Figure 19 where the methyl groups are close together.

Another view comparing the methyl groups of m^5dC in Z-DNA and B-DNA is shown in the van der Waals end view of the sequences CpG and GpC in the methylated form (Fig. 23). The upper base pair has different atoms distinguished by shading, but the lower base pair shows only the outline of the bases. The methyl carbon atom on the lower base pair is shaded gray. If one looks at the sequence CpG of Z-DNA, the methyl groups on the two base pairs are fairly close together. Further, there is a limited accessibility to the methyl group by solvent water molecules, which would be found atop of the base pair in the diagram. Accessibility to the gray methyl group of the bottom base pair is limited because of the presence of the amino group and carbonyl oxygen atom of the upper base pair. This situation is in marked contrast to the methyl group of the CpG sequence in B-DNA, shown in the lower part of Figure 23. It can be seen that the methyl group in B-DNA is thrust strongly into the solvent region so that water molecules have access to it as well as to contiguous sections of the cytosine ring. A difference in accessibility of the GpC residues is shown in the other two diagrams. In the Z conformation the methyl group of the upper base pair is in van der Waals contact with the imidazole group of the guanine below, which projects considerably further away from the center of the molecule than the methyl group. In the GpC sequence of B-DNA, the methyl is somewhat protected by the imidazole group of guanine below it but the protection is less than that seen in the Z-DNA structure.

There is also a reordering of solvent molecules around the outside of the Z-DNA helix due to the additional methyl group. Figure 24 shows the electron density found in a 3 Å thick section through a GC base pair in the Z-DNA helix. The water molecule W is hydrogen bonding to the amino group in the N4 position of cytosine. This water molecule is 2.9 Å away from the amino group. The water molecule forms an angle of 145° between W–N4 and the C4–N4 bond of the cytosine residue. The amino group is in the planar trigonal conformation, so one would expect this to exhibit an angle close to 120°. In the nonmethylated structure that same water molecule is found hydrogen bonded to the N4 position of cytosine, but in that structure the angle between the C4–N4 bond of the cytosine and the water–N4 hydrogen bond is 112°. Thus the presence of the methyl group has effectively forced the water molecule away from the position occupied by the methyl group. This demonstrates that the packing of water molecules is different in the primary hydration shell of the methylated form of Z-DNA and the nonmethylated form.

These observations strongly suggest that the Z-DNA methyl group making close van der Waals contact with both the imidazole group of guanine and the carbon atoms of the sugar residue of guanosine is stabilized by hydrophobic interactions with these residues and is somewhat shielded from surrounding water molecules. In contrast, the methyl group on C5 of cytosine in B-DNA has a much greater surface area accessible to solvent water molecules. It is likely that the stabilization of Z-DNA

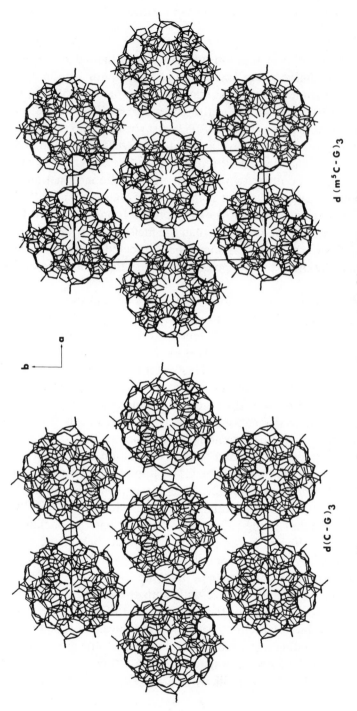

Figure 22. End views of the lattice of d(C-G)₃ and d(m⁵C-G)₃. Both lattices are orthorhombic and have similar but not identical dimensions. The methylated crystal has its cytosine residues somewhat farther away from the axis compared with the unmethylated crystal.

d(C - G)₃

d (m⁵C - G)₃

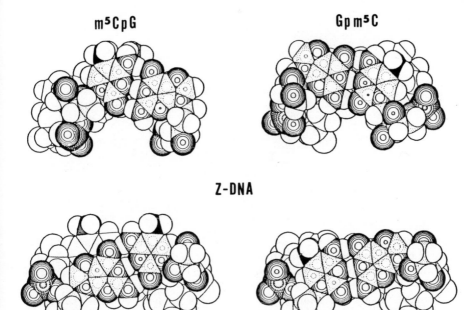

Figure 23. Fragments of Z-DNA and B-DNA are shown in van der Waals diagrams containing two base pairs with the sequences m^{5}CpG and Gpm^{5}C. The base pair closer to the reader has shaded atoms, while the base pair away from the reader has the atoms shown only in outline. The methyl group on 5-methylcytosine in the upper base pair closer to the reader is solid black, while the methyl group attached to the lower base pair is shaded gray. The two diagrams at the bottom show the methyl groups in B-DNA, which are more exposed to solvent water molecules than are the methyl groups in the upper two base pairs in Z-DNA.

on methylation relative to B-DNA is due to the destabilization of B-DNA from the methyl group in the major groove interacting with water molecules and, secondly, a stabilization of Z-DNA itself through the formation of a hydrophobic area on the surface of the molecule in which the methyl group fills a slight depression in the surface of the Z-DNA helix.

In vivo, especially for higher eukaryotes, the effects of methylation of CpG sequences in DNA are associated with an inhibition of RNA synthesis. Behe and Felsenfeld (1981) have shown that Z-DNA formation in poly(dG-dm^{5}C) is facilitated by small amounts of cations, especially the

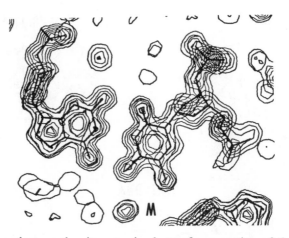

Figure 24. An electron density map is shown for a section of the methylated polymer that encloses one base pair. The electron density map covers a 3 Å-thick section of the map perpendicular to the *c* axis. The 5-methylcytosine of the base pair is on the left and the guanosine residue on the right. A water molecule is hydrogen bonded to the amino group on the 4 position of cytosine. The presence of the methyl group nearby forces that water molecule to occupy a position closer to the line of the cytosine C4–N4 bond than is the case in the structure of the nonmethylated polymer.

polyamines. Spermine stabilizes the formation of Z-DNA for the methylated polymer in submicromolar concentrations. This supports the contention that methylation of alternating dC-dG sequences induces the formation of Z-DNA, perhaps even in short segments of DNA. These structural studies provide a rationale for understanding the mechanism for this stabilization. What is not answered at present is the question of how short a segment of Z-DNA can be formed given the stimulus that methylation of cytosine residues produces for Z-DNA formation. We should like to be able to answer the question whether the methylation of CpG sequences, which occurs *in vivo*, actually results in the formation of small stretches of Z-DNA. Structural studies show us that the destabilization of B-DNA and the relative stabilization of Z-DNA are associated with interactions in the immediate vicinity of the methyl groups themselves. It is thus conceivable that small sections of Z-DNA could form in the middle of B-DNA. In order to study this question it will be necessary to carry out a different kind of experiment in which small segments of a

DNA oligomer are methylated in the hope of trapping, in a single crystal lattice, a segment containing a B-Z-B interface.

9. ADENINE-THYMINE BASE PAIRS IN Z-DNA

Recently, it has been shown that Z-DNA forms in deoxyoligomers containing A·T as well as CG base pairs (Wang et al., 1984). Through the use of antibodies specific for Z-DNA, segments of DNA in negatively supercoiled plasmids have been shown to form Z-DNA, and they contain A·T base pairs (Nordheim and Rich, 1983). However, A·T base pairs are less stable than CG base pairs. We have addressed the problem by solving the structure of Z-DNA with A·T base pairs, specifically two hexanucleoside pentaphosphates with the general self-conplementary sequence d(CGTACG) in which the cytosine residues have either methyl groups or bromine atoms on their 5 positions.

The structures of both the 5-methyl and 5-bromo derivatives of d(CGTACG) are similar to each other and likewise similar to the structure of (dC-dG)$_3$ as well as (m^{5}dC-dG)$_3$. The self-complementary molecules form a double helix with six base pairs. The left-handed Z-DNA double-helical fragments stack together to form an essentially continuous double helix running along the c axis of the unit cell. The crystal of the methylated hexamer diffracts to a resolution of 1.2 Å, and the brominated derivative to 1.5 Å. Figure 25 is a diagram of a van der Waals model of the brominated d(CGTACG), which shows the end-to-end organization of three of the double helical molecules. The overall form of the molecule is very similar to that of the unmethylated as well as the methylated (dC-dG)$_3$.

A van der Waals view of the 5-bromocytosine derivative of d(CGTACG) is shown in Figure 25. The large black atoms are the bromine atoms. The thymine methyl groups are also shaded black but have hydrogen atoms attached to them. The carbonyl oxygen 2 atoms of the thymine groups are cross-hatched in order to illustrate their position near the axis of the molecule at the bottom of the deep groove. Unlike the cytosine O2 atoms, these oxygens do not have amino groups hydrogen-bonding to them because of the absence of an N2 amino group in adenine. The bromine atoms in Figure 24 occur as pairs on the surface of the molecule and they are found at distances of 4.56 and 4.77 Å apart. The

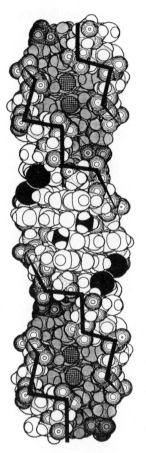

Figure 25. Stereo diagram of a van der Waals model of the methylated hexamer d(m⁵CGTAm⁵CG). Portions of the molecule that form the helical groove are shaded. The oxygen atoms are drawn with circles while the phosphorus atoms have spiked circles. The zigzag array of the phosphate groups bordering the helical groove is readily seen. This diagram shows three molecules stacked up as they are found in the crystal. The methyl groups are found on the outer convex portion of the molecule. The methyl group carbon atoms from the 5-methyl cytosine derivatives are crosshatched, while those from the thymine residues are solid.

two methyl groups on the thymine residues are separated by 4.85 Å in a manner similar to that seen in the methylated derivative described above. Homologous structural features are generally found in both the methylated and brominated derivatives.

In refining the crystal structure, a large number of water molecules and solvent ions were identified. In the 5-methyl derivative of d(CGTACG), 98 solvent atoms were located. The vast majority of these peaks are water molecules, although five of the peaks have been identified as cations due to the coordination geometry they exhibit. In order to visualize the water molecules in the lattice, we present in Figure 26 six sections of the electron density map. These composites each represent

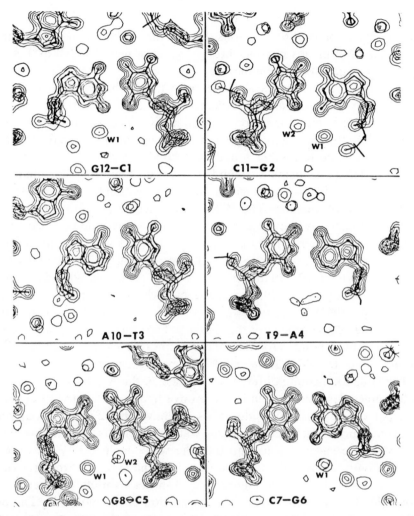

Figure 26. Sections of the electron density map are shown through the six base pairs in the double helix d(m^{5}CGTAm^{5}CG). The base pairs are oriented so that the groove is near the bottom of the diagram. Two different types of water molecules are seen in the helical groove. W_1 receives a hydrogen bond from the amino group of guanine and in turn donates a hydrogen bond to a phosphate oxygen. That water molecule lies near the plane of the bases. The second water molecule, W_2, lies halfway between the O_2 atoms of the stacked pyrimidine residues. In general it has fewer contours than W_1 due to the sectioning. The lack of ordered water molecules in the groove near the A·T base pairs is readily apparent. Water molecule W_2 does not appear with such intensity in the sections at the top and the bottom of the molecule because the occupancy of these water molecules is somewhat less in the intermolecular solvent channel.

the superposition of three sections through the six base pairs in the molecule, taken in planes passing through the base pairs as well as 1 Å on either side. The axis of the molecule passes near the O2 of the pyrimidine residue and the base pairs are oriented with the helical groove of the molecule on the lower part of the diagram. Two different types of water molecules are found in the groove, W1 and W2. The water molecule W1 is in position to accept a hydrogen bond from the amino group on position 2 of guanine, and it donates a hydrogen bond to a phosphate oxygen of the same deoxyguanosine nucleotide (except for the terminal guanine G6 or G12). This bridging water molecule is in a position to stabilize the *syn* conformation of guanine. The W2 water molecule fits into the groove, where it donates two hydrogen bonds to O2 of pyrimidine residues in successive base pairs. The oxygen atom of W2 is located between the base pairs, thus it usually has less electron density in the sections that are shown in Figure 26. In the structure of both the unmethylated hexamer $(dC-dG)_3$ (Wang et al., 1979) and the methylated hexamer $(m^5dC-dG)_3$ (Fujii et al., 1982), these two water molecules are found completely occupying the helical groove of the Z-DNA. Thus there are two water molecules hydrogen bonded to electronegative nitrogen and oxygen atoms at the bottom of the Z-DNA helical groove per base pair.

The situation is quite different in the case of the A·T base pairs, as no water molecules can be seen in the groove hydrogen bonded to the A·T base pairs. The adenine residue does not have an N2 amino group and so it is not surprising that the bridging water molecule is absent. However, it was a little surprising to find that the water molecule W2 hydrogen bonding onto the pyrimidine O2 atoms along the bottom of the groove is also missing in this region. Figure 26 also shows a variety of other water molecules found on the outer, convex portion of the molecule. The positions of some of these water molecules are also modified by the presence of the methyl group on the 5 position of cytosine (see Figure 24).

Another visualization of hydration around the Z-DNA molecule is seen in the stereo diagram of Figure 27. This shows a van der Waals model of $d(m^5CGTAm^5CG)$ (Wang et al., 1984) together with associated solvent molecules (stippled spheres). The convex outer portion of the structure is coated with a sheath of water molecules which are binding to most of the electronegative atoms in the molecule. There are small patches of hydrophobic residues which do not have water molecules covering them. Figure 27 also shows the helical groove of Z-DNA to be filled with water

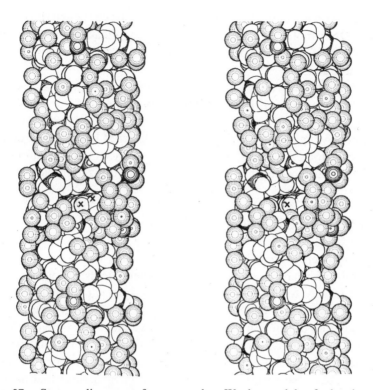

Figure 27. Stereo diagram of a van der Waals model of the hexamer d(m⁵CGTAm⁵CG) showing the first shell hydration. Only solvent molecules are included for which the temperature factor is lower than 45 Å². Water molecules are represented as spheres with stippled circles, magnesium metal ions are drawn as spheres with concentric circles, while phosphorus atoms are drawn with circles with cross striations. The helical groove is generally filled with ordered water molecules except in one region where the two A·T base pairs lie. In order to visualize this section of the groove, two water molecules were removed which are bound to the phosphate groups near the periphery of the molecule.

molecules except at the position occupied by the two base pairs. In this figure we are able to inspect the bottom of the helical groove where the base pairs themselves can be seen devoid of ordered hydration structure. The bottom of the groove does not show solvent even though there are water molecules at the outer part of the groove which are hydrogen bonded to the negatively charged phosphates. Two of these (at positions marked ×) have been removed in Figure 27 in order to make visible the disordered solvent at the bottom of the groove. Even though the geometry of the Z-DNA molecule is not modified to a significant extent through the introduction of A·T base pairs, hydration in the helical groove is significantly altered.

It is known that A·T base pairs in B-DNA are somewhat less stable than CG base pairs. In Z-DNA, they are probably considerably less stable, not only because of the absence of a third hydrogen bond between the bases, but owing as well to the paucity of ordered hydration structure in the helical groove. Because the Z-DNA sequences contain information, it is likely that these differences in stability are utilized in biological systems.

10. FUTURE STUDIES

There has been a great advance from a decade or so ago in available structural information regarding the nucleic acids. Progress can be ascribed largely to the fact that synthetic oligonucleotides have yielded crystalline fragments of DNA and RNA. These have in turn permitted high resolution diffraction studies which have taught us a great deal about the three-dimensional structure and conformation of the nucleic acids. Long helical fragments are now available comprising segments from four to over a dozen base pairs. Many of the crystal structures that have been solved are at atomic resolution. From this association of organic chemistry and X-ray crystallography has come an enormous broadening in our comprehension of the fine details of nucleic acid conformation as well as the manner by which they interact with water molecules and cations. These advances have not been confined to the traditional structures of DNA, the A and the B form, but now encompass the discovery of a novel left-handed Z form. If one were to extrapolate these studies into the future, it seems almost certain that increasingly more structural detail at high resolution will be obtained through the use of larger oligomers.

Are there other novel conformations yet to be discovered? Here again we think the answer is likely to be yes. The structural features characterizing Z-DNA are dependent on a particular preferred sequence of alternating purines and pyrimidines. Is it possible that there are other motifs involving repetitions of three or more nucleotides that produce an alternative conformational pattern? Is it possible that, through their interactions with other molecules, including proteins, the nucleic acids will be shown capable of maintaining conformations that exist in solution even in the absence of proteins? It is always difficult to predict the future, but we now have a methodology based on the use of synthetic oligonucleotides of known sequence that are capable of forming single crystals whose structures can be solved. This powerful tool will undoubtedly lead to a continued flowering of discovery well beyond the sketchy outlines available up to now through fiber diffraction studies. Among these new results are likely to be a proliferation of new conformational states. The discovery of the Z form of DNA, for all it has told us about the structural chemistry of nucleic acids, may ultimately be seen as an even more important event for awakening us to the unrealized possibilities inherent in the double helix.

Acknowledgments. This research was supported by grants from the National Institutes of Health, the American Cancer Society, the National Science Foundation, and the National Aeronautics and Space Administration.

REFERENCES

Arnott, S. (1970). *Progress in Biophysics and Molecular Biology*, J. A. V. Butler and D. Noble, Eds. Pergamon, New York.

Arnott, S., Chandrasekaran, D. L., Birdsall, D. L., Leslie, A. G. W., and Ratliff, R. L. (1980). *Nature* **283**, 743–745.

Behe, M., and Felsenfeld, G. (1981). *Proc. Natl. Acad. Sci. USA* **78**, 1619–1623.

Crawford, J. L., Kolpak, F. J., Wang, A. H.-J., and Quigley, G. J., van Boom, J. H., van der Marel, G., and Rich, A. (1980). *Proc. Natl. Acad. Sci. USA* **77**, 4016–4020.

Day, R. O., Seeman, N. C., Rosenberg, J. M., and Rich, A. (1973). *Proc. Natl. Acad. Sci. USA* **70**, 849–853.

Drew, H., Takano, T., Tanaka, S., Itakura, K., and Dickerson, R. E. (1980). *Nature* **286**, 567–573.

Ehrlich, M., and Wang, R. Y.-H. (1981). *Science* **212**, 1350–1357.

Franklin, R. E., and Gosling, R. G. (1951). *Nature* **171**, 740–741.

Fujii, S., Wang, A. H.-J., Quigley, G. J., Westerink, H., van Boom, J. H., van der Marel, G., and Rich, A. (1984). Submitted.

Fujii, S., Wang, A. H.-J., van der Marel, G., van Boom, J. H., and Rich, A. (1982). *Nucl. Acids Res.* **10**, 7879–7892.

Gupta, G., Bansal, M., and Sasiskeharan, V. (1980). *Proc. Natl. Acad. Sci. USA* **77**, 6486–6490.

Haschemeyer, A. E. V., and Rich, A. (1967). *J. Mol. Biol.* **27**, 369–384.

Hopkins, R. C. (1981). *Science* **211**, 289–291.

Lafer, E. M., Moller, A., Nordheim, A., Stollar, B. D., and Rich, A. (1981). *Proc. Natl. Acad. Sci. USA* **78**, 3546–3550.

Nordheim, A., and Rich, A. (1983). *Nature* **303**, 674–679.

Pohl, F. M., and Jovin, T. M. (1972). *J. Mol. Biol.* **67**, 375–396.

Pohl, F. M., Ranade, A., and Stockburger, M. (1973). *Biochim. Biophys. Acta* **335**, 85–92.

Razin, A., and Riggs, A. D. (1980). *Science* **210**, 604–610.

Rodley, G. A., Scobie, R. S., Bates, R. H., and Lewitt, R. M. (1976). *Proc. Natl. Acad. Sci. USA* **73**, 2959–2963.

Rosenberg, J. M., Seeman, N. C., Day, R. O., and Rich, A. (1976). *J. Mol. Biol.* **104**, 145–167.

Rosenberg, J. M., Seeman, N. C., Kim, J. J. P., Suddath, F. L., Nicholas, H. B., and Rich, A. (1973). *Nature* **243**, 150–154.

Sasiskeharan, V. (1982). *Cold Spring Harbor Symp. Quant. Biol.* **47**, 45–52.

Seeman, N. C., Rosenberg, J. M., Suddath, F. L., Kim, J. J. P., and Rich, A. (1976). *J. Mol. Biol.* **104**, 109–144.

Son, T.-D., Guschlbauer, W., and Gueron, M. (1972). *J. Am. Chem. Soc.* **94**, 7903–7911.

Sunderalingam, M. (1969). *Biopolymer* **7**, 821–833.

Thamann, T. J., Lord, R. C., Wang, A. H.-J., and Rich, A. (1981). *Nucl. Acids Res.* **9**, 5443–5457.

Voet, D., and Rich, A. (1970). *Prog. Nucl. Acid Res. Mol. Biol.* **10**, 183–265 (1970).

Wang, A. H.-J., Fujii, S., van Boom, J. H., and Rich, A. (1982). *Cold Spring Harbor Symp. Quant. Biol.* **47**, 33–44.

Wang, A. H.-J., Hakoshima, T., van der Marel, G., van Boom, J. H., and Rich, A. (1984). *Cell* (in press).

Wang, A. H.-J., Quigley, G. J., Kolpak, F. J., Crawford, J. L., van Boom, J. H., van der Marel, G., and Rich, A. (1979). *Nature* **282**, 680–686.

Wang, A. H.-J., Quigley, G. J., Kolpak, F. J., van der Marel, G., van Boom, J. H., and Rich, A. (1981). *Science* **211**, 171–176.

Watson, J. D., and Crick, F. H. (1953). *Nature* **171**, 737.

Kink-Antikink Bound States in DNA Structure

Kink-Antikink Bound States in DNA Structure

4

Kink-Antikink Bound States in DNA Structure

HENRY M. SOBELL
The University of Rochester School of Medicine and Dentistry
Rochester, New York

CONTENTS

1. INTRODUCTION

Mechanistically, how do drugs and dyes intercalate into DNA? Do they intercalate *directly* into the A, B, and Z structure, or is there some other structural form for DNA into which drugs and dyes intercalate?

We have proposed the existence of another DNA structure to be involved in the intercalation process. This structure—called β-DNA or β premelted DNA—forms spontaneously in the centers of soliton-antisoliton (i.e., kink-antikink) bound states in DNA structure. Such structural solitons contain a modulated β alternation in sugar puckering about the central β premelted core region that gradually merges into B-DNA (or A-DNA) on either side. We call such composite structures β premeltons. We are uncertain, as yet, whether the centers of β premeltons are open enough to accommodate an intercalator directly; however, structures such as these are known to have an intrinsic ability to undergo low frequency breather motions, and it is possible that such motions facilitate the intercalation process.

We envision intercalation to begin with the loose (external) binding by an intercalator to the B helix. A β premelton then appears in the region. This permits the drug molecule to intercalate either immediately or after a breathing motion of sufficient amplitude. Simple intercalators pin (immobilize) the β premelton. They also prevent soliton-antisoliton annihilation—a spontaneous event that returns DNA in the immediate region to its ground state structure, B-DNA. This is because simple intercalators bind the β premelted core region tightly and stabilize the structural soliton. Complex intercalators vary in this regard. Actinomycin binds the β premelted core within the β premelton, pinning the structure and also preventing its self-annihilation. Daunomycin, on the other hand, is unable to do this and therefore ends up binding to a conformation very similar to B-DNA on either side of the intercalated chromophore.

The β premelted core region within the soliton-antisoliton bound state structure can be thought of as a transition state intermediate that nucleates the DNA melting process (hence, the name β premelton). At lower temperatures, soliton-antisoliton pairs surround small β premelted cores. As the temperature rises, these bounding soliton-antisoliton pairs move apart, leaving the central region of each growing β premelted core more and more disordered. Finally, with increasing temperature, denaturation bubbles appear that contain regions of native DNA connected to denatured DNA through kink-antikink pairs. These kink and antikink structures, therefore, are phase boundaries that connect native DNA with denatured DNA during the helix-to-coil phase transition. They also act as (partial) phase boundaries in a variety of DNA structural phase transitions. In this chapter we describe the nature of one such boundary, the B to A boundary.

We begin by reviewing structural information obtained from X-ray crystallographic investigations of small RNA- and DNA-like oligonucleotide: intercalator complexes. These studies have allowed the direct visualization of drug–nucleic acid intercalative binding at atomic resolution. Two broad classes of intercalators can be distinguished—simple and complex. Simple intercalators bind to DNA through stacking interactions with the base pairs and electrostatic interactions between the positively charged chromophore and the sugar-phosphate backbone. Typically, simple intercalators bind to a C3′-*endo* (3′-5′) C2′-*endo* mixed puckered structure—called the β structural element—in both DNA-model and RNA-model dinucleotide systems. Examples include ethidium, acridine orange, ellipticine, and terpyridine platinum. Complex intercalators utilize additional types of interactions when binding to DNA or RNA, and these may (or may not) alter the final intercalative geometry that is observed. Examples include proflavine, daunomycin, and actinomycin. An important insight that emerges from these studies concerns the structural basis for neighbor exclusion, a phenomenon that is observed at saturating concentrations of simple intercalators such as ethidium. This reflects the formation of an organized helical structure in which drug molecules intercalate between every other base pair. Such a structure contains a dinucleotide (i.e., the β structural element) as the asymmetric unit and has an alternating pattern of sugar puckering down the polynucleotide backbone with alternate base pairs either partially or completely unstacked. It is a structure uniquely different from A, B, or Z-DNA. We have called this structure β-DNA or β premelted DNA.

As stated earlier, we think this structure plays a key role in a variety of DNA premelting phenomena. Later in Sections 3, 4, and 5, we explain how we have used soliton physics concepts to qualitatively undertand these and other aspects of DNA physical chemistry. Important biological implications are also described.

2. SIMPLE VERSUS COMPLEX INTERCALATORS

There is now a wealth of crystallographic information on drug–nucleic acid crystalline complexes contributed over the years by our laboratory and other laboratories around the world. From these data, it is possible to divide intercalators into two general classes, simple and complex.

Simple intercalators (summarized in Table 1 and Fig. 1) bind to the β structural element in both DNA and RNA model dinucleotide systems (i.e., the C3'-*endo*-(3'-5')-C2'-*endo* mixed puckered structure). A typical example is ethidium complexed to 5-iodocytidylyl-(3'-5')-guanosine, shown in Figure 2 (Jain et al., 1977). This structure contains an ethidium molecule [ethidium(1)] intercalated between guanine–cytosine base pairs in this self-complementary dinucleoside monophosphate structure. Another ethidium molecule [molecule(2)] is stacked on either side of the complex to maintain charge neutrality and to provide additional stabilization to the solid state complex. Base pairs above and below ethidium(1) are separated by 6.8 Å and are twisted 10°; this reflects the detailed nature

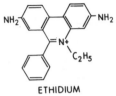

ETHIDIUM

ACRIDINE ORANGE

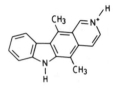

ELLIPTICINE

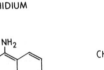

9-AMINOACRIDINE

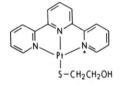

3,5,6,8-TETRAMETHYL-
N-METHYLPHENANTHROLINIUM

N,N-DIMETHYLPROFLAVINE

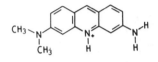

2-HYDROXYETHANETHIOLATO-2,2',2"-
TERPYRIDINE-PLATINUM (II)

Figure 1. Chemical structures of simple intercalators.

of the sugar-phosphate backbone geometry. Cytidine residues are C3'-*endo* in the low *anti* conformation (characteristic of the monomer unit in an A-type polynucleotide duplex), whereas guanosine residues are C2'-*endo* in the high *anti* conformation (characteristic of the monomer unit in a B-type polynucleotide duplex). Torsional angles that define the geometry of the phosphodiester linkages resemble both B-DNA and A-DNA (Table 2). The β structural element, therefore, has both A and B character and can be thought of as a structure intermediate between A and B polynucleotide conformations.

The β structural element has been observed in *15* separate crystallographic determinations. These involve *seven* different intercalators complexed to a variety of DNA-like and RNA-like self-complementary dinucleoside monophosphates. Four structures [5-iodocytidylyl-(3'-5')-

Table 1. Unit Cell Constants, Space Groups, and Structural Data for Simple Intercalators Complexed to DNA-like and RNA-like Self-complementary Dinucleoside Monophosphates

Complex	St[a]	Cell Constants	Space Group	Sugar Puckering	Twist Angle	Reference
Ethidium: 5-iodouridylyl-(3'-5')-adenosine	2:2	a = 28.45 Å b = 13.54 Å c = 34.13 Å β = 98.6°	C2	C3'-endo-(3'-5')-C2'-endo	8 ± 1°	Tsai et al., 1977
Ethidium: 5-iodocytidylyl-(3'-5')-guanosine	2:2	a = 14.06 Å b = 32.34 Å c = 16.53 Å β = 117.8°	P2$_1$	C3'-endo-(3'-5')-C2'-endo	8 ± 1°	Jain et al., 1977
Ethidium: uridylyl-(3'-5')-adenosine	2:2	a = 13.70 Å b = 31.67 Å c = 15.13 Å β = 113.9°	'P2$_1$	C3'-endo-(3'-5')-C2'-endo	9 ± 1°	Jain and Sobell, 1984a
Ethidium: cytidylyl-(3'-5')-guanosine	2:2	a = 13.79 Å b = 31.94 Å c = 15.66 Å β = 117.5°	P2$_1$	C3'-endo-(3'-5')-C2'-endo	9 ± 1°	Jain and Sobell, 1984b
Ethidium: cytidylyl-(3'-5')-guanosine	2:2	a = 13.64 Å b = 32.16 Å c = 14.93 Å β = 114.8°	P2$_1$	C3'-endo-(3'-5')-C2'-endo	8 ± 1°	Jain and Sobell, 1984b
Acridine orange: 5-iodocytidylyl-(3'-5')-guanosine	2:2	a = 14.33 Å b = 19.68 Å c = 20.67 Å β = 102.1°	P2$_1$	C3'-endo-(3'-5')-C2'-endo	10 ± 1°	Reddy et al., 1979
Acridine orange: cytidylyl-(3'-5')-guanosine	1:2	a = 12.65 Å b = 11.79 Å c = 15.74 Å	P1	C3'-endo-(3'-5')-C2'-endo	9 ± 1°	Wang et al., 1979

	Stoichiometry[a]		Space group	Sugar conformation	Angle	Reference
		$\alpha = 92.7°$				
		$\beta = 107.0°$				
		$\gamma = 90.0°$				
Ellipticine: 5-iodocytidylyl-(3'-5')-guanosine	2:2	$a = 13.88$ Å $b = 19.11$ Å $c = 21.42$ Å $\beta = 105.4°$	$P2_1$	C3'-endo-(3'-5')-C2'-endo	11 ± 1°	Jain et al., 1979
3,5,6,-Tetramethyl-N-methyl phenanthrolinium: 5-iodocytidylyl-(3'-5')-guanosine	2:2	$a = 13.99$ Å $b = 19.11$ Å $c = 21.31$ Å $\beta = 104.9°$	$P2_1$	C3'-endo-(3'-5')-C2'-endo	11 ± 1°	Jain et al., 1979
9-Aminoacridine: 5-iodocytidylyl-(3'-5')-guanosine	4:4	$a = 13.98$ Å $b = 30.58$ Å $c = 22.47$ Å $\beta = 113.9°$	$P2_1$	C3'-endo-(3'-5')-C2'-endo C3'-endo-(3'-5')-C2'-endo	8 ± 2° 10 ± 2°	Sakore et al., 1979
N,N-Dimethylproflavine: 5-iodocytidylyl-(3'-5')-guanosine	2:2	$a = 11.78$ Å $b = 14.55$ Å $c = 15.50$ Å $\alpha = 89.2°$ $\beta = 86.2°$ $\gamma = 96.4°$	P1	C3'-endo-(3'-5')-C2'-endo	9 ± 1°	Bhandary et al., 1984
N,N-Dimethylproflavine: 5-iodocytidylyl-(3'-5')-guanosine	2:2	$a = 14.20$ Å $b = 18.99$ Å $c = 20.73$ Å $\beta = 103.6°$	$P2_1$	C3'-endo-(3'-5')-C2'-endo	9 ± 1°	Bhandary et al., 1984
N,N-Dimethylproflavine: deoxycytidylyl-(3'-5')-deoxyguanosine	2:2	$a = 20.79$ Å $b = 33.82$ Å $c = 13.40$ Å	$P2_12_12$	C3'-endo-(3'-5')-C2'-endo	9 ± 1°	Sakore et al., 1984
Terpyridine platinum: deoxycytidylyl-(3'-5')-deoxyguanosine	2:2	$a = 22.25$ Å $b = 13.57$ Å $c = 33.12$ Å	$P2_12_12_1$	C3'-endo-(3'-5')-C2'-endo	12 ± 1°	Wang et al., 1978

[a] Stoichiometry.

177

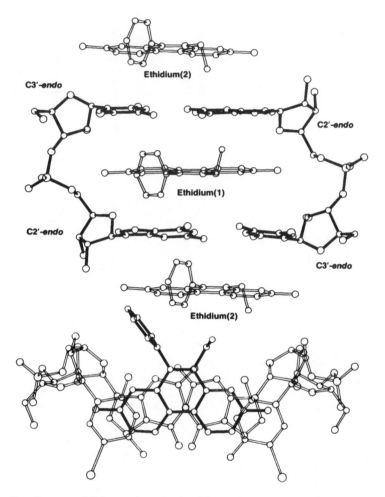

Figure 2. Structural information obtained from the ethidium: 5-iodocytidylyl-(3'-5')-guanosine crystalline complex.

guanosine complexed to ellipticine, acridine orange, tetramethyl-N-methyl-phenanthrolinium, and N,N-dimethylproflavine] are isomorphous and therefore demonstrate a host–guest relationship. The remaining 11 structures crystallize in different lattice environments that contain variable numbers of water molecules. The invariance of the β geometry in these studies with simple intercalators argues that (1) the β structural element is a particularly stable structure that can accommodate a large variety of

Table 2. Torsional Angles (°) Describing Conformations of Sugar–Phosphate Chains in A-DNA and B-DNA, Ethidium-IodoCpG Crystalline Complex and the Ethidium-DNA Neighbor-exclusion Structure.[a]

	χ	ε	ζ	α	β	γ	δ	ε	ζ	χ	References
A-DNA	27	47	83	175	315	270	211	47	83	27	Arnott et al., 1976
B-DNA	85	31	157	159	261	321	209	31	157	85	Arnott et al., 1976
Ethidium-iodoCpG	29	51	87	226	281	286	210	72	131	101	Jain et al., 1977
	24	90	84	225	291	291	224	55	134	109	
β-DNA	40	63	82	298	297	38	169	66	143	101	from Table 3

aromatic ring compounds without significant alterations in its geometry, and (2) the β structural element forms equally well with either DNA-like or RNA-like dinucleotides.

Two key predictions arise if one uses this information to understand how simple intercalators bind to DNA. The first is that simple intercalators unwind DNA by about 26° (i.e., 36° − 10° = 26°). The second is that simple intercalators bind to DNA in a neighbor-exclusion mode (i.e., they bind between every other base pair) at high drug–DNA binding ratios. We have called this structure β-DNA or, interchangeably, β premelted DNA (Fig. 3 and Table 3) (Sobell et al., 1977; Sobell et al., 1983a, 1983b; Banerjee and Sobell, 1983). β-DNA has a dinucleotide (the β structural element) as its asymmetric unit; the helical operation that takes one dinucleotide into the next is a rotation of 47.2° and a translation along the helix axis of 9.8 Å. Available evidence supports both predictions (Crothers, 1968; Wang, 1974; Bond et al., 1975; Stasiak et al., 1983). Similar predictions arise regarding drug binding to double-stranded RNA; however, there is limited information concerning this at the present time.

Important additional information concerning sequence binding preferences between simple intercalators and short-chain oligonucleotides has emerged from solution studies. Ethidium has been shown to bind tightly to dinucleotide sequences of the general type, *pyrimidine-purine* (Krugh and Reinhardt, 1975). This information has been confirmed and extended to include additional simple intercalators complexed to longer oligonucleotides (Patel, 1980). A similar specificity for binding the pyrimidine-purine sequence has been observed in the solid-state studies. Important

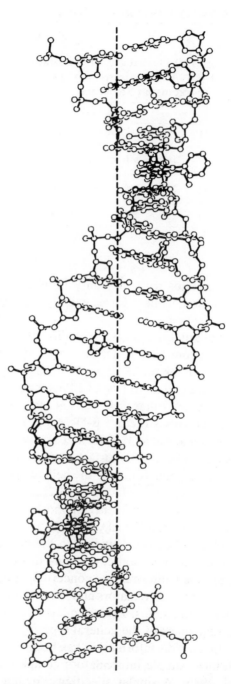

Figure 3. Ethidium-DNA neighbor exclusion structure. This DNA conformation corresponds to a higher energy β-DNA or β premelted DNA form.

Table 3. Coordinates for Ethidium–DNA Neighbor Exclusion Structure

	Deoxyadenosine 5'-Monophosphate					Thymidine 5'-Monophosphate					Ethidium		
		R (Å)	θ (°)	z (Å)			R (Å)	θ (°)	z (Å)			R (Å)	θ (°)	z (Å)
1	AN1	3.605	−5.094	−3.457	22	TN1	6.435	55.721	2.284	1	EC1	2.751	30.517	−0.320
2	AC2	4.855	0.342	−3.531	23	TC2	5.658	45.317	2.580	2	EC2	3.624	48.883	−0.630
3	AN3	5.472	12.886	−3.813	24	TO2	6.303	35.404	2.644	3	EC3	4.997	43.240	−0.793
4	AC4	4.816	25.813	−4.044	25	TN3	4.305	47.160	2.804	4	EC4	5.522	29.476	−0.630
5	AC5	3.515	31.508	−4.005	26	TC4	3.814	65.465	2.757	5	EN5	6.019	6.267	−0.153
6	AC6	2.661	11.360	−3.692	27	TO4	2.707	74.334	2.976	6	EC6	6.016	−6.298	0.150
7	AN6	1.336	5.490	−3.618	28	TC5	5.026	74.875	2.439	7	EC7	5.545	−29.695	0.637
8	AN7	3.749	52.727	−4.297	29	TC7	5.279	91.359	2.368	8	EC8	5.019	−43.466	0.803
9	AC8	5.021	50.717	−4.501	30	TC6	6.167	67.948	2.217	9	EC9	3.708	−49.112	0.657
10	AN9	5.628	37.454	−4.365	31	TC1'	7.862	53.027	2.037	10	EC10	2.800	−32.073	0.350
11	AC1'	7.096	35.887	−4.521	32	TO1'	8.728	59.585	2.631	11	EC11	3.633	−11.480	0.170
12	AO1'	7.675	35.562	−3.231	33	TC2'	8.244	53.097	0.562	12	EC12	3.623	10.448	−0.150
13	AC2'	7.854	44.536	−5.214	34	TC3'	8.914	62.153	0.389	13	EC13	4.975	15.662	−0.313
14	AC3'	9.257	42.563	−4.707	35	TO3'	9.768	61.801	−0.752	14	EC14	4.969	−16.159	0.320
15	AO3'	9.940	36.628	−5.468	36	TC4'	9.708	62.956	1.687	15	EC15	7.381	−10.674	0.320
16	AC4'	9.066	39.959	−3.250	37	TC5'	10.275	70.741	2.037	16	EC16	8.051	−15.711	−0.700
17	AC5'	9.258	46.810	−2.227	38	TO5'	9.320	77.005	2.252	17	EC17	9.416	−17.509	−0.530
18	AO5'	8.358	54.151	−2.380	39	P	10.023	84.967	2.749	18	EC18	10.028	−16.333	0.647
19	P	9.141	63.193	−2.199	40	PO1	9.389	92.512	2.341	19	EC19	9.385	−12.661	1.690
20	PO1	8.364	71.478	−2.177	41	PO2	11.434	84.627	2.319	20	EC20	8.057	−9.667	1.527
21	PO2	10.221	63.370	−3.204						21	EC21	7.425	10.531	−0.320
										22	EC22	8.006	8.042	−1.690
										23	EN23	6.029	52.298	−1.107
										24	EN24	6.020	−52.831	1.117

181

related information concerns the relative stabilities of double-helical dinucleotide complexes in solution containing self-complementary sequences. Young and Krugh (1975) have demonstrated by nuclear magnetic
resonance spectroscopy that purine-pyrimidine sequences form more stable structures than pyrimidine-purine sequences in aqueous solution. This
suggests that the sequence binding preferences demonstrated by ethidium
and other simple intercalators reflect the relative ease of unstacking pyrimidine-purine versus purine-pyrimidine sequences, rather than differences in stacking energies between the intercalators and base pairs in
these sequences (Pack and Loew, 1977). We return to this point when
we discuss the structures of poly[d(A-T)] and poly[d(G-C)] and the etiol-
ogy of their sequence-dependent nuclease sensitivities. In summary, then,
simple intercalators have the following properties:

1. They bind to the β structural element in both DNA and double-
    stranded RNA and unwind DNA by about 26°.
2. Their interactions with these polymers are "simple" in the sense
    that they involve stacking interactions with the base pairs and elec-
    trostatic interactions with the sugar-phosphate chains.
3. They demonstrate a sequence binding preference, CpG and TpA,
    in their interactions with short-chain oligonucleotides. This reflects
    the ease of unstacking pyrimidine-purine sequences versus purine-
    pyrimidine sequences—an important consideration in understand-
    ing how simple intercalators bind to synthetic DNA-like polymers
    such as poly[d(A-T)] and poly[d(G-C)]. Other physical effects may
    predominate, however, when simple intercalators bind to long
    chain naturally occurring DNA molecules (see Section 4).
4. They bind to DNA in a neighbor-exclusion mode at high drug–
    DNA binding ratios (i.e., a structure that corresponds to β-DNA,
    or β premelted DNA).

Complex intercalators (summarized in Fig. 4 and Table 4) utilize additional, "complex" types of interactions when binding to DNA or RNA.
As a result of these additional interactions, there may or may not be
alterations in the sugar-phosphate geometries (with concomitant variations in the magnitude of angular unwinding), sequence binding affinities,
and neighbor-exclusion volumes.

Table 4. Unit Cell Constants, Space Groups, and Structural Data for Complex Intercalators Complexed to DNA- and RNA-like, Self-Complementary Dinucleotides and Oligonucleotides

Complex	St[a]	Cell Constants	Space Group	Sugar Puckering	Twist Angle	Reference
Proflavine: 5-iodocytidylyl-(3'-5')-guanosine	2:2	$a = 32.11$ Å, $b = 22.23$ Å, $c = 18.45$ Å, $\beta = 123.2°$	C2	C3'-endo-(3'-5')-C3'-endo	$36 \pm 1°$	Reddy et al., 1979
Proflavine: cytidylyl-(3'-5')-guanosine	3:2	$a = 16.10$ Å, $b = 16.76$ Å, $c = 12.93$ Å, $\alpha = 122.7°$, $\beta = 113.7°$, $\gamma = 99.4°$	P1	C3'-endo-(3'-5')-C3'-endo	$36 \pm 1°$	Neidle et al., 1977
Daunomycin: d(CpGpTpApCpG)	2:2	$a = 27.92$ Å, $b = 27.92$ Å, $c = 52.89$ Å	$P4_12_12$	C2'-endo-(3'-5')-C2'-endo	$36 \pm 1°$	Quigley et al., 1980

[a] Stoichiometry.

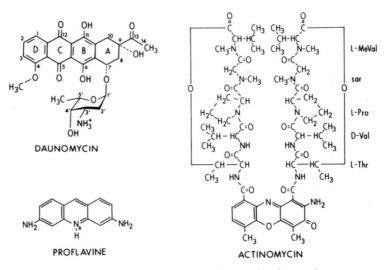

Figure 4.   Chemical structures of complex intercalators.

Proflavine behaves as a complex intercalator when binding to self-complementary dinucleoside monophosphates in these solid-state studies (Neidle et al., 1977; Reddy et al., 1979). This reflects its ability to form hydrogen bonds with its amino groups to phosphate oxygen atoms on opposite chains when intercalating into these model systems (see Figs. 5a and 6a).

Proflavine binds to a unique intercalative structure that has all sugar residues in the C3'-*endo* conformation. Although the torsional angles defining the phosphodiester geometries have values similar to the β structural element, it is important to realize that this structure is clearly *different* and should not be confused with the β element. Base pairs are separated by 6.8 Å; however, they are twisted 36°. The angular unwinding component present in the β structural element is therefore *absent* from this structure—this reflects the homogeneous sugar puckering that is observed. Phosphate oxygen atoms on opposing chains are separated by 15.3 Å, a distance that allows proflavine to hydrogen bond simultaneously (i.e., span) both phosphate oxygens. This gives rise to the symmetric interaction observed in the proflavine intercalative complexes.

Acridine orange, however, binds to the β element in these model studies (Reddy et al., 1979; Wang et al., 1979) (see Figs. 5b and 6b). Because

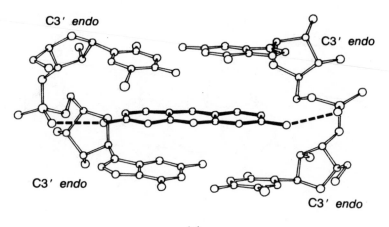

(a)

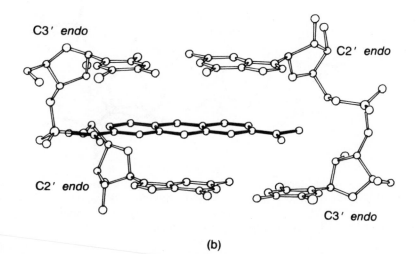

(b)

Figure 5. View taken at a skew angle to base pairs and drug molecules, showing differences between (a) proflavine–5-iodocytidylyl-(3′-5′)-guanosine crystalline complex, and (b) the acridine orange–5-iodocytidylyl-(3′-5′)-guanosine crystalline complex. Dashed lines denote hydrogen bonds.

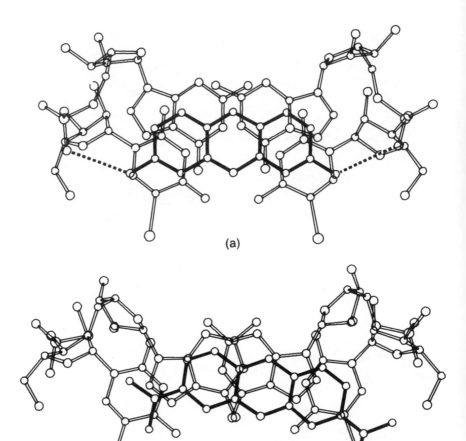

(a)

(b)

Figure 6. View taken perpendicular to base pairs and drug molecules to show differences between (*a*) the proflavine–5-iodocytidylyl-(3′-5′)-guanosine crystalline complex and (*b*) the acridine orange–5-iodocytidylyl-(3′-5′)-guanosine crystalline complex. Dashed lines denote hydrogen bonds.

it is a methylated proflavine derivative, acridine orange is unable to hydrogen bond to the phosphate oxygen atoms on opposite chains. Instead, acridine orange intercalates asymmetrically into the β structural element, forming tight stereo-specific stacking interactions with adjacent guanine rings. This asymmetry is a feature common to several different acridines in these solid state studies (see Sakore et al., 1979). It is possible for proflavine to intercalate into the β structural element; however, it would not be able to form hydrogen bonds simultaneously with phosphate oxygens on opposite chains (distance, 17.3 Å). One might expect such a complex to be asymmetric, with proflavine possibly forming one hydrogen bond with a neighboring phosphate oxygen.

To investigate this type of interaction, we have synthesized N,N-dimethylproflavine and have examined three different complexes between RNA-like and DNA-like self-complementary dinucleoside monophosphates (Bhandary et al., 1984; Sakore et al., 1984). N,N-Dimethylproflavine intercalates into the β structural element in all three cases. One structure, N,N-dimethylproflavine: deoxycytidylyl-(3'-5')-deoxyguanosine, is shown in Figure 7a and b. As expected, N,N-dimethylproflavine intercalates asymmetrically into the β structural element; however, unexpectedly, it does *not* form a hydrogen bond with the neighboring phosphate oxygen (distance, 3.5 Å). Instead, this amino group is hydrogen bonded to a water molecule. This probably reflects the position of the phosphate oxygen atom relative to the N,N-dimethylproflavine molecule, which is about 0.7 Å from the N,N-dimethylproflavine least-squares plane in this complex and in the other complexes we have studied.

Although proflavine behaves as a complex intercalator in these model studies by virtue of its ability to form hydrogen bonds to phosphate oxygen atoms with each of its amino groups, it may behave as a simple intercalator when binding to DNA. This is because proflavine is known to unwind DNA much the same as acridine orange, unwinding DNA about 20° on the ethidium scale (Waring, 1970). With DNA, proflavine could bind to the β structural element in an asymmetric fashion—analogous to the other acridine–DNA complexes we have studied. With RNA, however, proflavine could behave as a complex intercalator, binding symmetrically to the C3'-*endo* structure with two hydrogen bonds. This might be detected by altered spectroscopic properties of the complex along with decreased angular unwinding accompanying intercalation. Further information is needed to understand the interaction of proflavine with double-stranded RNA.

Another example of a complex intercalator is daunomycin. This drug molecule forms a complex with the DNA-like hexanucleotide d(CpGpTpApCpG), and this complex has been analyzed by X-ray crystallography (Quigley et al. 1980; see Fig. 8a, b). The daunomycin aglycone chromophore is oriented at right angles to the long dimension of the

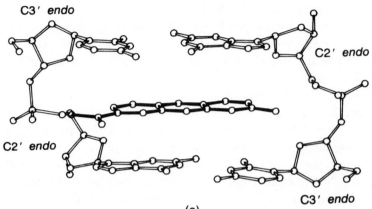

C3' endo

C2' endo

C2' endo

C3' endo

(a)

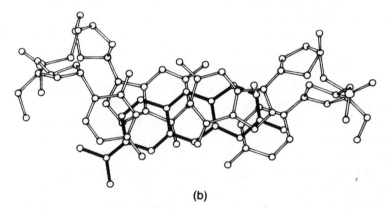

(b)

Figure 7. Structural information obtained from the *N,N*-dimethylproflavine–deoxycytidylyl-(3'-5')-deoxyguanosine crystalline complex. (*a*) Skew view to base pairs and drug molecule. (*b*) Perpendicular view to base pairs and drug molecule. This figure should be compared with the information shown in Figures 5 and 6.

DNA base pairs, and the cyclohexene ring rests in the minor groove. Substituents on this ring have hydrogen bonding interactions with the base pairs above and below the intercalation sites. The amino sugar lies in the minor groove, forming electrostatic interactions with phosphate groups. The intercalation geometry contains sugar residues in the C2'-*endo* conformation and involves the sequence d(CpG). Base pairs above and below the anthracycline ring are twisted about 36°; again, as in the

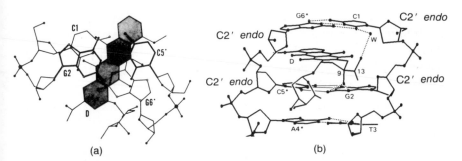

(a)   (b)

Figure 8. Structural information obtained from the daunomycin–d(CpGpTpApCpG) crystalline complex. (*a*) Complex viewed perpendicualr to the plane of the base pairs. (*b*) Complex viewed at a skew angle to base pairs and drug molecule.

proflavine studies, there is no unwinding at the immediate intercalation site. There is a small amount of unwinding in neighboring regions, however, which is estimated to be about 4° on either side. This decreased magnitude of angular unwinding is in good agreement with solution binding studies with superhelical DNA, where the angular unwinding has been estimated to be near 11° (Waring, 1970). Daunomycin, therefore, appears to behave as a complex intercalator when binding to DNA.

Finally, actinomycin D can be considered to be a complex intercalator when binding to DNA. Actinomycin demonstrates a d(GpX) (where X = C,T,A, or G) sequence-binding affinity with oligonucleotides and synthetic and naturally occurring DNA polymers (Reich and Goldberg, 1964; Wells and Larsen, 1970; Krugh, 1972; Patel, 1974). This sequence-binding preference reflects the presence of the pentapeptide chains on actinomycin, which form hydrogen bonds with guanine residues [on opposite chains, in the case of the d(GpC) sequence] in the three-dimensional DNA interaction (Sobell et al., 1971; Sobell and Jain, 1972). Reinhardt (1976) has investigated the binding of ethidium to DNA in the presence of actinomycin and, also, actinomine, an analog of actinomycin that lacks the pentapeptide chains. His studies clearly indicate that, whereas actinomycin binds noncompetitively with ethidium for sites on DNA, actinomine *competes* with ethidium for these sites. These data indicate that the pentapeptide chains on actinomycin play a key role in determining the sequence-binding guanine specificity in the DNA-binding reaction. Actinomycin behaves as a complex intercalator by virtue of its pentapeptide chains, whose interactions with DNA give rise to altered sequence-binding affinities. The enhanced neighbor-exclusion volume—estimated to be about five base pairs—also reflects the presence of the pentapeptide chains. Nevertheless, actinomycin unwinds DNA roughly the same as ethidium (Waring, 1970); this reflects the lock and key fit between actinomycin and DNA. Early model-building studies suggested that actinomycin binds to the β structural element in DNA (Sobell and Jain, 1972; Sobell et al., 1977, 1978), and this explains the magnitude of angular unwinding accompanying actinomycin-DNA binding (i.e., about 28°).

3. KINK-ANTIKINK BOUND STATES IN DNA STRUCTURE—THE β PREMELTON

How does the β structural element arise in DNA? And what are the surrounding structural features on either side of these elements?

We propose that β elements arise as part of more organized β structure, whose appearance reflects the presence of a soliton-antisoliton (i.e., kink-antikink) bound state in DNA structure (see Figures 9a and b). Such a structure contains a modulated β alternation in sugar puckering about the central β premelted core region which gradually merges into B-DNA on either side. Typically, a structure of this kind would give rise to an energy density profile similar to that shown in Figure 9c. We call this type of composite structure a β premelton (Banerjee and Sobell, 1983).

We have already described the methods used to construct this β premelton (Sobell et al., 1983a, 1983b). The B-DNA to β-DNA transition was first computed as a homogeneous transition involving the entire polymer length. This was accomplished in a series of steps in which the sugar puckering of alternate deoxyribose residues was altered and the structure then energy minimized, subject to a series of constraints and restraints using the method of linked-atom least-squares (coordinates for these structural intermediates are presented in Table 5, along with the A-DNA to β-DNA transition). To simulate the bound state structure, base paired dinucleotide elements from each structure in this sequence were then pieced together using a least-squares procedure. The decision to compute 12 intermediate structures and to then use these to construct the soliton-antisoliton bound state without a detailed knowledge of the total energies involved, is somewhat arbitrary but does not alter the basic conclusions

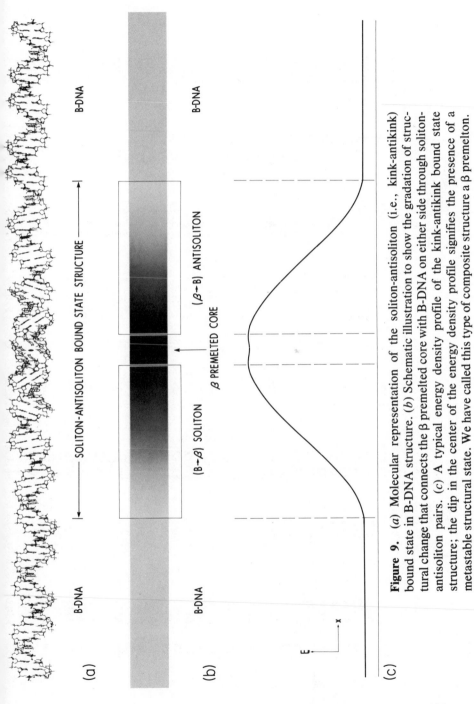

Figure 9. (a) Molecular representation of the soliton-antisoliton (i.e., kink-antikink) bound state in B-DNA structure. (b) Schematic illustration to show the gradation of structural change that connects the β premelted core with B-DNA on either side through soliton-antisoliton pairs. (c) A typical energy density profile of the kink-antikink bound state structure; the dip in the center of the energy density profile signifies the presence of a metastable structural state. We have called this type of composite structure a β premelton.

191

Table 5. Coordinates of Structural Intermediates Connecting B-DNA with A-DNA as Computed as a Uniform Transition Along the Polymer by the Method of Linked-Atom Least-Squares

| No. | Atom | $\Psi = (162,162)$ [5.00, 34.00 Å] | | | No. | Atom | $\Psi = (162,150)$ [5.02, 33.75 Å] | | |
		R (Å)	θ (°)	z (Å)			R (Å)	θ (°)	z (Å)
		Deoxyadenosine 5'-monophosphate					*Deoxyadenosine 5'-monophosphate*		
1	AN1	2.084	32.08	1.699	1	AN1	2.355	33.41	1.682
2	AC2	3.407	35.38	1.703	2	AC2	3.680	35.91	1.639
3	AH2	4.082	24.02	1.706	3	AH2	4.342	25.15	1.656
4	AN3	4.284	50.12	1.702	4	AN3	4.557	49.62	1.570
5	AC4	4.017	68.46	1.697	5	AC4	4.268	66.80	1.545
6	AC5	2.995	83.46	1.694	6	AC5	3.213	80.21	1.582
7	AC6	1.660	72.24	1.695	7	AC6	1.910	68.25	1.655
8	AN6	0.882	125.88	1.691	8	AN6	0.965	109.26	1.698
9	AH6	0.274	−126.38	1.693	9	AH6	0.041	−34.90	1.748
10	AH6	1.842	138.57	1.688	10	AH6	1.844	129.81	1.671
11	AN7	3.865	101.74	1.689	11	AN7	4.020	98.28	1.538
12	AC8	4.999	93.52	1.690	12	AC8	5.181	91.07	1.476
13	AH8	5.747	100.63	1.688	13	AH8	5.900	98.26	1.441
14	AN9	5.156	78.04	1.695	14	AN9	5.385	76.26	1.476
15	AC1'	6.536	72.57	1.698	15	AC1'	6.775	71.30	1.415
16	AH1'	6.643	63.99	1.567	16	AH1'	6.894	62.99	1.342
17	AC2'	7.455	77.79	0.664	17	AC2'	7.606	76.06	0.286
18	AH2'	7.397	85.49	0.627	18	AH2'	7.514	83.58	0.205
19	AH2'	7.366	74.02	−0.204	19	AH2'	7.475	71.95	−0.545
20	AC3'	8.822	75.73	1.261	20	AC3'	9.018	74.48	0.801
21	AH3'	9.531	79.63	0.932	21	AH3'	9.688	78.27	0.390
22	AO3'	9.361	67.63	0.950	22	AO3'	9.562	66.51	0.535
23	AC4'	8.621	77.00	2.762	23	AC4'	8.917	76.24	2.300
24	AH4'	9.139	72.50	3.257	24	AH4'	9.474	72.10	2.797
25	AO1'	7.122	75.18	2.948	25	AO1'	7.438	74.47	2.602
26	AC5'	9.123	85.52	3.307	26	AC5'	9.439	84.67	2.741
27	AH5'	10.107	85.55	3.130	27	AH5'	10.392	84.84	2.438
28	AH5'	8.962	85.80	4.293	28	AH5'	9.405	85.03	3.739
29	AO5'	8.588	93.20	2.666	29	AO5'	8.783	91.95	2.164
30	P1	9.553	101.17	2.820	30	P1	9.663	100.18	2.303
31	P101	9.219	108.22	1.950	31	P101	9.257	106.95	1.416
32	P102	10.933	98.46	2.620	32	P102	11.070	97.94	2.115

Thymidine 5'-monophosphate

33	TN1	5.157	42.07	−1.701
34	TC2	4.163	30.32	−1.701
35	TO2	4.665	15.85	−1.700
36	TN3	2.849	37.40	−1.702
37	TH3	2.314	18.46	−1.702
38	TC4	2.836	65.50	−1.703
39	TO4	2.130	89.24	−1.704
40	TC5	4.264	69.06	−1.703
41	TC6	5.195	57.27	−1.702
42	TH6	6.082	61.99	−1.702
43	TC7	4.992	85.41	−1.705
44	TCH7	4.515	96.03	−1.705
45	TCH7	5.569	85.25	−0.888
46	TCH7	5.569	85.23	−2.521
47	TC1'	6.536	36.58	−1.700
48	TH1'	6.643	28.00	−1.831
49	TC2'	7.455	41.80	−2.735
50	TH2'	7.365	38.03	−3.603
51	TH2'	7.397	49.51	−2.772
52	TC3'	8.822	39.73	−2.139
53	TH3'	9.531	43.63	−2.469
54	TO3'	9.360	31.63	−2.451
55	TC4'	8.623	41.01	−0.638
56	TH4'	9.141	36.51	−0.144
57	TO1'	7.124	39.20	−0.451
58	TC5'	9.126	49.52	−0.093
59	TH5'	10.110	49.55	−0.271
60	TH5'	8.965	49.80	0.893
61	TO5'	8.591	57.20	−0.734
62	P1	9.555	65.17	−0.579
63	PO1	9.222	72.22	−1.449
64	PO2	10.936	62.47	−0.778

Thymidine 5'-monophosphate

33	TN1	5.197	40.27	−1.931
34	TC2	4.274	27.90	−1.859
35	TO2	4.851	14.38	−1.819
36	TN3	2.926	32.78	−1.835
37	TH3	2.496	13.58	−1.783
38	TC4	2.747	60.70	−1.876
39	TO4	1.918	83.63	−1.848
40	TC5	4.145	66.65	−1.951
41	TC6	5.143	55.48	−1.976
42	TH6	5.999	60.80	−2.029
43	TC7	4.784	84.17	−1.999
44	TCH7	4.260	95.00	−1.971
45	TCH7	5.403	84.37	−1.214
46	TCH7	5.317	84.31	−2.845
47	TC1'	6.605	35.62	−1.959
48	TH1'	6.732	27.10	−1.999
49	TC2'	7.421	40.24	−3.116
50	TH2'	7.295	35.84	−3.932
51	TH2'	7.320	47.92	−3.240
52	TC3'	8.839	38.86	−2.605
53	TH3'	9.492	42.89	−3.004
54	TO3'	9.440	30.95	−2.897
55	TC4'	8.730	40.45	−1.101
56	TH4'	9.216	35.82	−0.613
57	TO1'	7.250	39.53	−0.792
58	TC5'	9.360	48.59	−0.612
59	TH5'	10.316	48.38	−0.903
60	TH5'	9.317	48.79	0.386
61	TO5'	8.815	56.33	−1.162
62	P1	9.807	63.95	−0.980
63	PO1	9.534	70.74	−1.880
64	PO2	11.183	61.11	−1.116

Table 5. (Continued)

No.	Atom	Ψ = (162,138) (5.04, 33.50 Å) R (Å)	θ (°)	z (Å)	No.	Atom	Ψ = (162,126) [5.06, 33.25 Å] R (Å)	θ (°)	z (Å)
		Deoxyadenosine 5'-monophosphate					*Deoxyadenosine 5'-monophosphate*		
1	AN1	2.606	33.19	1.612	1	AN1	2.765	33.69	1.551
2	AC2	3.931	35.34	1.542	2	AC2	4.089	35.42	1.456
3	AH2	4.588	25.18	1.592	3	AH2	4.739	25.56	1.534
4	AN3	4.800	48.26	1.407	4	AN3	4.957	47.76	1.259
5	AC4	4.483	64.48	1.336	5	AC4	4.628	63.40	1.149
6	AC5	3.397	76.62	1.393	6	AC5	3.530	74.87	1.224
7	AC6	2.130	64.13	1.539	7	AC6	2.284	62.54	1.437
8	AN6	1.074	95.75	1.608	8	AN6	1.189	89.53	1.532
9	AH6	0.268	29.94	1.711	9	AH6	0.428	37.09	1.683
10	AH6	1.852	121.34	1.540	10	AH6	1.900	116.28	1.428
11	AN7	4.139	94.55	1.288	11	AN7	4.233	92.52	1.065
12	AC8	5.320	88.17	1.174	12	AC8	5.421	86.60	0.902
13	AH8	6.010	95.47	1.092	13	AH8	6.091	93.91	0.778
14	AN9	5.572	73.89	1.194	14	AN9	5.698	72.67	0.940
15	AC1'	6.971	69.39	1.089	15	AC1'	7.099	68.39	0.793
16	AH1'	7.109	61.32	1.069	16	AH1'	7.220	60.44	0.808
17	AC2'	7.726	73.71	-0.113	17	AC2'	7.784	72.25	-0.472
18	AH2'	7.596	81.08	-0.235	18	AH2'	7.627	79.48	-0.644
19	AH2'	7.571	69.30	-0.906	19	AH2'	7.609	67.58	-1.231
20	AC3'	9.169	72.59	0.337	20	AC3'	9.251	71.46	-0.088
21	AH3'	9.800	76.28	-0.142	21	AH3'	9.858	74.97	-0.626
22	AO3'	9.729	64.75	0.117	22	AO3'	9.800	63.61	-0.259
23	AC4'	9.142	74.80	1.822	23	AC4'	9.294	74.11	1.377
24	AH4'	9.734	70.97	2.324	24	AH4'	9.914	70.52	1.880
25	AO1'	7.685	73.06	2.217	25	AO1'	7.861	72.49	1.858
26	AC5'	9.667	83.19	2.168	26	AC5'	9.824	82.47	1.630
27	AH5'	10.582	83.46	1.768	27	AH5'	10.706	82.71	1.160
28	AH5'	9.734	83.68	3.163	28	AH5'	9.964	83.15	2.613
29	AO5'	8.917	90.13	1.646	29	AO5'	9.020	89.16	1.134
30	P1	9.721	98.57	1.784	30	P1	9.759	97.74	1.297
31	P1O1	9.256	105.16	0.892	31	P1O1	9.271	104.30	0.416
32	P1O2	11.147	96.74	1.602	32	P1O2	11.202	96.36	1.116

Thymidine 5'-monophosphate

33	TN1	5.236	38.11	−2.130
34	TC2	4.372	25.41	−1.986
35	TO2	5.003	12.63	−1.922
36	TN3	3.009	28.75	−1.919
37	TH3	2.656	9.82	−1.815
38	TC4	2.705	56.01	−1.981
39	TO4	1.785	77.47	−1.909
40	TC5	4.067	63.85	−2.132
41	TC6	5.111	53.25	−2.201
42	TH6	5.938	59.01	−2.304
43	TC7	4.629	82.26	−2.208
44	TCH7	4.072	93.19	−2.139
45	TCH7	5.288	82.69	−1.457
46	TCH7	5.117	82.77	−3.081
47	TC1'	6.662	34.09	−2.207
48	TH1'	6.829	25.73	−2.134
49	TC2'	7.366	37.86	−3.476
50	TH2'	7.225	32.66	−4.212
51	TH2'	7.195	45.43	−3.694
52	TC3'	8.825	37.35	−3.057
53	TH3'	9.419	41.51	−3.515
54	TO3'	9.471	29.57	−3.342
55	TC4'	8.799	39.21	−1.557
56	TH4'	9.252	34.47	−1.069
57	TO1'	7.369	39.16	−1.151
58	TC5'	9.539	47.00	−1.125
59	TH5'	10.460	46.57	−1.508
60	TH5'	9.594	47.11	−0.127
61	TO5'	9.006	54.81	−1.598
62	P1	10.029	62.07	−1.380
63	PO1	9.830	68.66	−2.306
64	PO2	11.394	59.08	−1.447

Thymidine 5'-monophosphate

33	TN1	5.227	36.59	−2.316
34	TC2	4.410	23.62	−2.106
35	TO2	5.079	11.31	−2.018
36	TN3	3.041	25.95	−1.999
37	TH3	2.745	7.16	−1.849
38	TC4	2.651	52.86	−2.083
39	TO4	1.679	73.34	−1.972
40	TC5	3.980	62.03	−2.304
41	TC6	5.050	51.74	−2.412
42	TH6	5.853	57.83	−2.561
43	TC7	4.490	81.18	−2.408
44	TCH7	3.915	92.29	−2.302
45	TCH7	5.183	81.76	−1.688
46	TCH7	4.933	81.93	−3.302
47	TC1'	6.661	32.96	−2.436
48	TH1'	6.866	24.80	−2.257
49	TC2'	7.268	35.77	−3.791
50	TH2'	7.152	29.72	−4.429
51	TH2'	7.006	43.10	−4.107
52	TC3'	8.749	36.30	−3.455
53	TH3'	9.258	40.76	−3.956
54	TO3'	9.479	28.82	−3.750
55	TC4'	8.783	38.36	−1.960
56	TH4'	9.205	33.50	−1.470
57	TO1'	7.415	39.08	−1.503
58	TC5'	9.606	45.89	−1.572
59	TH5'	10.496	45.27	−2.015
60	TH5'	9.726	45.90	−0.580
61	TO5'	9.101	53.81	−1.991
62	P1	10.154	60.79	−1.734
63	PO1	10.055	67.23	−2.682
64	PO2	11.511	57.70	−1.734

Table 5. *(Continued)*

No.	Atom	Ψ = (165,114) [5.08, 33.00 Å]			No.	Atom	Ψ = (168,102) [5.10, 32.75 Å]		
		R (Å)	θ (°)	z (Å)			R (Å)	θ (°)	z (Å)
		Deoxyadenosine 5'-monophosphate					*Deoxyadenosine 5'-monophosphate*		
1	AN1	2.880	35.28	1.525	1	AN1	3.005	33.57	1.288
2	AC2	4.202	36.24	1.380	2	AC2	4.328	34.88	1.157
3	AH2	4.836	26.47	1.475	3	AH2	4.973	25.62	1.310
4	AN3	5.078	47.96	1.110	4	AN3	5.181	46.34	0.830
5	AC4	4.754	63.20	0.976	5	AC4	4.815	61.08	0.617
6	AC5	3.663	74.37	1.095	6	AC5	3.694	71.60	0.713
7	AC6	2.427	62.86	1.386	7	AC6	2.498	59.46	1.070
8	AN6	1.340	87.56	1.530	8	AN6	1.355	81.02	1.201
9	AH6	0.564	47.39	1.734	9	AH6	0.674	38.99	1.460
10	AH6	2.015	113.07	1.399	10	AH6	1.951	108.75	1.007
11	AN7	4.350	91.50	0.892	11	AN7	4.318	88.97	0.422
12	AC8	5.530	85.82	0.660	12	AC8	5.506	83.66	0.164
13	AH8	6.187	93.04	0.501	13	AH8	6.129	91.06	-0.059
14	AN9	5.812	72.17	0.694	14	AN9	5.837	70.16	0.261
15	AC1'	7.203	67.93	0.479	15	AC1'	7.237	66.20	0.040
16	AH1'	7.330	60.12	0.549	16	AH1'	7.362	58.43	0.103
17	AC2'	7.797	71.24	-0.855	17	AC2'	7.814	69.56	-1.300
18	AH2'	7.601	78.37	-1.068	18	AH2'	7.609	76.66	-1.509
19	AH2'	7.594	66.19	-1.562	19	AH2'	7.610	64.51	-2.005
20	AC3'	9.287	70.81	-0.562	20	AC3'	9.308	69.20	-1.021
21	AH3'	9.849	74.13	-1.177	21	AH3'	9.856	72.51	-1.647
22	AO3'	9.861	63.03	-0.672	22	AO3'	9.895	61.47	-1.128
23	AC4'	9.422	74.05	0.865	23	AC4'	9.455	72.50	0.400
24	AH4'	10.095	70.82	1.358	24	AH4'	10.155	69.42	0.881
25	AO1'	8.034	72.39	1.459	25	AO1'	8.083	70.58	1.012
26	AC5'	9.925	82.47	0.993	26	AC5'	9.914	81.00	0.514
27	AH5'	10.719	82.83	0.389	27	AH5'	10.704	81.46	-0.094
28	AH5'	10.213	83.29	1.939	28	AH5'	10.203	81.91	1.458
29	AO5'	9.005	88.88	0.623	29	AO5'	8.960	87.23	0.141
30	P1	9.671	97.74	0.776	30	P1	9.579	96.29	0.288
31	P101	9.069	104.27	-0.060	31	P101	8.907	102.87	-0.501
32	P102	11.113	96.66	0.534	32	P102	11.015	95.49	-0.018

33	TN1	5.145	35.28	−2.554
34	TC2	4.378	21.92	−2.268
35	TO2	5.090	9.99	−2.136
36	TN3	3.006	23.25	−2.138
37	TH3	2.773	4.31	−1.932
38	TC4	2.531	50.28	−2.267
39	TO4	1.520	70.60	−2.128
40	TC5	3.823	60.82	−2.567
41	TC6	4.915	50.62	−2.699
42	TH6	5.692	57.07	−2.905
43	TC7	4.289	80.97	−2.724
44	TCH7	3.709	92.54	−2.589
45	TCH7	5.020	81.79	−2.045
46	TCH7	4.680	81.74	−3.643
47	TC1'	6.586	32.01	−2.702
48	TH1'	6.829	24.00	−2.453
49	TC2'	7.129	34.16	−4.102
50	TH2'	7.083	27.46	−4.658
51	TH2'	6.765	41.13	−4.494
52	TC3'	8.606	35.87	−3.825
53	TH3'	9.021	40.81	−4.326
54	TO3'	9.436	28.85	−4.178
55	TC4'	8.670	37.75	−2.325
56	TH4'	9.073	32.67	−1.856
57	TO1'	7.355	38.91	−1.861
58	TC5'	9.539	45.12	−1.921
59	TH5'	10.435	44.25	−2.339
60	TH5'	9.632	45.16	−0.926
61	TO5'	9.106	53.18	−2.367
62	P1	10.216	59.82	−2.122
63	PO1	10.170	66.02	−3.113
64	PO2	11.533	56.32	−2.068

33	TN1	5.179	33.11	−2.577
34	TC2	4.447	19.85	−2.227
35	TO2	5.175	8.25	−2.100
36	TN3	3.082	20.89	−2.024
37	TH3	2.873	2.60	−1.775
38	TC4	2.566	47.08	−2.140
39	TO4	1.535	65.72	−1.934
40	TC5	3.822	58.24	−2.512
41	TC6	4.919	48.35	−2.715
42	TH6	5.673	54.95	−2.962
43	TC7	4.251	78.67	−2.660
44	TCH7	3.664	90.11	−2.468
45	TCH7	5.014	79.40	−2.016
46	TCH7	4.594	79.91	−3.594
47	TC1'	6.614	30.03	−2.804
48	TH1'	6.900	22.50	−2.441
49	TC2'	7.030	30.67	−4.267
50	TH2'	7.092	22.66	−4.413
51	TH2'	7.075	23.26	−4.675
52	TC3'	8.480	33.96	−4.163
53	TH3'	8.741	39.22	−4.719
54	TO3'	9.396	27.47	−4.577
55	TC4'	8.675	36.20	−2.685
56	TH4'	9.103	31.20	−2.220
57	O5	9.133	51.56	−2.806
58	P1	10.241	58.22	−2.577
59	P101	10.183	64.37	−3.575
60	P102	11.560	54.75	−2.527
61	O1	7.433	37.83	−2.155
62	TC5'	9.592	43.52	−2.394
63	TH5'	10.441	42.56	−2.895
64	TH5'	9.783	43.64	−1.413

Table 5. *(Continued)*

No.	Atom	$\Psi = (174,90)$ [5.12, 32.50 Å]			No.	Atom	$\Psi = (168,78)$ [5.14, 32.25 Å]		
		R (Å)	θ (°)	z (Å)			R (Å)	θ (°)	z (Å)
		Deoxyadenosine 5'-monophosphate					*Deoxyadenosine 5'-monophosphate*		
1	AN1	2.979	34.72	1.098	1	AN1	3.275	32.04	1.058
2	AC2	4.301	35.70	0.950	2	AC2	4.584	33.63	0.836
3	AH2	4.941	26.41	1.133	3	AH2	5.254	25.22	1.020
4	AN3	5.156	46.97	0.571	4	AN3	5.377	44.51	0.373
5	AC4	4.791	61.69	0.317	5	AC4	4.932	58.42	0.115
6	AC5	3.679	72.42	0.423	6	AC5	3.779	67.85	0.291
7	AC6	2.487	60.74	0.838	7	AC6	2.679	54.96	0.791
8	AN6	1.365	83.27	0.985	8	AN6	1.474	70.75	1.013
9	AH6	0.657	43.94	1.287	9	AH6	0.955	32.52	1.368
10	AH6	1.972	110.10	0.755	10	AH6	1.906	100.77	0.768
11	AN7	4.299	89.68	0.080	11	AN7	4.281	85.57	-0.082
12	AC8	5.473	84.20	-0.217	12	AC8	5.459	81.02	-0.465
13	AH8	6.089	91.58	-0.479	13	AH8	6.017	88.80	-0.756
14	AN9	5.805	70.65	-0.097	14	AN9	5.869	67.71	-0.375
15	AC1'	7.196	66.58	-0.350	15	AC1'	7.259	64.23	-0.720
16	AH1'	7.361	58.82	-0.294	16	AH1'	7.449	56.60	-0.633
17	AC2'	7.763	70.20	-1.683	17	AC2'	7.689	67.52	-2.117
18	AH2'	7.526	77.31	-1.897	18	AH2'	7.407	74.64	-2.325
19	AH2'	7.586	65.06	-2.386	19	AH2'	7.470	62.12	-2.782
20	AC3'	9.256	70.16	-1.396	20	AC3'	9.202	67.78	-1.971
21	AH3'	9.753	73.74	-2.028	21	AH3'	9.664	71.20	-2.655
22	AO3'	9.879	62.50	-1.495	22	AO3'	9.844	60.15	-2.088
23	AC4'	9.380	73.54	0.024	23	AC4'	9.447	71.37	-0.581
24	AH4'	10.123	70.74	0.496	24	AH4'	10.210	68.55	-0.153
25	AO1'	8.039	70.91	0.631	25	AO1'	8.151	69.14	0.158
26	AC5'	9.736	82.32	0.136	26	AC5'	9.850	80.02	-0.528
27	AH5'	10.562	82.98	-0.415	27	AH5'	10.621	80.54	-1.158
28	AH5'	9.944	83.51	1.093	28	AH5'	10.155	81.19	0.402
29	AO5'	8.747	88.21	-0.339	29	AO5'	8.844	85.99	-0.909
30	P1	9.256	97.80	-0.206	30	P1	9.364	95.41	-0.714
31	P101	8.451	104.59	-0.880	31	P101	8.535	102.28	-1.317
32	P102	10.661	97.77	-0.657	32	P102	10.760	95.37	-1.191

33	TN1	5.088	33.44	−2.618	33	TN1	5.310	29.05	−2.821
34	TC2	4.382	19.99	−2.212	34	TC2	4.665	16.25	−2.330
35	TO2	5.125	8.40	−2.080	35	TO2	5.440	5.82	−2.118
36	TN3	3.026	20.91	−1.958	36	TN3	3.305	16.27	−2.088
37	TH3	2.839	2.47	−1.670	37	TH3	3.157	−0.17	−1.736
38	TC4	2.493	47.60	−2.074	38	TC4	2.657	39.66	−2.291
39	TO4	1.473	66.94	−1.819	39	TO4	1.553	53.41	−2.037
40	TC5	3.729	59.06	−2.506	40	TC5	3.804	52.45	−2.808
41	TC6	4.816	48.94	−2.760	41	TC6	4.944	43.62	−3.053
42	TH6	5.556	55.69	−3.045	42	TH6	5.627	50.58	−3.404
43	TC7	4.159	79.96	−2.659	43	TC7	4.084	73.68	−3.058
44	TCH7	3.591	91.68	−2.424	44	TCH7	3.457	84.99	−2.815
45	TCH7	4.951	80.47	−2.048	45	TCH7	4.904	75.38	−2.501
46	TCH7	4.463	81.43	−3.605	46	TCH7	4.320	74.95	−4.026
47	TC1′	6.515	30.36	−2.901	47	TC1′	6.749	26.62	−3.095
48	TH1′	6.833	23.05	−2.484	48	TH1′	7.115	20.30	−2.563
49	TC2′	6.850	30.13	−4.387	49	TC2′	7.068	24.82	−4.568
50	TH2′	7.009	22.39	−4.701	50	TH2′	6.310	27.58	−5.135
51	TH2′	6.214	35.21	−4.898	51	TH2′	7.419	17.51	−4.724
52	TC3′	8.237	34.89	−4.416	52	TC3′	8.303	31.28	−4.770
53	TH3′	8.350	40.65	−4.956	53	TH3′	8.212	36.95	−5.340
54	TO3′	9.206	29.02	−4.964	54	TO3′	9.345	26.38	−5.388
55	TC4′	8.549	36.97	−2.952	55	TC4′	8.713	34.09	−3.359
56	TH4′	9.009	31.85	−2.536	56	TH4′	9.248	29.43	−2.935
57	TO1′	7.357	38.69	−2.373	57	TO1′	7.544	35.58	−2.742
58	TC5′	9.492	44.31	−2.701	58	TC5′	9.607	41.70	−3.255
59	TH5′	10.327	43.20	−3.218	59	TH5′	10.385	40.77	−3.862
60	TH5′	9.705	44.50	−1.725	60	TH5′	9.931	42.12	−2.312
61	TO5′	9.047	52.44	−3.123	61	TO5′	9.037	49.56	−3.640
62	P1	10.182	59.04	−2.939	62	P1	10.105	56.63	−3.538
63	PO1	10.145	65.14	−3.954	63	PO1	9.926	62.67	−4.564
64	PO2	11.498	55.51	−2.907	64	PO2	11.451	53.40	−3.558

Table 5. (Continued)

No.	Atom	Ψ = (165,66) [5.16, 32.00 Å]			No.	Atom	Ψ = (162,54) [5.18, 31.75 Å]		
		R (Å)	θ (°)	z (Å)			R (Å)	θ (°)	z (Å)
		Deoxyadenosine 5'-monophosphate					*Deoxyadenosine 5'-monophosphate*		
1	AN1	3.526	30.94	1.021	1	AN1	3.860	28.68	0.997
2	AC2	4.818	32.82	0.724	2	AC2	5.127	30.83	0.619
3	AH2	5.511	25.04	0.899	3	AH2	5.846	23.73	0.771
4	AN3	5.554	43.36	0.192	4	AN3	5.796	40.98	0.026
5	AC4	5.048	56.66	−0.055	5	AC4	5.222	53.52	−0.193
6	AC5	3.874	65.11	0.193	6	AC5	4.031	60.76	0.136
7	AC6	2.857	51.62	0.764	7	AC6	3.114	46.87	0.767
8	AN6	1.619	62.86	1.060	8	AN6	1.859	53.63	1.145
9	AH6	1.228	28.59	1.456	9	AH6	1.598	24.52	1.571
10	AH6	1.900	93.51	0.812	10	AH6	1.960	82.94	0.907
11	AN7	4.279	83.06	−0.192	11	AN7	4.320	78.73	−0.243
12	AC8	5.448	79.27	−0.654	12	AC8	5.477	75.97	−0.780
13	AH8	5.957	87.37	−0.958	13	AH8	5.927	84.41	−1.086
14	AN9	5.915	66.18	−0.604	14	AN9	6.008	63.26	−0.788
15	AC1'	7.293	63.13	−1.039	15	AC1'	7.368	60.78	−1.314
16	AH1'	7.528	55.66	−0.927	16	AH1'	7.654	53.52	−1.197
17	AC2'	7.603	66.21	−2.476	17	AC2'	7.556	63.73	−2.776
18	AH2'	7.271	73.31	−2.682	18	AH2'	7.178	70.79	−2.963
19	AH2'	7.363	60.51	−3.100	19	AH2'	7.296	57.84	−3.368
20	AC3'	9.120	66.81	−2.447	20	AC3'	9.068	64.65	−2.864
21	AH3'	9.511	70.27	−3.176	21	AH3'	9.387	68.21	−3.618
22	AO3'	9.790	59.22	−2.579	22	AO3'	9.741	57.06	−3.032
23	AC4'	9.455	70.63	−1.092	23	AC4'	9.489	68.51	−1.535
24	AH4'	10.251	67.91	−0.707	24	AH4'	10.303	65.78	−1.198
25	AO1'	8.221	68.45	−0.249	25	AO1'	8.318	66.43	−0.604
26	AC5'	9.847	79.31	−1.090	26	AC5'	9.891	77.13	−1.558
27	AH5'	10.569	79.83	−1.775	27	AH5'	10.561	77.66	−2.294
28	AH5'	10.217	80.53	−0.186	28	AH5'	10.326	78.28	−0.680
29	AO5'	8.807	85.22	−1.402	29	AO5'	8.838	83.09	−1.781
30	P1	9.309	94.71	−1.188	30	P1	9.354	92.48	−1.534
31	P1O1	8.414	101.72	−1.667	31	P1O1	8.448	99.68	−1.898
32	P1O2	10.661	94.98	−1.775	32	P1O2	10.664	93.05	−2.206

Thymidine 5'-monophosphate

33	TN1	5.533	26.26	−2.980
34	TC2	4.941	14.11	−2.421
35	TO2	5.743	4.69	−2.144
36	TN3	3.580	13.45	−2.190
37	TH3	3.462	−1.35	−1.785
38	TC4	2.839	33.89	−2.465
39	TO4	1.681	42.83	−2.213
40	TC5	3.900	47.45	−3.049
41	TC6	5.084	39.88	−3.285
42	TH6	5.714	46.94	−3.689
43	TC7	4.031	68.63	−3.379
44	TCH7	3.347	79.33	−3.130
45	TCH7	4.862	71.51	−2.868
46	TCH7	4.214	69.64	−4.360
47	TC1'	6.982	24.41	−3.244
48	TH1'	7.375	18.81	−2.648
49	TC2'	7.311	21.78	−4.695
50	TH2'	6.492	22.48	−5.263
51	TH2'	7.829	15.31	−4.752
52	TC3'	8.358	29.43	−5.048
53	TH3'	8.076	34.94	−5.593
54	TO3'	9.413	25.48	−5.788
55	TC4'	8.871	32.67	−3.694
56	TH4'	9.489	28.40	−3.306
57	TO1'	7.728	33.60	−3.003
58	TC5'	9.692	40.52	−3.698
59	TH5'	10.423	39.76	−4.367
60	TH5'	10.092	41.17	−2.788
61	TO5'	9.011	48.07	−4.047
62	P1	10.006	55.59	−4.029
63	PO1	9.708	61.53	−5.062
64	PO2	11.380	52.73	−4.110

Thymidine 5'-monophosphate

33	TN1	5.857	22.45	−3.087
34	TC2	5.303	11.15	−2.473
35	TO2	6.119	2.78	−2.117
36	TN3	3.939	9.95	−2.281
37	TH3	3.835	−3.16	−1.830
38	TC4	3.123	27.20	−2.643
39	TO4	1.926	31.74	−2.418
40	TC5	4.099	40.88	−3.280
41	TC6	5.331	34.80	−3.480
42	TH6	5.908	41.77	−3.930
43	TC7	4.061	61.17	−3.706
44	TCH7	3.301	70.67	−3.471
45	TCH7	4.876	65.53	−3.237
46	TCH7	4.208	61.62	−4.695
47	TC1'	7.320	21.18	−3.312
48	TH1'	7.730	16.42	−2.648
49	TC2'	7.698	17.85	−4.721
50	TH2'	6.880	16.59	−5.274
51	TH2'	8.365	12.54	−4.684
52	TC3'	8.522	26.15	−5.230
53	TH3'	8.059	31.09	−5.754
54	TO3'	9.563	23.03	−6.072
55	TC4'	9.112	30.10	−3.960
56	TH4'	9.819	26.41	−3.601
57	P102	11.276	51.08	−4.690
58	P1	9.873	53.31	−4.486
59	P101	9.416	59.13	−5.503
60	O5	8.993	45.25	−4.382
61	O1	8.005	30.39	−3.172
62	TC5'	9.815	38.17	−4.098
63	TH5'	10.473	37.62	−4.845
64	TH5'	10.307	39.15	−3.244

Table 5. (Continued)

No.	Atom	Ψ = (162,42) [5.20, 31.50 Å]			No.	Atom	Ψ = (162,30) [5.22, 31.25 Å]		
		R (Å)	θ (°)	z (Å)			R (Å)	θ (°)	z (Å)
		Deoxyadenosine 5'-monophosphate					*Deoxyadenosine 5'-monophosphate*		
1	AN1	4.181	24.90	0.764	1	AN1	4.384	24.11	0.741
2	AC2	5.426	27.96	0.365	2	AC2	5.606	27.39	0.292
3	AH2	6.192	21.87	0.551	3	AH2	6.397	21.81	0.482
4	AN3	6.006	37.81	-0.292	4	AN3	6.123	36.93	-0.424
5	AC4	5.317	49.27	-0.558	5	AC4	5.369	47.79	-0.697
6	AC5	4.076	54.85	-0.216	6	AC5	4.120	52.53	-0.308
7	AC6	3.306	40.11	0.486	7	AC6	3.448	37.76	0.455
8	AN6	2.027	41.58	0.885	8	AN6	2.185	37.25	0.903
9	AH6	1.960	16.20	1.363	9	AH6	2.203	14.75	1.420
10	AH6	1.869	69.73	0.605	10	AH6	1.897	63.12	0.610
11	AN7	4.176	73.16	-0.656	11	AN7	4.106	70.82	-0.774
12	AC8	5.326	71.69	-1.242	12	AC8	5.230	70.10	-1.419
13	AH8	5.676	80.74	-1.595	13	AH8	5.515	79.52	-1.794
14	AN9	5.983	59.39	-1.224	14	AN9	5.950	58.03	-1.418
15	AC1'	7.345	57.73	-1.790	15	AC1'	7.290	56.81	-2.047
16	AH1'	7.682	50.57	-1.707	16	AH1'	7.649	49.65	-2.005
17	AC2'	7.478	61.11	-3.243	17	AC2'	7.354	60.59	-3.490
18	AH2'	7.057	68.17	-3.393	18	AH2'	6.917	67.77	-3.592
19	AH2'	7.238	55.22	-3.852	19	AH2'	7.098	54.71	-4.111
20	AC3'	8.981	62.49	-3.359	20	AC3'	8.849	62.15	-3.661
21	AH3'	9.250	66.36	-4.100	21	AH3'	9.084	66.26	-4.390
22	AO3'	9.710	55.11	-3.592	22	AO3'	9.579	54.76	-3.961
23	AC4'	9.407	66.22	-2.020	23	AC4'	9.321	65.66	-2.320
24	AH4'	10.257	63.58	-1.750	24	AH4'	10.187	63.02	-2.099
25	AO1'	8.271	63.60	-1.077	25	AO1'	8.229	62.68	-1.347
26	AC5'	9.753	75.02	-1.998	26	AC5'	9.650	74.57	-2.253
27	AH5'	10.418	75.82	-2.732	27	AH5'	10.327	75.52	-2.969
28	AH5'	10.185	76.12	-1.117	28	AH5'	10.062	75.58	-1.358
29	AO5'	8.665	80.87	-2.191	29	AO5'	8.556	80.45	-2.444
30	P1	9.143	90.55	-1.951	30	P1	9.029	90.27	-2.208
31	P1O1	8.140	97.73	-2.116	31	P1O1	7.998	97.46	-2.238
32	P1O2	10.345	91.60	-2.788	32	P1O2	10.146	91.63	-3.144

33	TN1	6.268	20.00	-3.115		33	TN1	6.519	18.11	-3.248
34	TC2	5.741	9.78	-2.431		34	TC2	6.015	8.55	-2.509
35	TO2	6.557	2.09	-2.055		35	TO2	6.838	1.55	-2.066
36	TN3	4.382	8.94	-2.194		36	TN3	4.654	7.47	-2.296
37	TH3	4.267	-2.46	-1.694		37	TH3	4.544	-2.91	-1.752
38	TC4	3.547	23.96	-2.574		38	TC4	3.780	20.70	-2.749
39	TO4	2.354	27.12	-2.304		39	TO4	2.577	22.30	-2.494
40	TC5	4.458	36.45	-3.286		40	TC5	4.632	32.75	-3.515
41	TC6	5.698	31.28	-3.531		41	TC6	5.898	28.46	-3.737
42	TH6	6.227	37.89	-4.030		42	TH6	6.385	34.84	-4.282
43	TC7	4.306	55.14	-3.736		43	TC7	4.361	50.36	-4.046
44	TCH7	3.510	63.09	-3.460		44	TCH7	3.524	57.31	-3.774
45	TCH7	5.121	60.02	-3.318		45	TCH7	5.152	56.24	-3.674
46	TCH7	4.394	55.54	-4.732		46	TCH7	4.423	50.12	-5.044
47	TC1'	7.722	18.86	-3.391		47	TC1'	7.980	17.29	-3.500
48	TH1'	8.163	14.66	-2.708		48	TH1'	8.419	13.16	-2.823
49	TC2'	8.055	15.14	-4.786		49	TC2'	8.344	13.95	-4.900
50	TH2'	7.238	12.56	-5.250		50	TH2'	7.574	10.60	-5.338
51	TH2'	8.826	10.82	-4.736		51	TH2'	9.193	10.49	-4.866
52	TC3'	8.643	23.58	-5.463		52	TC3'	8.740	22.61	-5.602
53	TH3'	8.018	28.01	-5.903		53	TH3'	8.015	26.64	-5.957
54	TO3'	9.571	20.95	-6.463		54	TO3'	9.597	20.67	-6.699
55	TC4'	9.378	28.03	-4.322		55	TC4'	9.516	27.23	-4.511
56	TH4'	10.177	24.83	-4.069		56	TH4'	10.379	24.47	-4.355
57	P1	9.738	51.23	-5.044		57	TO1'	8.624	25.98	-3.418
58	P101	9.075	56.77	-6.011		58	TC5'	9.932	35.67	-4.761
59	P102	11.138	49.48	-5.386		59	TH5'	10.534	35.61	-5.560
60	O5	8.998	42.69	-4.814		60	TH5'	10.452	37.36	-3.961
61	O1	8.391	27.73	-3.347		61	TO5'	8.872	41.53	-5.001
62	TC5'	9.946	36.19	-4.584		62	P1	9.453	50.51	-5.386
63	TH5'	10.533	35.77	-5.391		63	PO1	8.587	55.73	-6.268
64	TH5'	10.504	37.57	-3.792		64	PO2	10.812	49.21	-5.913

Table 5. (Continued)

No.	Atom	Ψ = (162,18) [5.25, 31.00 Å] R (Å)	θ (°)	z (Å)	No.	Atom	Ψ = (150,18) [5.28, 30.75 Å] R (Å)	θ (°)	z (Å)
		Deoxyadenosine 5'-monophosphate					*Deoxyadenosine 5'-monophosphate*		
1	AN1	4.802	21.45	0.633	1	AN1	4.767	22.45	0.967
2	AC2	5.990	24.88	0.124	2	AC2	5.951	25.62	0.433
3	AH2	6.810	20.01	0.308	3	AH2	6.756	20.40	0.564
4	AN3	6.423	33.85	-0.651	4	AN3	6.403	34.81	-0.307
5	AC4	5.586	43.66	-0.917	5	AC4	5.613	45.10	-0.502
6	AC5	4.330	47.05	-0.469	6	AC5	4.380	48.98	-0.021
7	AC6	3.793	32.83	0.353	7	AC6	3.796	34.72	0.757
8	AN6	2.561	30.28	0.863	8	AN6	2.569	33.23	1.292
9	AH6	2.674	12.11	1.415	9	AH6	2.639	14.29	1.803
10	AH6	2.109	51.13	0.566	10	AH6	2.187	54.90	1.048
11	AN7	4.151	64.57	-0.947	11	AN7	4.280	66.57	-0.425
12	AC8	5.235	65.17	-1.657	12	AC8	5.371	66.87	-1.126
13	AH8	5.420	74.90	-2.045	13	AH8	5.610	76.39	-1.463
14	AN9	6.049	53.86	-1.688	14	AN9	6.123	55.36	-1.220
15	AC1'	7.358	53.34	-2.390	15	AC1'	7.425	54.62	-1.932
16	AH1'	7.763	46.40	-2.375	16	AH1'	7.859	47.90	-1.831
17	AC2'	7.320	57.25	-3.828	17	AC2'	7.345	57.38	-3.411
18	AH2'	6.836	64.32	-3.890	18	AH2'	6.835	64.26	-3.540
19	AH2'	7.066	51.30	-4.443	19	AH2'	7.101	50.95	-3.944
20	AC3'	8.793	59.29	-4.074	20	AC3'	8.803	59.49	-3.728
21	AH3'	8.965	63.58	-4.803	21	AH3'	8.934	63.61	-4.488
22	AO3'	9.549	52.07	-4.429	22	AO3'	9.574	52.33	-4.076
23	AC4'	9.314	62.80	-2.750	23	AC4'	9.373	63.33	-2.449
24	AH4'	10.206	60.34	-2.582	24	AH4'	10.220	60.47	-2.239
25	AO1'	8.295	59.43	-1.730	25	AO1'	8.348	61.29	-1.375
26	AC5'	9.589	71.81	-2.665	26	AC5'	9.793	72.01	-2.488
27	AH5'	10.226	73.01	-3.407	27	AH5'	10.399	72.59	-3.277
28	AH5'	10.036	72.78	-1.786	28	AH5'	10.306	73.03	-1.648
29	AO5'	8.452	77.48	-2.780	29	AO5'	8.742	78.16	-2.597
30	P1	8.894	87.55	-2.592	30	P1	9.336	87.54	-2.505
31	P101	7.830	94.59	-2.464	31	P101	8.384	94.91	-2.559
32	P102	9.876	89.27	-3.657	32	P102	10.411	88.03	-3.512

Thymidine 5'-monophosphate

33	TN1	6.972	15.61	−3.394
34	TC2	6.503	7.00	−2.590
35	TO2	7.330	0.53	−2.151
36	TN3	5.153	6.37	−2.306
37	TH3	5.039	−2.65	−1.719
38	TC4	4.265	18.13	−2.748
39	TO4	3.079	19.57	−2.427
40	TC5	5.056	28.88	−3.586
41	TC6	6.314	25.12	−3.878
42	TH6	6.752	31.09	−4.464
43	TC7	4.697	44.91	−4.110
44	TCH7	3.851	50.58	−3.782
45	TCH7	5.484	50.85	−3.788
46	TCH7	4.695	44.69	−5.110
47	TC1'	8.418	14.78	−3.723
48	TH1'	8.863	10.77	−3.062
49	TC2'	8.704	11.48	−5.136
50	TH2'	7.944	7.74	−5.494
51	TH2'	9.591	8.58	−5.154
52	TC3'	8.932	19.91	−5.897
53	TH3'	8.124	23.53	−6.137
54	TO3'	9.649	18.21	−7.099
55	TC4'	9.807	24.68	−4.918
56	TH4'	10.715	22.34	−4.903
57	TO1'	9.103	22.88	−3.668
58	TC5'	10.080	33.13	−5.168
59	TH5'	10.634	33.43	−5.999
60	TH5'	10.615	34.97	−4.391
61	TO5'	8.922	38.32	−5.329
62	P1	9.326	47.52	−5.818
63	PO1	8.296	52.18	−6.609
64	PO2	10.633	46.73	−6.489

Thymidine 5'-monophosphate

33	TN1	6.970	15.62	−3.449
34	TC2	6.463	6.81	−2.696
35	TO2	7.274	0.46	−2.196
36	TN3	5.096	5.59	−2.538
37	TH3	4.970	−3.75	−1.979
38	TC4	4.207	16.97	−3.058
39	TO4	2.992	17.27	−2.849
40	TC5	5.030	28.18	−3.834
41	TC6	6.322	24.89	−4.003
42	TH6	6.788	30.88	−4.563
43	TC7	4.655	43.88	−4.436
44	TCH7	3.765	49.20	−4.196
45	TCH7	5.391	50.49	−4.082
46	TCH7	4.736	43.14	−5.431
47	TC1'	8.442	15.11	−3.642
48	TH1'	8.839	11.09	−2.953
49	TC2'	8.871	12.06	−5.029
50	TH2'	8.158	8.37	−5.468
51	TH2'	9.764	9.31	−4.975
52	TC3'	9.141	20.41	−5.751
53	TH3'	8.347	23.82	−6.065
54	TO3'	9.975	18.89	−6.881
55	TC4'	9.902	25.20	−4.692
56	TH4'	10.813	22.98	−4.591
57	TO1'	9.090	23.30	−3.515
58	TC5'	10.167	33.59	−4.930
59	TH5'	10.696	33.96	−5.776
60	TH5'	10.724	35.34	−4.164
61	TO5'	9.004	38.73	−5.043
62	P1	9.396	47.96	−5.486
63	PO1	8.376	52.60	−6.287
64	PO2	10.721	47.36	−6.129

Table 5. *(Continued)*

No.	Atom	Ψ = (138,18) [5.30, 30.50 Å] R (Å)	Θ (°)	Z (Å)	No.	Atom	Ψ = (126,18) [5.32, 30.25 Å] R (Å)	Θ (°)	Z (Å)
		Deoxyadenosine 5'-monophosphate					*Deoxyadenosine 5'-monophosphate*		
1	AN1	4.788	22.68	1.166	1	AN1	4.799	21.67	1.178
2	AC2	5.966	25.93	0.624	2	AC2	5.974	25.39	0.654
3	AH2	6.767	20.59	0.719	3	AH2	6.785	20.19	0.738
4	AN3	6.424	35.31	−0.079	4	AN3	6.419	35.02	−0.017
5	AC4	5.651	45.75	−0.222	5	AC4	5.628	45.38	−0.146
6	AC5	4.424	49.60	0.272	6	AC5	4.386	48.82	0.334
7	AC6	3.827	35.14	1.008	7	AC6	3.809	33.76	1.035
8	AN6	2.604	33.54	1.552	8	AN6	2.585	30.87	1.559
9	AH6	2.673	14.39	2.026	9	AH6	2.705	11.61	2.010
10	AH6	2.232	55.19	1.347	10	AH6	2.163	52.47	1.370
11	AN7	4.351	67.19	−0.076	11	AN7	4.289	66.67	0.009
12	AC8	5.453	67.58	−0.759	12	AC8	5.406	67.42	−0.647
13	AH8	5.714	77.02	−1.055	13	AH8	5.663	77.01	−0.924
14	AN9	6.183	56.12	−0.894	14	AN9	6.157	56.02	−0.784
15	AC1'	7.488	55.42	−1.601	15	AC1'	7.477	55.55	−1.464
16	AH1'	7.977	49.13	−1.396	16	AH1'	8.016	49.71	−1.170
17	AC2'	7.379	56.75	−3.110	17	AC2'	7.383	55.69	−2.984
18	AH2'	6.787	63.05	−3.326	18	AH2'	6.701	61.14	−3.277
19	AH2'	7.217	49.79	−3.543	19	AH2'	7.344	48.40	−3.333
20	AC3'	8.804	59.45	−3.482	20	AC3'	8.763	59.27	−3.379
21	AH3'	8.886	63.41	−4.269	21	AH3'	8.782	63.32	−4.163
22	AO3'	9.621	52.45	−3.796	22	AO3'	9.657	52.68	−3.713
23	AC4'	9.376	63.77	−2.242	23	AC4'	9.302	63.84	−2.145
24	AH4'	10.187	60.74	−1.967	24	AH4'	10.066	60.73	−1.771
25	AO1'	8.358	62.87	−1.167	25	AO1'	8.274	63.72	−1.128
26	AC5'	9.882	72.20	−2.400	26	AC5'	9.874	72.12	−2.352
27	AH5'	10.430	72.33	−3.236	27	AH5'	10.374	72.05	−3.218
28	AH5'	10.470	73.25	−1.612	28	AH5'	10.514	73.04	−1.601
29	AO5'	8.891	78.65	−2.478	29	AO5'	8.940	78.88	−2.384
30	P1	9.581	87.59	−2.482	30	P1	9.706	87.49	−2.424
31	P1O1	8.705	95.03	−2.649	31	P1O1	8.906	95.08	−2.633
32	P1O2	10.682	87.26	−3.463	32	P1O2	10.807	86.79	−3.399

Thymidine 5'-monophosphate

33	TN1	7.046	16.02	−3.468
34	TC2	6.514	7.13	−2.757
35	TO2	7.312	0.73	−2.252
36	TN3	5.144	5.86	−2.650
37	TH3	5.004	−3.60	−2.119
38	TC4	4.270	17.21	−3.180
39	TO4	3.048	17.34	−3.017
40	TC5	5.119	28.42	−3.908
41	TC6	6.418	25.23	−4.029
42	TH6	6.905	31.19	−4.561
43	TC7	4.762	43.82	−4.518
44	TCH7	3.858	48.88	−4.318
45	TCH7	5.477	50.39	−4.135
46	TCH7	4.886	43.10	−5.509
47	TC1'	8.525	15.54	−3.606
48	TH1'	8.897	11.45	−2.917
49	TC2'	9.004	12.73	−4.987
50	TH2'	8.305	9.20	−5.465
51	TH2'	9.893	9.99	−4.912
52	TC3'	9.306	21.03	−5.670
53	TH3'	8.527	24.43	−6.007
54	TO3'	10.182	19.65	−6.771
55	TC4'	10.024	25.64	−4.568
56	TH4'	10.930	23.44	−4.437
57	TO1'	9.165	23.65	−3.430
58	TC5'	10.297	33.94	−4.777
59	TH5'	10.805	34.37	−5.635
60	TH5'	10.872	35.55	−4.015
61	TO5'	9.139	39.08	−4.844
62	P1	9.540	48.27	−5.227
63	PO1	8.561	52.84	−6.077
64	PO2	10.902	47.82	−5.791

Thymidine 5'-monophosphate

33	TN1	7.172	16.52	−3.355
34	TC2	6.606	7.74	−2.678
35	TO2	7.381	1.14	−2.188
36	TN3	5.232	6.84	−2.585
37	TH3	5.061	−2.58	−2.079
38	TC4	4.397	18.46	−3.101
39	TO4	3.174	19.33	−2.954
40	TC5	5.288	29.18	−3.794
41	TC6	6.580	25.78	−3.899
42	TH6	7.093	31.58	−4.407
43	TC7	4.977	44.22	−4.386
44	TCH7	4.078	49.26	−4.201
45	TCH7	5.693	50.33	−3.977
46	TCH7	5.124	43.69	−5.374
47	TC1'	8.651	15.84	−3.476
48	TH1'	9.029	11.87	−2.781
49	TC2'	9.138	13.04	−4.852
50	TH2'	8.448	9.48	−5.327
51	TH2'	10.032	10.40	−4.772
52	TC3'	9.423	21.21	−5.545
53	TH3'	8.645	24.52	−5.898
54	TO3'	10.304	19.87	−6.643
55	TC4'	10.129	25.88	−4.449
56	TH4'	11.040	23.77	−4.314
57	TO1'	9.274	23.90	−3.309
58	TC5'	10.380	34.12	−4.670
59	TH5'	10.859	34.61	−5.544
60	TH5'	10.981	35.73	−3.930
61	TO5'	9.212	39.15	−3.691 $\frac{7}{32}$
62	P1	9.571	48.27	−5.123
63	PO1	8.570	52.56	−5.981
64	PO2	10.924	47.98	−5.710

Table 5. (*Continued*)

No.	Atom	Ψ = (114,16) [5.34, 30.00 Å]			No.	Atom	Ψ = (102,12) [5.36, 29.75 Å]		
		R (Å)	θ (°)	z (Å)			R (Å)	θ (°)	z (Å)
		Deoxyadenosine 5'-monophosphate					*Deoxyadenosine 5'-monophosphate*		
1	AN1	4.766	21.05	1.247	1	AN1	4.600	18.76	1.265
2	AC2	5.936	25.01	0.727	2	AC2	5.778	23.88	0.847
3	AH2	6.748	19.75	0.783	3	AH2	6.621	19.06	0.986
4	AN3	6.383	34.92	0.094	4	AN3	6.205	34.21	0.227
5	AC4	5.600	45.42	0.004	5	AC4	5.360	44.48	0.025
6	AC5	4.356	48.79	0.485	6	AC5	4.063	47.08	0.393
7	AC6	3.773	33.27	1.143	7	AC6	3.534	30.05	1.051
8	AN6	2.552	29.87	1.665	8	AN6	2.306	23.19	1.468
9	AH6	2.688	10.12	2.082	9	AH6	2.562	2.77	1.907
10	AH6	2.125	51.92	1.507	10	AH6	1.753	46.14	1.237
11	AN7	4.276	66.89	0.206	11	AN7	3.902	66.27	0.020
12	AC8	5.407	67.71	-0.425	12	AC8	5.062	67.50	-0.553
13	AH8	5.682	77.34	-0.668	13	AH8	5.300	77.72	-0.858
14	AN9	6.145	56.26	-0.589	14	AN9	5.876	55.83	-0.590
15	AC1'	7.474	55.86	-1.254	15	AC1'	7.249	55.78	-1.160
16	AH1'	8.046	50.46	-0.881	16	AH1'	7.844	51.02	-0.657
17	AC2'	7.397	54.85	-2.770	17	AC2'	7.301	53.20	-2.646
18	AH2'	6.630	59.25	-3.120	18	AH2'	6.464	55.79	-3.096
19	AH2'	7.488	47.45	-3.032	19	AH2'	7.597	45.91	-2.777
20	AC3'	8.708	59.36	-3.211	20	AC3'	8.500	59.08	-3.116
21	AH3'	8.644	63.57	-3.980	21	AH3'	8.341	63.23	-3.892
22	AO3'	9.664	53.25	-3.607	22	AO3'	9.627	54.77	-3.661
23	AC4'	9.240	64.05	-1.982	23	AC4'	8.950	64.43	-1.906
24	AH4'	9.974	60.81	-1.574	24	AH4'	9.666	61.30	-1.426
25	AO1'	8.215	64.53	-1.026	25	AO1'	7.882	65.35	-1.059
26	AC5'	9.868	72.18	-2.215	26	AC5'	9.576	72.76	-2.196
27	AH5'	10.340	71.64	-3.091	27	AH5'	10.065	72.29	-3.065
28	AH5'	10.498	73.11	-1.456	28	AH5'	10.232	73.80	-1.463
29	AO5'	8.977	79.13	-2.301	29	AO5'	8.692	79.96	-2.280
30	P1	9.795	87.49	-2.380	30	P1	9.508	88.39	-2.577
31	P101	9.027	95.07	-2.636	31	P101	8.740	96.08	-2.917
32	P102	10.899	86.34	-3.336	32	P102	10.544	86.60	-3.580

Thymidine 5'-monophosphate

33	TN1	7.216	17.51	-3.269
34	TC2	6.622	8.69	-2.632
35	TO2	7.379	1.92	-2.150
36	TN3	5.245	7.95	-2.569
37	TH3	5.054	-1.59	-2.090
38	TC4	4.439	19.83	-3.078
39	TO4	3.214	21.06	-2.961
40	TC5	5.364	30.39	-3.728
41	TC6	6.653	26.85	-3.804
42	TH6	7.189	32.59	-4.285
43	TC7	5.093	45.29	-4.311
44	TCH7	4.194	50.34	-4.152
45	TCH7	5.801	51.14	-3.874
46	TCH7	5.270	44.85	-5.294
47	TC1'	8.696	16.72	-3.357
48	TH1'	9.054	12.84	-2.643
49	TC2'	9.210	13.75	-4.714
50	TH2'	8.535	10.10	-5.187
51	TH2'	10.108	11.21	-4.607
52	TC3'	9.487	21.78	-5.437
53	TH3'	8.695	24.90	-5.796
54	TO3'	10.378	20.36	-6.523
55	TC4'	10.179	26.62	-4.357
56	TH4'	11.095	24.61	-4.213
57	TO1'	9.325	24.73	-3.203
58	TC5'	10.408	34.82	-4.611
59	TH5'	10.821	34.99	-5.521
60	TH5'	11.031	36.51	-3.896
61	TO5'	9.231	39.77	-4.602
62	P1	9.565	48.93	-5.024
63	PO1	8.534	53.17	-5.854
64	PO2	10.912	48.69	05.626

Thymidine 5'-monophosphate

33	TN1	7.209	19.25	-3.355
34	TC2	6.581	10.52	-2.726
35	TO2	7.311	3.26	-2.303
36	TN3	5.205	10.66	-2.601
37	TH3	4.974	1.04	-2.133
38	TC4	4.466	23.50	-3.041
39	TO4	3.262	26.60	-2.872
40	TC5	5.442	33.33	-3.689
41	TC6	6.698	29.00	-3.824
42	TH6	7.263	34.58	-4.293
43	TC7	5.269	48.35	-4.199
44	TCH7	4.410	53.99	-4.002
45	TCH7	6.022	53.22	-3.747
46	TCH7	5.425	48.31	-5.186
47	TC1'	8.676	17.92	-3.510
48	TH1'	9.033	13.69	-2.843
49	TC2'	9.109	15.23	-4.910
50	TH2'	8.397	11.74	-5.365
51	TH2'	10.002	12.54	-4.868
52	TC3'	9.382	23.49	-5.590
53	TH3'	8.589	26.92	-5.879
54	TO3'	10.201	22.20	-6.737
55	TC4'	10.155	27.93	-4.511
56	TH4'	11.074	25.83	-4.449
57	TO1'	9.371	25.66	-3.313
58	TC5'	10.389	36.21	-4.680
59	TH5'	10.845	36.91	-5.560
60	TH5'	11.005	37.68	-3.942
61	TO5'	9.214	41.18	-4.636
62	P1	9.538	50.35	-5.071
63	PO1	8.513	54.55	-5.916
64	PO2	10.884	50.21	-5.675

Table 5. (Continued)

No.	Atom	Ψ = (90,6) [5.38, 29.50 Å] R (Å)	θ (°)	z (Å)	No.	Atom	Ψ = (78,12) [5.40, 29.25 Å] R (Å)	θ (°)	z (Å)
		Deoxyadenosine 5'-monophosphate					*Deoxyadenosine 5'-monophosphate*		
1	AN1	4.876	17.51	1.171	1	AN1	5.051	17.17	1.171
2	AC2	6.042	22.86	0.772	2	AC2	6.237	22.40	0.849
3	AH2	6.896	18.39	0.900	3	AH2	7.083	18.03	1.017
4	AN3	6.442	33.04	0.184	4	AN3	6.668	32.30	0.296
5	AC4	5.563	42.68	-0.002	5	AC4	5.793	41.52	0.057
6	AC5	4.252	44.33	0.352	6	AC5	4.461	42.90	0.331
7	AC6	3.780	27.60	0.974	7	AC6	3.966	26.87	0.922
8	AN6	2.573	19.74	1.374	8	AN6	2.740	19.27	1.244
9	AH6	2.872	1.55	1.788	9	AH6	3.011	1.80	1.651
10	AH6	1.944	38.97	1.159	10	AH6	2.115	36.99	0.998
11	AN7	4.012	62.75	0.003	11	AN7	4.208	60.28	-0.054
12	AC8	5.175	64.98	-0.542	12	AC8	5.393	62.72	-0.540
13	AH8	5.370	75.19	-0.828	13	AH8	5.570	72.52	-0.846
14	AN9	6.042	54.01	-0.583	14	AN9	6.287	52.42	-0.508
15	AC1'	7.425	54.50	-1.126	15	AC1'	7.698	53.03	-0.972
16	AH1'	8.029	49.87	-0.630	16	AH1'	8.292	49.17	-0.374
17	AC2'	7.511	52.30	-2.620	17	AC2'	7.901	49.84	-2.417
18	AH2'	6.637	53.96	-3.061	18	AH2'	7.026	49.59	-2.900
19	AH2'	7.921	45.62	-2.773	19	AH2'	8.472	44.09	-2.449
20	AC3'	8.602	59.13	-3.074	20	AC3'	8.816	57.28	-2.964
21	AH3'	8.347	63.85	-3.742	21	AH3'	8.422	61.56	-3.622
22	AO3'	9.677	54.58	-3.666	22	AO3'	9.930	53.60	-3.618
23	AC4'	9.088	63.51	-1.793	23	AC4'	9.299	62.10	-1.731
24	AH4'	9.800	60.08	-1.376	24	AH4'	10.037	59.01	-1.301
25	AO1'	8.016	63.99	-0.981	25	AO1'	8.218	62.52	-0.944
26	AC5'	9.738	71.78	-1.947	26	AC5'	9.894	70.28	-1.967
27	AH5'	10.308	71.55	-2.767	27	AH5'	10.323	69.89	-2.868
28	AH5'	10.318	72.58	-1.145	28	AH5'	10.559	71.15	-1.236
29	AO5'	8.865	78.88	-2.066	29	AO5'	8.983	77.13	-1.979
30	P1	9.658	86.89	-2.560	30	P1	9.651	85.18	-2.605
31	P101	8.848	94.02	-3.031	31	P101	8.715	91.55	-3.137
32	P102	10.677	84.48	-3.537	32	P102	10.685	82.71	-3.562

Thymidine 5'-monophosphate

33	TN1	7.556	19.52	−3.193		33	TN1	7.606	18.83	−3.175
34	TC2	6.927	11.10	−2.591		34	TC2	6.997	10.27	−2.598
35	TO2	7.654	3.99	−2.215		35	TO2	7.738	3.11	−2.294
36	TN3	5.556	11.38	−2.435		36	TN3	5.634	10.50	−2.380
37	TH3	5.317	2.26	−1.990		37	TH3	5.414	1.36	−1.960
38	TC4	4.821	23.60	−2.822		38	TC4	4.889	22.77	−2.685
39	TO4	3.621	26.39	−2.629		39	TO4	3.696	25.37	−2.443
40	TC5	5.793	33.02	−3.445		40	TC5	5.847	32.39	−3.287
41	TC6	7.047	28.97	−3.609		41	TC6	7.093	28.39	−3.510
42	TH6	7.610	34.42	−4.054		42	TH6	7.650	33.97	−3.930
43	TC7	5.608	47.32	−3.896		43	TC7	5.665	46.81	−3.650
44	TCH7	4.740	52.31	−3.686		44	TCH7	4.802	51.66	−3.406
45	TCH7	6.350	51.86	−3.421		45	TCH7	6.414	50.98	−3.152
46	TCH7	5.768	47.71	−4.883		46	TCH7	5.817	47.77	−4.634
47	TC1'	9.019	18.21	−3.379		47	TC1'	9.060	17.54	−3.426
48	TH1'	9.385	14.09	−2.724		48	TH1'	9.456	13.44	−2.788
49	TC2'	9.420	15.67	−4.790		49	TC2'	9.400	15.02	−4.854
50	TH2'	8.715	12.09	−5.217		50	TH2'	8.654	11.69	−5.265
51	TH2'	10.330	13.26	−4.776		51	TH2'	10.290	12.36	−4.873
52	TC3'	9.613	23.73	−5.486		52	TC3'	9.636	23.08	−5.534
53	TH3'	8.787	26.87	−5.737		53	TH3'	8.829	26.46	−5.763
54	TO3'	10.387	22.62	−6.668		54	TO3'	10.380	21.91	−6.734
55	TC4'	10.405	28.21	−4.441		55	TC4'	10.476	27.27	−4.491
56	TH4'	11.347	26.44	−4.442		56	TH4'	11.397	25.22	−4.493
57	TO1'	9.721	25.64	−3.184		57	TO1'	9.768	24.93	−3.254
58	TC5'	10.539	36.42	−4.582		58	TC5'	10.698	35.32	−4.634
59	TH5'	10.998	37.34	−5.454		59	TH5'	11.036	35.93	−5.568
60	TH5'	11.126	38.02	−3.831		60	TH5'	11.377	36.61	−3.944
61	TO5'	9.304	40.81	−4.544		61	TO5'	9.531	40.10	−4.453
62	P1	9.491	49.97	−5.049		62	P1	9.750	48.94	−4.985
63	PO1	8.349	53.68	−5.798		63	PO1	8.614	52.46	−5.753
64	PO2	10.772	50.16	−5.782		64	PO2	11.044	48.90	−5.697

Thymidine 5'-monophosphate

Table 5. *(Continued)*

No.	Atom	$\Psi = (66,16)$ [5.42, 29.00 Å] R (Å)	θ (°)	z (Å)	No.	Atom	$\Psi = (54,18)$ [5.44, 28.75 Å] R (Å)	θ (°)	z (Å)
		Deoxyadenosine 5'-monophosphate					*Deoxyadenosine 5'-monophosphate*		
1	AN1	5.140	18.03	1.186	1	AN1	5.280	18.86	1.154
2	AC2	6.345	23.13	0.935	2	AC2	6.503	23.52	0.929
3	AH2	7.178	18.73	1.134	3	AH2	7.319	18.99	1.123
4	AN3	6.812	32.89	0.425	4	AN3	7.005	33.02	0.451
5	AC4	5.955	41.93	0.145	5	AC4	6.176	41.99	0.179
6	AC5	4.611	43.30	0.345	6	AC5	4.831	43.60	0.355
7	AC6	4.077	27.72	0.898	7	AC6	4.250	28.79	0.873
8	AN6	2.830	20.53	1.148	8	AN6	2.980	22.83	1.098
9	AH6	3.066	3.16	1.538	9	AH6	3.156	5.72	1.466
10	AH6	2.230	37.80	0.881	10	AH6	2.430	39.81	0.841
11	AN7	4.375	60.02	-0.061	11	AN7	4.631	59.60	-0.035
12	AC8	5.582	62.43	-0.487	12	AC8	5.850	61.79	-0.429
13	AH8	5.762	71.84	-0.804	13	AH8	6.044	70.79	-0.732
14	AN9	6.479	52.51	-0.394	14	AN9	6.732	52.17	-0.329
15	AC1'	7.913	53.12	-0.780	15	AC1'	8.175	52.65	-0.683
16	AH1'	8.478	50.09	-0.078	16	AH1'	8.721	49.43	0.008
17	AC2'	8.236	48.76	-2.137	17	AC2'	8.509	48.85	-2.062
18	AH2'	7.406	46.65	-2.615	18	AH2'	7.695	46.27	-2.514
19	AH2'	8.944	44.11	-2.028	19	AH2'	9.283	44.79	-1.992
20	AC3'	8.946	56.35	-2.854	20	AC3'	9.080	56.76	-2.779
21	AH3'	8.407	59.83	-3.512	21	AH3'	8.447	60.42	-3.315
22	AO3'	10.068	53.22	-3.565	22	AO3'	10.110	54.22	-3.667
23	AC4'	9.444	62.03	-1.734	23	AC4'	9.683	61.85	-1.650
24	AH4'	10.227	59.46	-1.296	24	AH4'	10.501	59.20	-1.315
25	AO1'	8.378	62.48	-0.920	25	AO1'	8.667	61.68	-0.747
26	AC5'	9.949	70.11	-2.139	26	AC5'	10.157	69.91	-1.989
27	AH5'	10.253	69.48	-3.085	27	AH5'	10.518	69.81	-2.922
28	AH5'	10.702	71.32	-1.518	28	AH5'	10.898	71.05	-1.350
29	AO5'	9.001	76.74	-2.075	29	AO5'	9.179	76.23	-1.891
30	P1	9.615	85.11	-2.645	30	P1	9.683	84.50	-2.557
31	P101	8.643	91.59	-3.087	31	P101	8.633	90.16	-3.087
32	P102	10.625	83.08	-3.659	32	P102	10.731	82.62	-3.540

Thymidine 5'-monophosphate

33	TN1	7.599	18.26	−3.107
34	TC2	7.003	9.50	−2.566
35	TO2	7.753	2.27	−2.324
36	TN3	5.646	9.62	−2.309
37	TH3	5.443	0.32	−1.919
38	TC4	4.893	21.98	−2.542
39	TO4	3.703	24.36	−2.271
40	TC5	5.844	31.91	−3.111
41	TC6	7.085	27.97	−3.371
42	TH6	7.643	33.68	−3.759
43	TC7	5.670	46.50	−3.397
44	TCH7	4.808	51.27	−3.135
45	TCH7	6.418	50.38	−2.872
46	TCH7	5.828	47.97	−4.373
47	TC1'	9.046	17.03	−3.398
48	TH1'	9.458	12.82	−2.790
49	TC2'	9.349	14.75	−4.846
50	TH2'	8.575	11.70	−5.262
51	TH2'	10.220	11.90	−4.897
52	TC3'	9.631	22.93	−5.478
53	TH3'	8.846	26.55	−5.687
54	TO3'	10.366	21.84	−6.686
55	TC4'	10.495	26.73	−4.409
56	TH4'	11.399	24.49	−4.414
57	TO1'	9.750	25.29	−3.120
58	TC5'	10.710	35.53	−4.580
59	TH5'	11.052	36.20	−5.510
60	TH5'	11.401	36.77	−3.900
61	TO5'	9.557	40.39	−4.384
62	P1	9.760	49.09	−4.974
63	PO1	8.594	52.57	−5.699
64	PO2	11.011	48.01	−5.754

Thymidine 5'-monophosphate

33	TN1	7.614	18.93	−2.983
34	TC2	7.055	9.95	−2.474
35	TO2	7.838	2.95	−2.257
36	TN3	5.699	9.48	−2.220
37	TH3	5.540	0.12	−1.855
38	TC4	4.889	21.42	−2.428
39	TO4	3.689	22.68	−2.162
40	TC5	5.799	32.03	−2.964
41	TC6	7.059	28.56	−3.220
42	TH6	7.597	34.51	−3.584
43	TC7	5.568	46.79	−3.220
44	TCH7	4.683	51.18	−2.966
45	TCH7	6.290	50.92	−2.674
46	TCH7	5.737	48.54	−4.191
47	TC1'	9.067	18.06	−3.270
48	TH1'	9.504	14.15	−2.632
49	TC2'	9.386	15.46	−4.699
50	TH2'	8.630	12.16	−5.100
51	TH2'	10.271	12.75	−4.725
52	TC3'	9.629	23.50	−5.389
53	TH3'	8.833	26.93	−5.638
54	TO3'	10.379	22.25	−6.583
55	TC4'	10.467	27.72	−4.349
56	TH4'	11.380	25.58	−4.326
57	TO1'	9.731	25.64	−3.137
58	TC5'	10.721	35.73	−4.526
59	TH5'	11.068	36.47	−5.452
60	TH5'	11.425	36.93	−3.855
61	TO5'	9.583	40.67	−4.314
62	P1	9.770	49.23	−4.963
63	PO1	8.574	52.68	−5.644
64	PO2	10.977	48.91	−5.811

Table 5. (*Continued*)

No.	Atom	$\Psi = (42,18)$ [5.46, 28.50 Å] R (Å)	θ (°)	z (Å)	No.	Atom	$\Psi = (30,18)$ [5.48, 28.25 Å] R (Å)	θ (°)	z (Å)
		Deoxyadenosine 5'-monophosphate					Deoxyadenosine 5'-monophosphate		
1	AN1	5.336	18.35	1.084	1	AN1	5.077	21.15	1.107
2	AC2	6.575	22.75	0.904	2	AC2	6.352	24.65	0.944
3	AH2	7.375	18.13	1.117	3	AH2	7.107	19.28	1.131
4	AN3	7.109	32.09	0.452	4	AN3	6.982	33.97	0.544
5	AC4	6.306	41.01	0.155	5	AC4	6.280	43.69	0.284
6	AC5	4.958	42.79	0.286	6	AC5	4.958	46.93	0.403
7	AC6	4.335	28.45	0.780	7	AC6	4.183	33.24	0.840
8	AN6	3.048	23.22	0.959	8	AN6	2.858	31.06	1.002
9	AH6	3.178	6.13	1.318	9	AH6	2.827	11.87	1.319
10	AH6	2.537	40.09	0.690	10	AH6	2.535	50.66	0.765
11	AN7	4.788	58.33	-0.117	11	AN7	4.962	62.48	0.048
12	AC8	6.021	60.32	-0.473	12	AC8	6.218	63.10	-0.270
13	AH8	6.227	69.01	-0.783	13	AH8	6.521	71.32	-0.542
14	AN9	6.892	50.90	-0.337	14	AN9	6.981	53.17	-0.152
15	AC1'	8.346	51.27	-0.642	15	AC1'	8.441	52.71	-0.424
16	AH1'	8.861	47.70	0.027	16	AH1'	8.959	48.79	0.190
17	AC2'	8.703	48.16	-2.049	17	AC2'	8.787	50.47	-1.869
18	AH2'	7.919	45.22	-2.499	18	AH2'	8.035	47.13	-2.311
19	AH2'	9.533	44.65	-2.020	19	AH2'	9.670	47.56	-1.906
20	AC3'	9.145	56.45	-2.723	20	AC3'	9.064	59.24	-2.477
21	AH3'	8.443	60.20	-3.143	21	AH3'	8.288	62.94	-2.769
22	AO3'	10.089	54.59	-3.745	22	AO3'	9.905	58.23	-3.616
23	AC4'	9.825	61.04	-1.586	23	AC4'	9.811	63.52	-1.346
24	AH4'	10.678	58.43	-1.355	24	AH4'	10.711	61.15	-1.246
25	AO1'	8.868	60.03	-0.595	25	AO1'	8.967	61.30	-0.258
26	AC5'	10.243	69.22	-1.838	26	AC5'	10.111	72.01	-1.480
27	AH5'	10.596	69.21	-2.773	27	AH5'	10.616	72.36	-2.341
28	AH5'	10.947	70.36	-1.160	28	AH5'	10.668	73.41	-0.689
29	AO5'	9.208	75.18	-1.734	29	AO5'	8.974	77.40	-1.545
30	P1	9.578	83.45	-2.492	30	P1	9.301	85.97	-2.302
31	P101	8.424	88.31	-3.029	31	P101	8.129	90.59	-2.880
32	P102	10.621	81.66	-2.477	32	P102	10.306	84.36	-3.252

33	TN1	7.560	19.18	-2.853
34	TC2	7.010	10.03	-2.367
35	TO2	7.798	2.95	-2.189
36	TN3	5.658	9.47	-2.090
37	TH3	5.510	-0.05	-1.744
38	TC4	4.844	21.55	-2.254
39	TO4	3.647	22.69	-1.970
40	TC5	5.750	32.41	-2.768
41	TC6	7.007	28.94	-3.047
42	TH6	7.546	35.00	-3.391
43	TC7	5.527	47.38	-2.977
44	TCH7	4.645	51.78	-2.713
45	TCH7	6.250	51.31	-2.416
46	TCH7	5.700	49.44	-3.941
47	TC1'	9.009	18.34	-3.164
48	TH1'	9.462	14.52	-2.520
49	TC2'	9.306	15.53	-4.589
50	TH2'	8.551	12.08	-4.964
51	TH2'	10.196	12.85	-4.619
52	TC3'	9.521	23.57	-5.313
53	TH3'	8.704	26.90	-5.544
54	TO3'	10.249	22.22	-6.517
55	TC4'	10.373	28.01	-4.305
56	TH4'	11.289	25.89	-4.295
57	TO1'	9.666	26.01	-3.071
58	TC5'	10.611	36.08	-4.513
59	TH5'	10.941	36.46	-5.455
60	TH5'	11.301	37.48	-3.841
61	TO5'	9.458	41.02	-4.354
62	P1	9.652	49.64	-5.020
63	PO1	8.430	53.34	-5.625
64	PO2	10.812	49.06	-5.927

33	TN1	7.075	21.26	-2.691
34	TC2	6.557	11.17	-2.262
35	TO2	7.385	3.91	-2.106
36	TN3	5.206	9.78	-2.013
37	TH3	5.118	-0.75	-1.707
38	TC4	4.337	22.62	-2.156
39	TO4	3.130	22.76	-1.902
40	TC5	5.227	35.18	-2.611
41	TC6	6.498	31.68	-2.860
42	TH6	7.039	38.33	-3.164
43	TC7	5.006	51.73	-2.793
44	TCH7	4.120	56.77	-2.557
45	TCH7	5.711	55.85	-2.197
46	TCH7	5.220	54.15	-3.746
47	TC1'	8.533	20.72	-2.970
48	TH1'	8.963	16.52	-2.333
49	TC2'	8.874	18.09	-4.399
50	TH2'	8.126	14.63	-4.818
51	TH2'	9.761	15.24	-4.418
52	TC3'	9.134	26.63	-5.069
53	TH3'	8.339	30.27	-5.314
54	TO3'	9.902	25.37	-6.253
55	TC4'	9.960	30.96	-4.006
56	TH4'	10.868	28.67	-3.972
57	TO1'	9.200	28.73	-2.811
58	TC5'	10.236	39.35	-4.165
59	TH5'	10.704	39.70	-5.046
60	TH5'	10.821	40.80	-3.399
61	TO5'	9.083	44.58	-4.187
62	P1	9.392	53.30	-4.876
63	PO1	8.222	57.73	-5.484
64	PO2	10.535	52.11	-5.786

Table 5. (*Continued*)

$\Psi = (18,18)$
[5.50, 28.00 Å]

No.	Atom	R (Å)	θ (°)	z (Å)	No.	Atom	R (Å)	θ (°)	z (Å)
		Deoxyadenosine 5'-monophosphate					*Thymidine 5'-monophosphate*		
1	AN1	5.191	21.74	1.118	33	TN1	7.139	20.38	-2.595
2	AC2	6.471	24.96	0.945	34	TC2	6.619	10.29	-2.202
3	AH2	7.217	19.53	1.102	35	TO2	7.445	3.05	-2.073
4	AN3	7.114	34.13	0.571	36	TN3	5.267	8.86	-1.957
5	AC4	6.427	43.80	0.352	37	TH3	5.179	-1.65	-1.679
6	AC5	5.109	47.11	0.486	38	TC4	4.398	21.57	-2.071
7	AC6	4.316	33.84	0.892	39	TO4	3.190	21.55	-1.825
8	AN6	2.990	32.24	1.064	40	TC5	5.291	34.13	-2.489
9	AH6	2.928	13.67	1.355	41	TC6	6.565	30.74	-2.733
10	AH6	2.692	51.15	0.859	42	TH6	7.109	37.39	-3.009
11	AN7	5.131	62.29	0.175	43	TC7	5.068	50.51	-2.638
12	AC8	6.389	62.88	-0.132	44	TCH7	4.176	55.36	-2.411
13	AH8	6.701	70.94	-0.372	45	TCH7	5.759	54.50	-2.019
14	AN9	7.140	53.11	-0.048	46	TCH7	5.300	53.13	-3.581
15	AC1'	8.600	52.61	-0.321	47	TC1'	8.599	19.88	-2.867
16	AH1'	9.029	48.42	0.313	48	TH1'	9.027	15.70	-2.233
17	AC2'	8.947	50.06	-1.752	49	TC2'	8.947	17.33	-4.298
18	AH2'	8.200	46.64	-2.176	50	TH2'	8.199	13.91	-4.722
19	AH2'	9.834	47.24	-1.771	51	TH2'	9.833	14.51	-4.316
20	AC3'	9.207	58.55	-2.413	52	TC3'	9.207	25.83	-4.958
21	AH3'	8.412	62.17	-2.659	53	TH3'	8.412	29.44	-5.205
22	AO3'	9.980	57.35	-3.593	54	TC4'	9.981	24.62	-6.140
23	AC4'	10.027	62.83	-1.343	55	TH4'	10.026	30.11	-3.888
24	AH4'	10.937	60.57	-1.306	56	TO1'	10.936	27.84	-3.851
25	AO1'	9.262	60.57	-0.153	57	TO1'	9.261	27.84	-2.699
26	AC5'	10.300	71.18	-1.493	58	TC5'	10.299	38.45	-4.039
27	AH5'	10.752	71.58	-2.382	59	TH5'	10.751	38.85	-4.928
28	AH5'	10.899	72.57	-0.735	60	TH5'	10.898	39.84	-3.280
29	AO5'	9.147	76.36	-1.485	61	TO5'	9.146	43.63	-4.030
30	P1	9.402	84.88	-2.247	62	P1	9.401	52.15	-4.792
31	P101	8.189	89.08	-2.808	63	PO1	8.188	56.36	-5.352
32	P102	10.495	83.56	-3.211	64	PO2	10.493	50.83	-5.758

216

presented here. More complete computations are in progress to obtain a family of minimum energy bound-state structural forms.

We cannot say at present whether the center of the β premelton is open enough to accommodate an intercalator such as ethidium. However, structures such as these have an intrinsic ability to undergo low frequency breather motions (one such possible breather motion is shown schematically in Figure 10) and it is possible that such motions facilitate the intercalation process.

As we have said earlier, we envision intercalation to begin with the loose (external) binding by an intercalator to the B helix. A β premelton then appears in the region—this permits the drug molecule to intercalate either immediately or after a breathing motion of sufficient amplitude. Simple intercalators pin (immobilize) the β premelton and also prevent soliton-antisoliton annihilation. This is because simple intercalators bind the β premelted core region tightly and stabilize this structure (this would lower the position of the energy minimum shown in Figure 9c, and separate kink from antikink in the β premelton structure). Complex intercalators, however, vary in this regard. Actinomycin binds the β premelted core within the β premelton, pinning the structure and preventing its self-annihilation. Daunomycin, on the other hand, is unable to do this and ends up binding to a conformation very similar to B-DNA on either side of the intercalated chromophore.

Our model for understanding the intercalation of drugs and dyes into A-DNA or A' double-helical RNA is very similar. Intercalation begins with the loose (external) binding by an intercalator to the A (or A') helix. A β premelton then appears in the region. This permits the drug molecule to intercalate either immediately or after a breathing motion of sufficient amplitude (of the kind shown in Fig. 10). The model predicts that simple intercalators bind to the β structural element when they intercalate into DNA and RNA. Evidence supporting this prediction has already been presented in the preceding section.

In Figure 9a we show a β premelton that contains a small β premelted core. β premeltons having larger cores can occur. Bifunctional intercalators, such as echinomycin, require β premeltons having larger cores. Intercalators that necessitate the transient rupture of hydrogen bonds connecting base pairs to gain entrance into (and exit from) DNA [i.e., meso-tetra-(4-N-methylpyridyl)porphine] (Fiel and Munson, 1980) require β premeltons with still larger cores. We term DNA breathing distortions

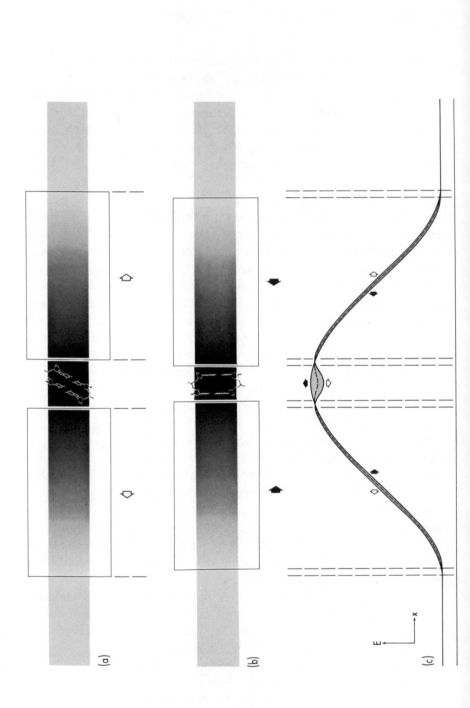

218

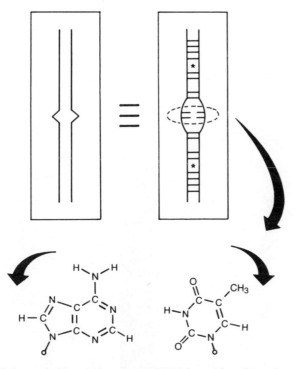

Figure 11. Schematic illustration shows DNA breathing distortions that combine base unstacking with the transient rupture of hydrogen bonds connecting base pairs. Such dynamic distortions take place in the centers of larger β premelton structures. Intercalators that necessitate the transient rupture of hydrogen bonds connecting base pairs to gain entrance into (and exit from) DNA [i.e., meso-tetra-(4-N-methylpyridyl)porphine] constitute convincing evidence that DNA breathing and drug intercalation are related phenomena.

those distortions that combine base unstacking with the transient rupture of hydrogen bonds connecting base pairs. We expect such dynamic dis-

Figure 10. (*Opposite*) A possible low frequency breather motion that could facilitate the intercalation of a drug molecule into the center of the β premelton structure. (*a*) and *b*) Kink and antikink pulsate in and out with a characteristic frequency determined by the process shown within (i.e., the molecular structures shown represent lower and higher energy forms of the central β structural element). (*c*) Energy density profile accompanying (*a*) and (*b*); notice that the center of the β premelton experiences larger energy fluctuations than the sides. The area indicated by the stippled pattern equals the sum of the areas indicated by slashed lines.

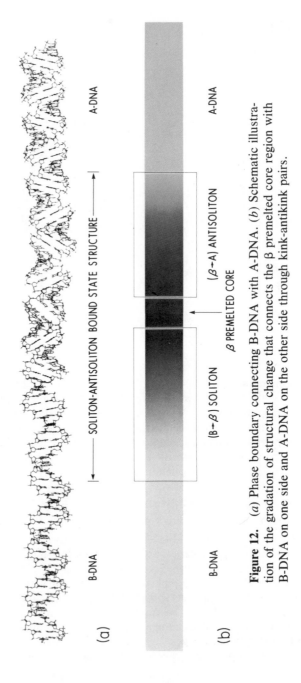

Figure 12. (a) Phase boundary connecting B-DNA with A-DNA. (b) Schematic illustration of the gradation of structural change that connects the β premelted core region with B-DNA on one side and A-DNA on the other side through kink-antikink pairs.

tortions to take place in the centers of these larger β premelton structures (see Fig. 11).

The term "soliton" is synonomous with "kink" and refers to a gradation of structural change either within a single structural type (i.e., a nontopological soliton) or connecting two different structures (i.e., a topological soliton). The structure shown in Figure 9a consists of *two* topological solitons, a (B → β) soliton and a (β → B) antisoliton. These two structures surround the β premelted core region and are related by twofold symmetry.

Soliton concepts allow us, in addition, to propose the molecular nature of the junction that connects B-DNA with A-DNA during the B to A structural phase transition. Such a junction (shown in Figs. 12a and b) has been constructed in a manner similar to that already described. A homogeneous β-DNA to A-DNA change was computed as a sequence of transitions involving the entire polymer length. Dinucleotide elements from this sequence and the B to β sequence were then connected using the least-squares joining procedure. The final structure consists of two *different* topological solitons, a (B → β) soliton and a (β → A) antisoliton, and vice versa. Such a junction constitutes a phase boundary transforming B-DNA to A-DNA. In the global sense, this kink-antikink bound state is topological because it alters the structure of the polymer as it moves along its length.

How does the B to A transition begin? We envision the B to A transition to begin at the centers of β premeltons within B-DNA structure. Nucleation could be site specific and involve growing regions of A-DNA joined to B-DNA on either side as [(B → β): (β → A)] and [(A → β): (β → B)] β premelton pairs (see Figs. 13a–d). Each β premelton contains a small β premelted core and moves along the polymer with minimal activation energy. Its high free energy and ease of movement could account for the cooperative nature of the phase transition.

4. PHYSICAL ASPECTS OF SOLITARY EXCITATIONS

The possibility that nonlinear excitations exist in biopolymers and play a central role in energy transfer was first advanced by Davydov in his class series of papers (Davydov and Kislukha, 1976; Davydov, 1979, 1981). These types of solitary excitations are known as vibrational solitons. In addition, a different class of solitons that gives rise to localized conformational changes in DNA structure has been proposed by Eng-

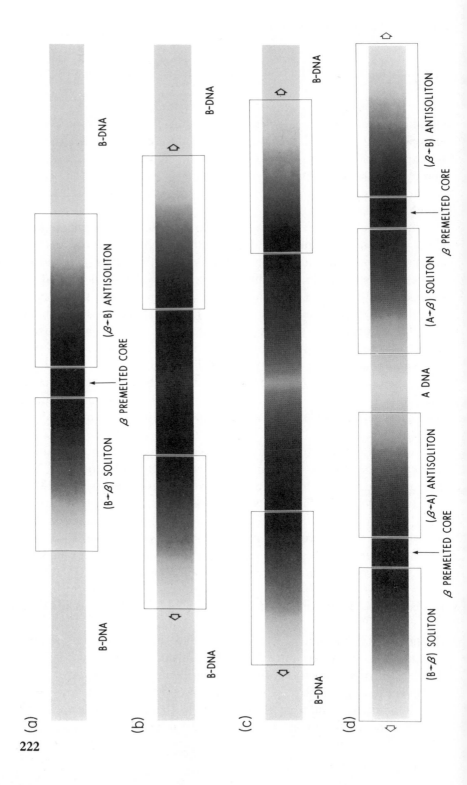

(a)

B-DNA   (B→β) SOLITON β PREMELTED CORE   (β→B) ANTISOLITON   B-DNA

(b)

B-DNA   B-DNA

(c)

B-DNA   B-DNA

(d)

(B→β) SOLITON β PREMELTED CORE   (β→A) ANTISOLITON   A DNA   (A→β) SOLITON β PREMELTED CORE   (β→B) ANTISOLITON lander et al. (1980) to explain DNA breathing phenomena; these are called structural solitons.

Solitons are intrinsic locally coherent excitations that move along a polymer chain with a velocity significantly less than the speed of sound (they may even be stationary). They are combinations of intramolecular and deformational excitations that appear as a consequence of an intrinsic nonlinear instability in the polymer structure. Extensive research on solitons in many physical systems has shown that this nonlinearity gives the spacially localized conformational excitation a robust character (Scott et al., 1973; Barone et al., 1971). Solitons do not significantly interact with conventional normal mode excitations (i.e., phonons). They have their own identity and can be treated by Newtonian dynamics as heavy Brownian-like particles, each having an "effective mass." Solitary structures— as sites for biochemical activity—behave like independent species and can be treated by statistical mechanics and chemical thermodynamics. They can arise from equilibrium or nonequilibrium processes.

Here we propose that collectively localized nonlinear excitations (in the form of *both* vibrational and structural solitons) exist in DNA structure. These arise (in B and A structure) as a consequence of an intrinsic nonlinear ribose inversion instability that results in a modulated β alternation in sugar puckering along the polymer backbone. Vibrational solitons resemble structural solitons in possessing this modulated β alternation in sugar puckering; however, they lack well defined structure in their centers and have no internal dynamic motion. Vibrational solitons can coalesce to form structural solitons. Structural solitons (also called soliton-antisoliton or kink-antikink bound states) possess well defined structure in their centers (i.e., the β premelted core regions) and can undergo low frequency breather motions. We have called such structural solitons β premeltons.

The stability of a β premelton is expected to reflect the collective properties of extended DNA regions and to be sensitive to temperature, pH, ionic strength, and other thermodynamic factors. It is also expected to

Figure 13. (*Opposite*) Schematic illustration of the B to A structural phase transition. (*a*) Nucleation begins at the centers of β premeltons within B-DNA structure. (*b*) and (*c*) (B → β) solitons and (β → B) antisolitons move apart, leaving growing regions of β premelted cores whose centers begin to form A-DNA. (*d*) Completion of the [(B → β): (β → A)] and [(A → β): (β → B)] β premelton pairs with intervening A-DNA structure.

reflect the detailed nucleotide base sequence. This is partly because the ease with which β premelted cores form primarily reflects the magnitude of localized base-stacking energies. Nucleotide sequences with minimal base overlap (as occur in alternating purine-pyrimidine sequences) may be favored to form β premelted core regions along with sequences that contain high A-T/G-C base ratios. Equally important are the energetics in the kink and antikink regions. As seen in Figure 9c, the stability of a kink-antikink bound state reflects the depth in the energy minimum within the β premelted core coupled with the height and separation of the domain walls on either side.

It is evident that sequentially homogeneous DNA polymers are not adequate models for understanding the properties of naturally occurring DNA because they lack sufficient information in their nucleotide base sequence to give rise to this site specificity.* In the context of soliton models, this poses the problem of describing soliton behavior in the presence of locally altered potentials. A general theory for this has been developed (Fogel et al., 1976). The theory shows that vibrational solitons either move nonuniformly or are trapped by locally favorable potentials. It remains to extend this theory to DNA structure to predict the localization of kink-antikink bound states at specific nucleotide sequences. Because the kink-antikink bound state is multiple base pairs in extent, it may be that the effective trapping potential involves the recognition of an extended sequence rather than being determined by any single base pair energetics or its immediate neighbors.

We note experimental evidence that indicates the presence of nuclease hypersensitive sites in eukaryotic DNA, many of these located at 5′ ends of genes (Jessee et al., 1982; Cartwright and Elgin, 1982). These same sites are sensitive to cleavage by a 1,10-phenanthroline–copper(I) complex, a known intercalating agent. It is possible that these sites correspond to the centers of β premeltons localized in these regions.

The presence of β premelted core regions within kink-antikink bound states could provide a key component in the recognition of the promoter by the RNA polymerase enzyme. RNA polymerase-promoter recognition

---

* However, as we have already described, we predict poly[d(A-T)] and poly[d(G-C)] to contain β premeltons that favor TpA and CpG sequences as β structural elements. Simple intercalators, therefore, are expected to preferentially intercalate into these sequences. This could also explain the sequence-dependent nuclease sensitivities that have been observed with these polymers (Sobell et al., 1983b).

could begin with the loose binding to β premelted nucleation sites within promoter regions, followed by a sequence of conformational transitions that progressively lead to a more open tight binding complex. Such an autocatalytic process can be viewed as one that creates "an avalanche of kinks" whose energies come from the stepwise interactions between the RNA polymerase and the promoter. This results in the formation and propagation of a new DNA phase. Such a phase could resemble the β structure but be more completely unwound and single-stranded.

The process of RNA transcription could begin within this more open β premelton structure. The RNA polymerase enzyme then acts to catalyze the stepwise polymerization of ribonucleoside triphosphates into RNA, using one of two β denatured DNA chains as template. Normal chain termination either occurs spontaneously or requires other factors. Actinomycin and other related antibiotics are known to cause premature chain termination. We suggest that these antibiotics intercalate tightly into β premelted DNA regions within the transcriptional complex. This tends to pin the complex, increasing the probability of premature chain termination.

5.   HIGHER ENERGY KINK-ANTIKINK BOUND STATE STRUCTURES

Over the years, increasing attention has focused on understanding the full range of conformational flexibility in DNA structure. DNA molecules in the process of genetic recombination are known to exhibit a phenomenon known as branch migration, a zipperlike process that allows homologous DNA molecules to exchange polynucleotide chains. Two forms of branch migration are known, single-strand and double-strand (see Figures 14a and b). Single-strand branch migration has been postulated to play an important role in the initial (asymmetric) phase of genetic recombination (synapsis), while double-strand branch migration occurs in the latter (symmetric) phase (see, for example, Radding, 1978). The double-strand branch migration junction has been termed the Holliday junction (Holliday, 1964).

Very similar dynamic processes are known to take place in negatively superhelical DNA molecules. Chain slippage in regions of DNA having repetitive sequences has been inferred by the presence of S1 nuclease

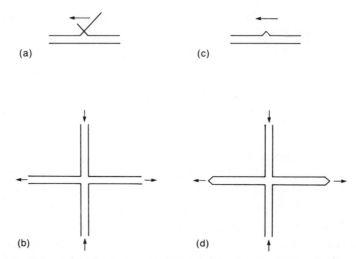

Figure 14. Schematic illustration showing (*a*) single-strand branch migration, (*b*) double-strand branch migration, (*c*) moving dislocations that give rise to chain slippage in regions of DNA having repetitive base sequences, (*d*) hairpin or cruciform structures that form at sequences that contain inverted repeats.

sensitivities in these regions (see Figure 14*c*) (Hentschel, 1982; Glikin et al., 1983). These types of moving dislocations resemble single-strand branch migration. Hairpin or cruciform structures can form at sequences that contain inverted repeats (Gellert et al., 1978; Lilly, 1980). Their formation involves processes analogous to double-strand branch migration (see Figure 14*d*).

Finally, the discovery of Z-DNA, a left-handed conformational state radically different from either A or B structures (see Chapters 1, 2, and 3, this volume) points to still wider conformational flexibility in DNA structure. What is the nature of the B-Z junction, and how does this structural phase transition take place?

We propose that these processes are related to one another and involve higher energy kink-antikink bound states in DNA structure. Such structural solitons can originate within β premeltons having higher energy nonlinear instabilities in their central premelted core regions. These instabilities arise because of particularly energetic breathing motions that distort and eventually disrupt hydrogen bonds connecting base pairs (refer to Figure 11). Two types of events, schematically illustrated in Figures 15*a* and *b*, can then take place.

The first involves polynucleotide chains snapping back at regions of DNA having inverted repeats to form (weakly) hydrogen bonded hairpin structures. This is then followed by a series of double-strand branch migration steps in which hydrogen bonds connecting base pairs within dinucleotide elements are broken and reformed in a concerted twofold symmetric fashion. Eventually, a Holliday-type junction is formed that contains a modulated β alternation in sugar puckering along the polymer backbone that merges into B-DNA on all four sides. Such a structure corresponds to a higher energy kink-antikink bound state and, more generally, serves as a mobile junction connecting homologous DNA molecules during the process of genetic recombination. We call this structure a β *branch migration* (see Figure 16). β branch migratons are expected to have dynamic breather modes, and the presence of such modes could facilitate the branch migration process.

The second arises by slippage of polynucleotide chains at regions of DNA having repetitive base sequences e.g., poly[d(A-G)]: poly[d(T-C)].

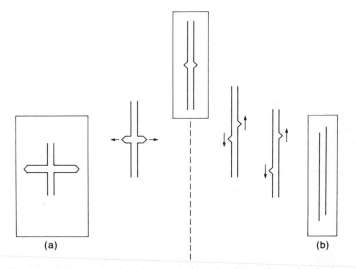

(a)

(b)

Figure 15. Possible mechanism interrelating the formation of hairpin or cruciform structures at sequences containing inverted repeats (*a*) with the origin of moving dislocations in regions of DNA having repetitive base sequences to give chain slippage (*b*). These processes could arise as a consequence of particularly energetic breathing distortions within β premeltons that occasionally result in the sequence of events shown.

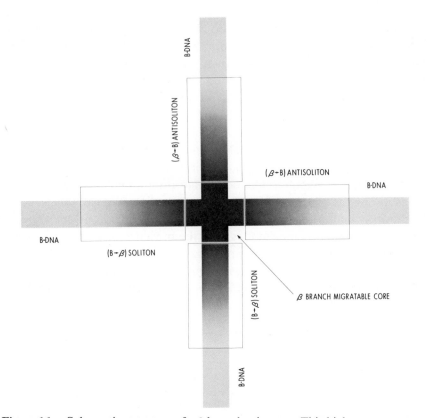

Figure 16. Schematic structure of a β branch migraton. This higher energy structural soliton contains a modulated β alternation in sugar puckering along the polymer backbone [in the form of two (B → β) and (β → B) soliton-antisoliton pairs] that merges into B-DNA on all four sides. Double-strand branch migration occurs stepwise in which hydrogen bonds connecting base pairs within dinucleotide elements are broken and re-formed in a concerted twofold symmetric fashion. This occurs in the center of the structural soliton, termed the β branch migratable core region.

These types of moving dislocations occur in pairs and, again, can be considered to be higher energy kink-antikink bound states. They contain dislocatable cores containing an even number of unpaired bases. These core regions are surrounded by a modulated β alternation in sugar puckering, which merges into B-DNA on either side (see Figure 17). We call these structures—β *dislocations*. Again, each β dislocation is expected

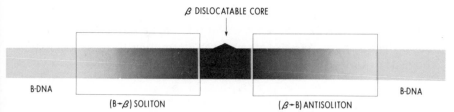

Figure 17. Schematic structure of a β dislocation. This higher energy structural soliton contains a modulated β alternation in sugar puckering along the polymer backbone [(B → β) and (β → B) soliton-antisoliton pairs] surrounding the β dislocatable core. The β dislocatable core contains an even number of unpaired bases. Movement of this structural soliton along DNA gives rise to polynucleotide chain slippage.

to possess characteristic breather motions and it is possible that these motions facilitate its movement along the polymer.

We have wondered whether the B to Z transition involves a similar slippage mechanism. In the presence of high salt (and/or negative superhelical strain energy), β premeltons appearing within B poly[d(G-C)] could give rise to β dislocaton pairs, which, as they move apart, could leave Z structure behind (see Figure 18). This would predict that hydrogen bonds connecting base pairs are transiently broken during the transition. It also predicts the slippage of polynucleotide chains. We note the presence of S1 sensitive sites on both sides of d(G-C) oligomers when these

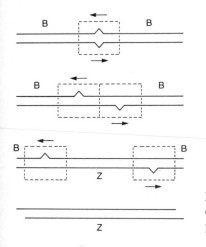

Figure 18. A possible mechanism for understanding the B to Z transition in the poly[d(G-C)] synthetic copolymer.

are inserted into negatively superhelical plasmids and held in the Z-form (Singleton et al., 1982). These observations are in agreement with the model proposed here. We are, therefore, attempting to model the stereochemistry of the B-Z junction with these concepts in mind.

6. CONCLUDING REMARKS

In this chapter, we have described how soliton physics concepts can be used to qualitatively understand many aspects of DNA structure, dynamics, and physical chemistry. Nonlinear dynamics, DNA premelting and melting, structural phase transitions, and moving dislocations that give rise to chain slippage and to branch migration are topics that lend themselves to soliton physics descriptions. More quantitative theories to describe these processes remain to be developed.

REFERENCES

Arnott, S., Smith, P. J. C., and Chandrasekaran, R. (1976). In *Handbook of Biochemistry and Molecular Biology*, G. D. Fasman, Ed., 3rd ed. Vol. 2, Chemical Rubber Company, Cleveland, Ohio, Sect B, pp. 411–422.

Banerjee, A., and Sobell, H. M. (1983). *J. Biomol. Struct. Dyn.* 1, 253–262.

Barone, A., Esposito, F., Magee, C. J., and Scott, A. C. (1971). *Riv. Nuovo Cimento* 1, 227–267.

Bhandary, K. K., Sakore, T. D., Sobell, H. M., King, D. and Gabbay, E. J., (1984). *J. Biomol. Struct. Dyn.* 1, No. 5, 1195–1217.

Bond, P. J., Langridge, R., Jennette, K. W., and Lippard, S. J. (1975). *Proc. Natl. Acad. Sci. USA* 72, 4825–4829.

Cartwright, I. L., and Elgin, S. C. R. (1982). *Nucl. Acids Res.* 10 (19), 5835–5852.

Crothers, D. M. (1968). *Biopolymers* 6, 575–583.

Davydov, A. S. (1979). *Phys. Scr.* 20, 387–394.

Davydov, A. S. (1981). *Physica 3D* (North Holland) 1, 1–22.

Davydov, A. S., and Kislukha, N. I. (1976). *Phys. Stat. Sol. B* 75, 735–742.

Englander, S. W., Kallenbach, N. R., Heeger, A. J., Krumhansl, J. A., and Litwin, S. (1980). *Proc. Nat. Acad. Sci. USA* 77, No. 12, 7222–7226.

Fiel, R. J., and Munson, B. R. (1980). *Nucl. Acids Res.* 8, 2835–2842.

Fogel, M. B., Trullinger, S. E., Bishop, A. R., and Krumhansl, J. A. (1976). *Phys. Rev. Lett.* **36**, 1411–1414.

Gellert, M., Mizuuchi, K., O'Dea, M. H., Ohmori, H., and Tomizawa, J. (1978). *Cold Spring Harbor Symp. Quant. Biol.* **43**, 35–40.

Glikin, G. C., Gargiulo, G., Rena-Descalzi, L., and Worcel, A. (1983). *Nature* **303** (5920), 770–774.

Hentschel, C. C. (1982). *Nature* **295**, 714–716.

Holliday, R. (1964). *Genet. Res.* **5**, 282–304.

Jain, S. C., Tsai, C.-C., and Sobell, H. M. (1977). *J. Mol. Biol.* **114**, 317–331.

Jain, S. C., Bhandary, K. K., and Sobell, H. M. (1979). *J. Mol. Biol.* **135**, 813–840.

Jain, S. C., and Sobell, H. M. (1984a). *J. Biomol. Struct. Dyn.* **1**, No. 5, 1161–1177.

Jain, S. C., and Sobell, H. M. (1984b). *J. Biomol. Struct. Dyn.* **1**, No. 5, 1179–1194.

Jessee, B., Gargiulo, G., Razvi, F., and Worcel, A. (1982). *Nucl. Acids Res.* **10** (19), 5823–5834.

Krugh. T. R. (1972). *Proc. Natl. Acad. Sci. USA* **69**, 1911–1914.

Krugh, T. R., and Reinhardt, C. (1975). *J. Mol. Biol.* **97**, 133–162.

Lilly, D. M. (1980). *Proc. Natl. Acad. Sci. USA* **77**, 6468–6472.

Neidle, S., Achari, A., Taylor, G. L., Berman, H. M., Carrell, H. L., Glusker, J. P., and Stallings, W. C. (1977). *Nature* **269**, 304–307.

Pack, G., and Loew, G. H. (1977). *Int. J. Quantum Chem. Symp.* **4**, 87–90.

Patel, D. J. (1974). *Biochemistry* **13**, 2396–2402.

Patel, D. J. (1980). In *Nucleic Acid Geometry and Dynamics,* R. H. Sarma, Ed. Pergamon, New York, Oxford, pp. 185–232.

Quigley, G. J., Wang, A. H.-J., Ughetto, G., van der Marel, G., van Boom, M. J. H., and Rich, A. (1980). *Proc. Natl. Acad. Sci. USA* **77**, 7204–7208.

Radding, C. (1978). *Ann. Rev. Biochem.* **47**, 847–880.

Reddy, B. S., Seshadri, T. P., Sakore, T. D., and Sobell, H. M. (1979). *J. Mol Biol.* **135**, 787–812.

Reich, E., and Goldberg, I. H. (1964). *Prog. Nucl. Acid. Res. Mol. Biol.* **3**, 183–234.

Reinhardt, C. (1976). Ph.D. Thesis, "Spectroscopic evidence for sequence preferences in the intercalative binding of ethidium bromide to nucleic acids" University of Rochester, Department of Chemistry.

Sakore, T. D., Bhandary, K. K., and Sobell, H. M. (1984). *J. Biomol. Struct. Dyn.* **1**, No. 5, 1219–1227.

Sakore, T. D., Reddy, B. S., and Sobell, H. M. (1979). *J. Mol. Biol.* **135**, 763–785.

Scott, A. C., Chu, F. Y. F., and McLaughlin, D. W. (1973). *Proc. IEEE* **61,** 1443–1483.

Singleton, C. K., Klysik, J., Stirdivant, S. M., and Wells, R. D. (1982). *Nature* **299,** 312–316.

Sobell, H. M., Banerjee, A., Lozansky, E. D., Zhou, G.-P., and Chou, K.-C. (1983a). In *Structure and Dynamics: Nucleic Acids and Proteins*, E. Clementi and R. H. Sarma, Eds. Adenine, Guilderland, New York, pp. 181–195.

Sobell, H. M., Lozansky, E. D., and Lessen, M. (1978). *Cold Spring Harbor Symp. Quant. Biol.* **43,** 11–19.

Sobell, H. M., Jain, S. C., Sakore, T. D., and Nordman, C. E. (1971). *Nature, New Biol.* **231,** 200–205.

Sobell, H. M., and Jain, S. C. (1972). *J. Mol. Biol.* **68,** 21–34.

Sobell, H. M., Sakore, T. D., Jain, S. C., Banerjee, A., Bhandary, K. K., Reddy, B. S., and Lozansky, E. D. (1983b). *Cold Spring Harbor Symp. Quant. Biol.* **47,** 293–314.

Sobell, H. M., Tsai, C.-C., Jain, S. C., and Gilbert, S. G. (1977). *J. Mol. Biol.* **114,** 333–365.

Stasiak, A., DiCapua, E., and Koller, T. (1983). *Cold Spring Harbor Symp. Quant. Biol.* **47,** 811–820.

Tsai, C.-C., Jain, S. C., and Sobell, H. M. (1977). *J. Mol. Biol.* **114,** 301–315.

Wang, J. C. (1974). *J. Mol. Biol.* **89,** 783–801.

Wang, A. H.-J., Nathans, J., van der Marel, G., van Boom, J. H., and Rich, A. (1978). *Nature* **276,** 471–474.

Wang, A. H.-J., Quigley, G. J., and Rich, A. (1979). *Nucleic Acids Res.* **6,** 3879–3890.

Waring, M. J. (1970). *J. Mol. Biol.* **54,** 247–279.

Wells, R. D., and Larson, J. E. (1970). *J. Mol. Biol.* **49,** 319–342.

Young, M. A., and Krugh, T. R. (1975). *Biochemistry* **14,** 4841–4847.

The Structure of Cro Repressor Protein

5

Y. TAKEDA
University of Maryland
Catonsville, Maryland

DOUGLAS H. OHLENDORF
Institute of Molecular Biology
University of Oregon
Eugene, Oregon

WAYNE F. ANDERSON
MRC Group on Protein Structure and Function
University of Alberta
Edmonton, Alberta, Canada

BRIAN W. MATTHEWS
Institute of Molecular Biology
University of Oregon
Eugene, Oregon

CONTENTS

1. INTRODUCTION

Bacteriophage lambda (λ) has long been a favorite system for study in genetics and molecular biology. The large body of information relating to bacteriophage λ and the regulation of its development makes this an excellent system for studying the role of protein–DNA interactions and recognition and their role in the control of cellular processes. The cro repressor may also be viewed as a model system for studying the interaction of proteins with specific base sequences of double helical DNA. In this chapter we review recent structural studies on the λ cro repressor. These studies have provided a detailed understanding of the relationship between the structure and function of this protein.

As a temperate bacteriophage, λ has two developmental pathways available. One is a lytic pathway in which the bacteriophage's regulatory and structural proteins are made, the DNA is replicated, and the host cell is lysed, releasing approximately 100 progeny phage particles. In the alternative lysogenic pathway, the phage DNA is circularized, inserted into the host genome, and transmitted to successive generations as part of the *E. coli* genome. The phage genome codes for two repressor proteins, the products of the *cI* and *cro* genes. The two repressors form a molecular switch between the two developmental pathways. In lytic development the cro repressor is synthesized in sufficient quantities to inhibit transcription of the phage *cI* gene, and, later in the infection, repress the transcription of all of the early genes, including its own. On the other hand, after the establishment of the lysogenic state, the only gene of the bacteriophage (now a prophage) that is expressed is the *cI* gene. The product of this gene (the cI or λ repressor) prevents the transcription of all of the other phage genes.

The λ cro repressor has been well characterized (Folkmanis et al., 1976; Takeda et al., 1977). It is a small, basic protein of 66 amino acid residues and has a monomeric molecular weight of 7351 daltons. Both the amino acid sequence of the protein (Hsiang et al., 1977) and the nucleotide sequence of the *cro* gene (Roberts et al., 1977) have been determined. In addition to the biochemical characterization, the role of the cro repressor in the control of the *cI* gene and the early genes of bacteriophage λ has been elucidated (Johnson et al., 1981; Ptashne et al., 1980; Takeda, 1979).

2. STRUCTURE DETERMINATION

The structure was determined by the method of isomorphous replacement (Green et al., 1954; Matthews, 1977; Blundell and Johnson, 1976). The purification of the protein and its crystallization from 1.1 to 1.3 *M* phosphate have been described (Takeda et al., 1977; Anderson et al., 1979). The space group of the crystals used in the determination of the structure is R32 and the cell dimensions in the hexagonal system are $a = b = 91.6$ Å, $c = 268.5$ Å. Within each asymmetric unit of the crystal there are four subunits related in pairs by mutually perpendicular intersecting (or nearly intersecting) twofold axes. In other words, the four monomers are arranged in the crystal with local 222-point symmetry (Anderson et al., 1981).

An initial electron density map was calculated at 5 Å resolution and extended to a nominal resolution of 2.8 Å. To improve the accuracy of the electron density map and facilitate its initial interpretation, the density of the four monomers in the asymmetric unit was averaged (Fig. 1). Upon inspection of the averaged map in an optical comparator (Richards, 1968), several helices and regions of β sheet were immediately apparent. The course of the polypeptide backbone could be readily followed from the

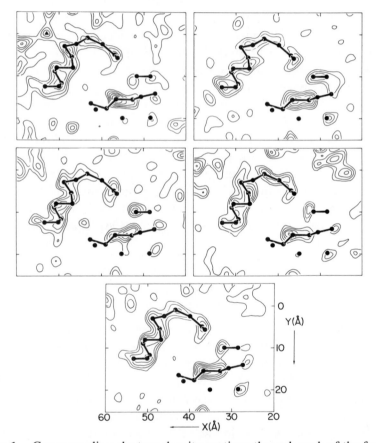

Figure 1. Corresponding electron density sections through each of the four cro molecules in the crystallographic asymmetric unit, transformed into a common coordinate frame. Clockwise from top left the density is for molecules O, A, B, and C, with the averaged density below. α-carbon atoms within ±2.5 Å of the section are indicated by the solid circles. Reprinted, with permission, from Anderson et al. (1981), copyright 1981 by MacMillan Journals, Ltd.

N-terminus to approximately residue Asn 61, but at this point the density becomes very weak, indicating that the five or six C-terminal residues are disordered in the crystals. This appears to be the case for each of the four cro monomers.

Utilizing this map, the averaged electron density map allowed preliminary interpretation of the structure of the monomer. The structure of the cro tetramer was then built from that of the monomer using an MMSX graphics facility (Molnar et al., 1976). Refinement of the coordinates using the Jack–Levitt program EREF (Jack and Levitt, 1971) was performed first at 3.5 Å resolution and then at 2.8 Å, 2.5 Å, and finally 2.2 Å resolution. The partially refined coordinates were checked against an improved map obtained by the phase refinement procedure of Bricogne (1974). Currently the crystallographic residual is 0.19 with an estimated error in the coordinates of about 0.35 Å (Ohlendorf and Matthews, unpublished results).

3. CONFORMATION OF CRO REPRESSOR

The conformation of a cro monomer is illustrated in Figure 2. The structure is very simple, consisting of three strands of antiparallel β sheet (residues 2–6, 39–45, 48–55) and three α helices (residues 7–14, 15–23, and 27–36) (Anderson et al., 1981). The detailed arrangement of the four repressor monomers in the crystals is shown in Figure 3. The respective local twofold symmetry axes relating the O ("original") monomer with the A, B, and C monomers are designated P, Q, and R, respectively.

It has been known for some time that the cro repressor exists as a dimer in physiological conditions (Takeda et al., 1977). This means that only one of the three local twofold axes observed in the crystal structure persists in solution. Careful analysis of the interactions between monomer O and monomers A, B, and C suggested that the O–B combination was the dimer present in solution. In this dimer, residues 54–59 form an extended "arm" (Fig. 2) that interacts with the other monomer. The stability of the O–B dimer is due to the formation of an antiparallel β ribbon between portions of these two arms and to the placing of the aromatic side chain of the Phe 58 of one monomer against the hydrophobic core of the other monomer (Anderson et al., 1981).

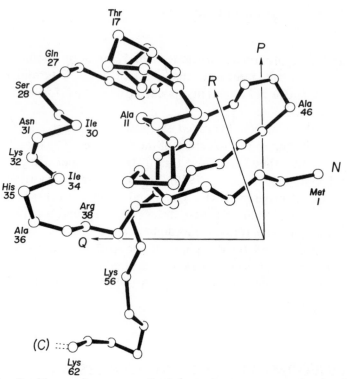

Figure 2. Backbone of monomer O of the cro repressor protein. As discussed in the text, the four C-terminal residues 63–66 are disordered. The local orthogonal twofold symmetry axes *P, Q,* and *R* relate the molecule shown to monomers A, B, and C, respectively. Reprinted, with permission, from Anderson et al. (1981), copyright 1981 by MacMillan Journals, Ltd.

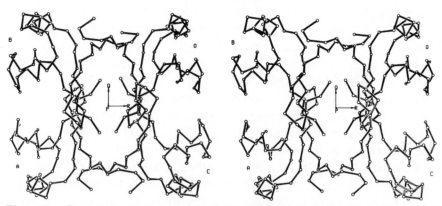

Figure 3. Stereo view of a tetramer of cro. Monomers O and B are drawn with solid bonds; monomers A and C are drawn with open bonds. The directions of the local twofold axes *P, Q,* and *R,* are given. The view of tetramer is along the *P* axis.

4.   MODE OF INTERACTION OF CRO WITH DNA

Both cI and cro repressors compete for three sites within both the left and the right operators of bacteriophage λ (Fig. 4). These 17-base-pair sites are called O_R1, O_R2, and O_R3 in the right operator and O_L1, O_L2, and O_L3 in the left operator. The three sites within each operator are separated by spacers of three to seven base pairs (Maniatis et al., 1975). The sequence within each site is an approximate inverted repeat or palindrome. Although the sequences of these six sites are similar to each other, they are not identical. This allows the cI and cro repressors to bind to these six sites with different relative affinities. Mutations within the 17-base-pair sites reduce the binding affinity of both cro and cI, whereas mutations in the spacer regions have no effect on binding (Ptashne et al., 1980; Takeda, 1979; Flashman, 1978; Meyer et al., 1980 and references therein).

Chemical probe experiments (Ptashne et al., 1980; Johnson et al., 1978; Johnson, 1980) indicate that both cro and cI repressors bind primarily along one face of the DNA double helix and protect many of the same groups. The results of such experiments for cro binding to O_R3 are illustrated in Figure 5. Methylation in the major groove by dimethylsulfate of the N7 of six guanines is prevented by the presence of cro, but methylation in the minor groove of N3 of adenine is not prevented. Furthermore, ethylation of any of the six phosphates shown in Figure 5 interferes

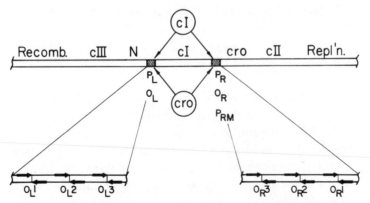

Figure 4.   Summary of sites within bacteriophage λ at which cI and cro act. P_L, P_R, and P_{rm} are the leftward, rightward, and repressor maintenance promoters. O_L and O_R are the left and right operator region, each consisting of three binding sites with approximate twofold symmetry.

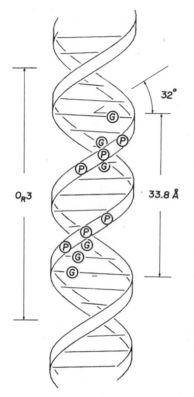

Figure 5. DNA operator O_R3, drawn in the B form (after Ptashne et al., 1980). Ethylation of the phosphates, labeled ⓟ, hinders cro repressor binding; binding of cro protects the N7 positions of guanines, labeled ⓖ, from methylation by dimethyl sulphate. The figure is based on experiments of Johnson, Ptashne, and collaborators (Ptashne et al., 1980; Johnson et al., 1978; Johnson, 1980). Reprinted, with permission, from Anderson et al. (1981), copyright 1981 by Macmillan Journals, Ltd.

with cro binding (Johnson, 1980). The pattern for cI is very similar, but it is slightly more extended. That is, four additional phosphates are protected. Such experiments suggest that both cro and cI contact the DNA in the major groove but not the minor (Ptashne et al., 1980; Johnson et al., 1978).

Because both cro and the DNA have approximate local twofold symmetry, related subunits of the protein contact the two symmetry-related halves of the DNA binding site. In this way the complex of bound repressor and DNA would have a common twofold symmetry axis. There would, of course, be deviations from exact twofold symmetry because of the variations in the base sequence of the DNA (Anderson et al., 1981; Ohlendorf et al., 1982).

Figure 6 shows a cro dimer viewed down its local twofold axis. In this view, one immediately notices that the cro dimer contains a pair of pro-

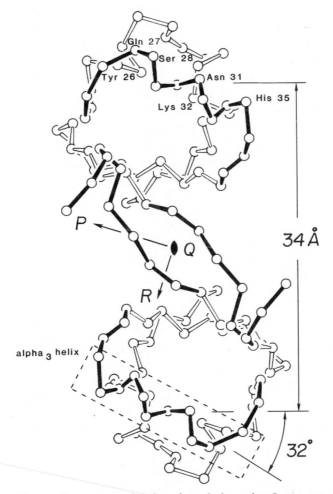

Figure 6. Dimer of two cro molecules viewed along the Q symmetry axis. The P and R symmetry axes are also shown. Regions of the protein backbone closest to the viewer, and presumed to interact with DNA, are drawn solid. These regions include two α helices, 34 Å apart and inclined at 32°, and also a pair of extended strands close to the Q axis. Reprinted, with permission, from Anderson et al. (1981), copyright 1981 by MacMillan Journals, Ltd.

truding twofold-related α helices (α₃) positioned so as to precisely complement the geometry of successive major grooves of right-handed B-DNA (see Figure 5). This suggested arrangement is shown in Figure 7.

The complementarity between protein and DNA is striking. In addition to the overall correspondence between the respective backbones of the DNA and protein, residues Tyr 26, Gln 27, Ser 28, Asn 31, Lys 32, and His 35 of the α₃ helix are in a position to interact either with the hydrogen bonding groups of the base pairs that are exposed in the major groove, conferring specificity, or with the negatively charged phosphates, promoting generalized binding. Nearby residues, including Arg 38 and Lys 39, are also suitably located to contribute to binding and recognition. Close to the twofold symmetry axis, the sequences Glu 54–Val 55–Lys

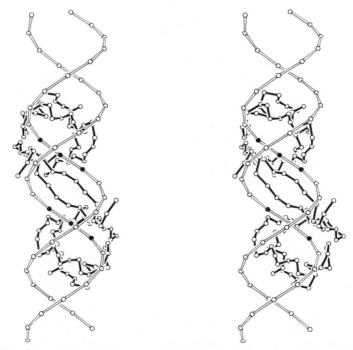

Figure 7. Stereo view of the presumed interaction between two molecules of cro repressor and DNA, viewed along their common twofold symmetry axis. The open circles along the DNA backbone indicate the positions of the phosphates: those phosphates whose ethylation hinders cro binding (Fig. 5) are drawn solid. Reprinted, with permission, from Anderson et al. (1981), copyright 1981 Macmillan Journals, Ltd.).

56 of the two monomers form a pair of antiparallel β sheet strands parallel to the minor groove (Anderson et al., 1981; Ohlendorf et al., 1982).

5. MODELING THE CRO–DNA COMPLEX

Starting with the preliminary protein–DNA alignment suggested above, an effort was made to build a detailed model of the cro–DNA complex (Ohlendorf et al., 1982). The general approach was to carry out model building on an MMSX computer graphics facility (Molnar et al., 1976) and to supplement this with energy minimization using the program EREF of Jack and Levitt (1971). Use of computer graphics permitted the screening of a large number of potential interactions between the protein and the DNA while energy minimization ensured that the stereochemistry of a proposed DNA–protein complex was reasonable.

There has been considerable speculation that the free energy of stabilization of DNA–protein complexes comes primarily from sequence-nonspecific interactions (von Hippel, 1979) while specificity derives largely from hydrogen bonds between protein side chains and the parts of the base pairs exposed in the grooves of the DNA (Yarus, 1969; von Hippel and McGhee, 1972; Seeman et al., 1976; von Hippel, 1979; Hélène and Lancelot, 1982). In model building, therefore, we attempted to maximize the number of hydrogen bonds and other interactions between cro and the DNA, and we assumed that each "mispositioned" (i.e., unsatisfied) hydrogen bond decreased the affinity of the DNA–protein complex. It is, of course, understood that other factors such as shape complementarity, hydrophobic and dipolar interactions, and the role of solvent are also important. The binding constant K for the cro–DNA interaction is related to the free energy of the interaction ΔG, by the formula $K = \exp(-\Delta G/RT)$, therefore each improperly positioned hydrogen bond donor or acceptor, which reduces the free energy by about 1 kcal/mole (von Hippel, 1979; Levitt, 1974), results in a reduction of the binding constant by a factor of approximately seven.

Because both cro and the DNA binding site have approximate twofold symmetry, it is in the first instance sufficient to consider the interaction between one monomer of cro and half of the binding site. We have numbered the base pairs within each half-site from 1 to 9, starting at the distal base pair. Bases $+1$ and -1 constitute a base pair, and so on. The "$+$"

base is the one for which the inward direction of numbering is from 5' to 3'.

The initial alignment of cro (Anderson et al., 1981) (Fig. 8) suggested that the amino acid side chains most likely to participate in interactions within the major groove of the DNA were Tyr 26, Gln 27, Ser 28, Asn 31, Lys 32, His 35, and Arg 38. Based upon the framework of the potential hydrogen bonds and the cro:DNA geometry suggested by the energy calculations, it is possible to construct a set of rules that can be used to predict the hydrogen bonds expected when cro interacts with a given sequence of DNA. The proposed rules for hydrogen bonding (Ohlendorf et al., 1981) are:

1. Arg 38 forms two hydrogen bonds to O6 and N7 of guanine −6. If base −6 is an adenine then Arg 38 makes a hydrogen bond with either the N7 of this adenine or the N7 of a purine at −7. If base −6 is a thymine then Arg 38 can make no hydrogen bonds to this

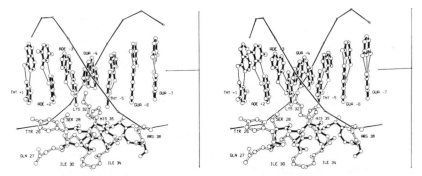

Figure 8. Stereo drawing illustrating the initial alignment of residues 26–38 of c pressor within the major groove of the DNA. Only the left half of the 17-base-pair b site is shown. The vertical line at the right of the figure is the common protein and twofold symmetry axis which relates the other half of the binding site to that shown The horizontal line at the right coincides with the long axis of the DNA. The conform of the individual protein side chains are those determined from the crystal structure repressor protein. The base sequence is that for O_R3, with numbering as defined text. For clarity, the details of the DNA backbone have been omitted and replac lines connecting the phosphates. Reprinted, with permission, from Ohlendorf et al. (copyright 1982 by Macmillan Journals, Ltd.

base because of the blocking methyl group of C6 but can form one hydrogen bond to N7 or a purine at -7 (i.e., a net loss of one hydrogen bond).

2. Lys 32 N^ε forms one hydrogen bond with either a purine N7 or an O4 of thymine at position -5. In addition, it forms a second hydrogen bond with O6 of guanine -4 and possibly a third hydrogen bond with N7 of guanine -4.

3. Ser 28 O^γ forms two hydrogen bonds with N6 and N7 of adenine -3 or, alternatively, one hydrogen bond to N7 of guanine, N4 of cytosine, or O4 of thymine at -3. In the case of a thymine at position -3 there is a second hydrogen bond from the N6 of the adenine at $+3$ to the carbonyl oxygen of the serine.

4. Gln 27 forms two hydrogen bonds to N6 and N7 of an adenine at $+2$. In all other cases it forms a single hydrogen bond.

5. Tyr 26 O^η forms a hydrogen bond with either O4 of thymine $+1$ or N7 of a purine at $+1$, but it cannot be placed in a stereochemically acceptable position to accept a hydrogen bond from N6 of adenine or N4 of cytosine.

As shown in Table 1, this model for the interaction between cro and DNA is consistent with the known relative affinities of cro for its six binding sites and for mutant sites as well (Johnson et al., 1979; Johnson, 1980).

The presumed sequence-specific hydrogen bonding between cro and O_R3 is shown in Figure 9 in a schematic representation based on that proposed by Woodbury et al., (1980); it is shown in stereo in Figure 10. The apparent interactions between the protein and the DNA backbone, which are not sequence specific, and, presumably, provide much of the overall energy of interaction, are summarized diagrammatically in Figure 11.

In the proposed cro–O_R3 complex, there are about 20 sequence-specific hydrogen bonds (Fig. 9) as well as many sequence-independent interactions (Fig. 11). These interactions could easily account for the observed dissociation constant of about 10^{-10} M (Takeda et al., 1977; Johnson et al., 1980). In the crystal, the five or six C-terminal residues of cro are "disordered" and appear to be freely moving about. It is proposed that in the cro–DNA complex these residues take up defined po-

Table 1. Cro Binding to Different DNA Sites[a]

Site	Sequence	Mutant	Affinity Relative to O_R3	Predicted H-bonds Relative to O_R3
| $O_R3$ | ```
1     5       9       13      17
T A T C A C C G C A A G G G A T A
A T A G T G G C G T T C C C T A T
                        G
                        C
                        C
                        G
          G
          C
``` | | 1 | 0 |
| | | v3C | 1/10 | −1 |
| | ```
    T
    A
``` | r1 | 1/3 | 0 |
| | ```
        T
        A
``` | r2 | <1 | 0 |
| | ```
        A
        T
``` | r3 | <1 | 0 |
| | | c12 | 1/10 | −1 |
| | | c10 | 1/10 | −1 |
| $O_R2$ | ```
T A A C A C C G T G G C G T G T T G
A T T G T G G C A C G C A C A A A C
          G
          C
``` | | 1/8 | −1 |
| | | v3C | 1/80 | −2 |
| | ```
        A
        T
``` | vi or vH | | −2 |
| | ```
      A A
      T T
``` | virC23 | 1/400 | −3 |
| | | vN or vC34 | <1/8 | −3 |
| $O_R1$ | ```
                  T
T A C C T C T G G C G G T G A T A
A T G G A G A C C G C C A C T A T
G
``` | | 1/8 | −1 |
| | ```
                  A
``` | vR18 | | −2 |
```

	Sequence	Variant base(s)	Mutation	Relative affinity	Change in H-bonds
		C			
		A / T (C)			
$O_{L}1$	T A C C A C T G G C G G T G A T A		vs326	1/40	−2
	A T G G T G A C C G C C A C T A T	A / T	v3		−1
		G / C	C G T A	vs387	<1/8	−2
			A	vC1	1/40	−2
			v101	~1/2	−1
					−2
$O_{L}2$	T A T C T G G C G G T G T T G	T A A T	v2		−2
	A T A G A C C G C C A C A A C	A / T	v003		−2
				v305	~1/2	−1
					−1
$O_{L}3$	A A C C A T C T G C G G T G A T A			~1/10	−2
	T T G G T A G A C G C C A C T A T				

Position: 1 5 9 13 17

Amino Acid: 26 (−28), Lys (32), Arg (38), Arg (38), Lys (32), Lys (26), (−28)

*The table lists the DNA sequences and relative affinities for the six 17-base-pair sites to which cro binds specifically in phage λ. Mutations in these sites with known relative affinities for cro binding are also shown (Data taken from Johnson, 1980, and Johnson et al., 1979). The relative binding affinities are to be compared with the change in the number of sequence-specific hydrogen bonds between cro and the DNA site as predicted from the model for cro binding proposed in the text. If the model is correct, and if hydrogen bonding is a major determinant in the specificity of cro:DNA recognition, then the loss of successive hydrogen bonds ought to correspond to successive reductions in the affinity of binding by about a factor of seven (see text).

Reprinted, with permission, from Ohlendorf et al. (1982), copyright 1982 by MacMillan Journals, Ltd.

247

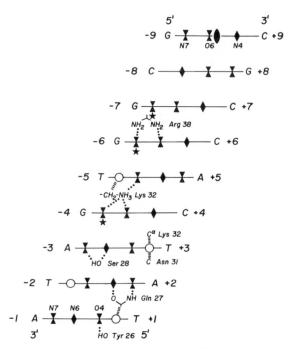

Figure 9. Schematic representation, following Woodbury et al. (1980), of the presumed sequence-specific interactions between cro and the parts of the base pairs exposed within the major groove of the DNA. The direction of view is imagined to be directly into the major groove of the DNA with the base pairs seen edge-on. The dyad symbol within the topmost base pair indicates the center of the overall 17-base-pair binding region. $\mathbf{I}$ hydrogen bond acceptor; ◆ hydrogen bond donor; ○ methyl group of thymine; ★ guanine N7, which is protected from methylation when cro is bound (Takeda et al. 1977; Johnson et al., 1980). Dots (···) indicate presumed hydrogen bonds between cro side chains and the bases. Dashes (||||) show apparent van der Waals contacts between cro and the thymine methyl groups. Other van der Waals contacts are not shown. Reprinted, with permission, from Ohlendorf et al. (1982), copyright 1982 by Macmillan Journals, Ltd.

sitions along the minor groove of the DNA (Ohlendorf et al., 1982) as shown in Figure 12. Because these presumed interactions between the C-terminal hexapeptides and the DNA are sequence-independent, it is possible to shift the hexapeptides one nucleotide along the DNA backbone toward the center of the DNA. When this is done the DNA-binding

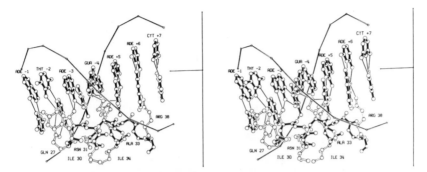

Figure 10. Stereo drawing (*cf*. Fig. 8) showing the presumed hydrogen bonding of cro to base pairs in the major groove of O_R3, following model building and energy refinement. Reprinted, with permission, from Ohlendorf et al. (1982), copypright 1982 by Macmillan Journals, Ltd.

helix (α_3) is pulled approximately 4 Å away from the major groove, as shown in Figure 13. This is a conformation that the cro dimer might use for sliding along the DNA while "searching" for its specific binding site. Such a mechanism has been proposed for the binding of the *lac* repressor (Richter and Eigen, 1974; Berg and Blomberg, 1976; Berg et al., 1981; Winter and von Hippel, 1981; Winter et al., 1981; Berg et al., 1982).

The proposed model (see Figs. 9–12) is consistent with solution studies of cro–DNA complexes. Removal with carboxypeptidase of four or five residues from the C-terminus of cro reduces its affinity for binding to λ DNA. The presumed involvement in DNA binding of the lysine residues listed in Figures 9 and 11 has been confirmed by chemical protection and modification experiments of cro protein (Y. Takeda, unpublished observations).

The model-building studies that have been described predict which base pairs are recognized by the protein and which specific interactions mediate this recognition. One way to test this model and to provide additional information is to study complexes of cro with specific DNA oligomers.

Crystals of the complex of cro with the complementary hexamer of ATCACC and GGTGAT have been obtained (Anderson et al., 1983). The space group is $C222_1$ with cell dimensions $a = 81.1$ Å, $b = 89.2$ Å, $c = 80.0$ Å. Diffraction spectra extend to at least 3.0 Å resolution. Analysis

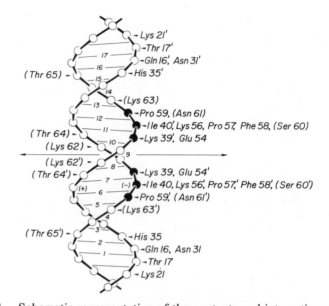

Figure 11. Schematic representation of the contacts and interactions presumed to occur between cro and the phosphates in the DNA backbone. In the presumed cro–DNA complex, at least one close approach occurs between a given phosphate group and the named amino acids. Those phosphates for which ethylation interferes with cro binding (Johnson, 1980) are drawn solid. Numbering of the base pairs and the signs $(+)$ and $(-)$ correspond to the identification used in the text. Amino acids belonging to "lower" and "upper" cro monomers related by the horizontal dyad axis are nonprimed and primed, respectively. Amino acids in parentheses are those at the C-terminus of the molecule, which are disordered in the crystal (residues 62–66) or for which the backbone conformation has been adjusted (residues 60, 61) to improve the contact with the DNA. Reprinted, with permission, from Ohlendorf et al. (1982), copyright 1982 by Macmillan Journals, Ltd.

of dissolved crystals reveals that they contain two six-base-pair duplexes per cro dimer. The volume of the unit cell is 581,000 Å³/ dalton, and each duplex corresponds to base pairs 2–7 of the left half of the O_R3 operator of bacteriophage λ (Fig. 9). In addition, crystals have been obtained with the complementary monomer ACCGCAAGG and CCTTGCGGT (Anderson et al., 1983). These crystals are isomorphous with those obtained with the hexamer. Cocrystallization experiments with half of the O_R3 operator of bacteriophage λ (Fig. 9) as well as with whole O_R3 operator fragments are in progress.

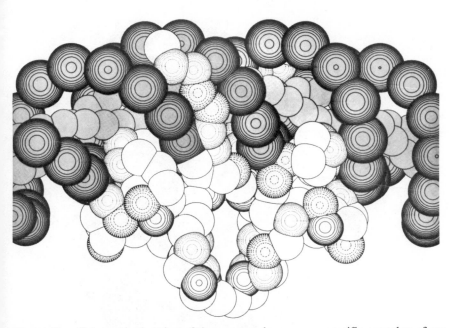

Figure 12. Schematic drawing of the proposed sequence-specific complex of cro with DNA. For the stylized DNA (largest spheres), the large concentric circles indicate the positions of the phosphate groups and the smaller circles follow the bottom of the major and minor grooves. For the cro dimer (smaller spheres), one circle is drawn for each amino acid; continuous concentric circles indicate acidic residues, broken circles indicate basic residues, dotted circles indicate non-charged hydrophilic residues, and blank spheres show hydrophobic residues. Reprinted, with permission, from Ohlendorf et al., (1982), copyright 1982 by Macmillan Journals, Ltd.

6. SEQUENCE HOMOLOGY IN GENE-REGULATORY PROTEINS

The amino acid sequence of cro was initially compared with the sequences of four related gene-regulatory proteins (Anderson et al., 1982). One of these proteins was the cI repressor from bacteriophage λ. As previously mentioned, both cI and cro recognize the same six DNA sequences within the λ genome but with different affinities (Johnson et al., 1978, 1979; Johnson, 1980). Another of the proteins, the product of the *cII* gene from bacteriophage λ (hereafter cII), is an activator of transcription (Schwarz et al., 1978; Shimatake and Rosenberg, 1981). A third protein was the

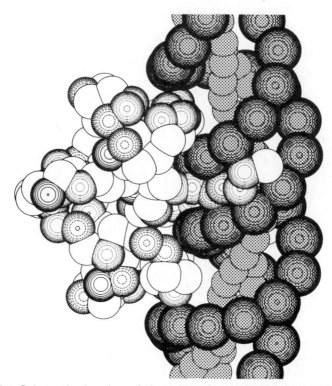

Figure 13. Schematic drawing of the proposed sequence-nonspecific complex of cro with DNA. For the stylized DNA (shaded spheres) the large concentric circles indicate the positions of the phosphate groups and the smaller circles follow the bottom of the major and minor grooves. For the cro dimer, one circle is drawn for each amino acid; continuous concentric circles indicate acidic residues, broken circles indicate basic residues, dotted circles indicate noncharged hydrophilic residues, and blank spheres show hydrophobic residues.

product of the *cro* gene from the related coliphage 434 (hereafter cro 434) which had been previously shown to be homologous with λ cro (Grosschedl and Schwartz, 1979). The fourth protein, the product of the *c2* gene from the related *Salmonella* phage P22 (hereafter c2 P22), functions in a manner analogous to that of cI in phage λ. The C-terminal domains of these two proteins (λ cI repressor and P22 c2 repressor) are homologous (Sauer et al., 1981).

Amino acid sequences of the respective proteins were compared, initially using segment lengths of 10 and 20 residues (data not shown). Two comparisons stood out: (1) the alignment of residues 13–32 of cro with

23–42 of cII (MBC = 0.75; significance 4.6 σ); (2) the alignment of 9–28 of cro 434 with 11–30 of P22 (MBC = 0.75; significance 4.6 σ). (MBC is the minimum base change per codon. The significance is estimated by comparing the best alignments with all other possible alignments.) Another unusually good alignment was between residues 11 and 30 of cro

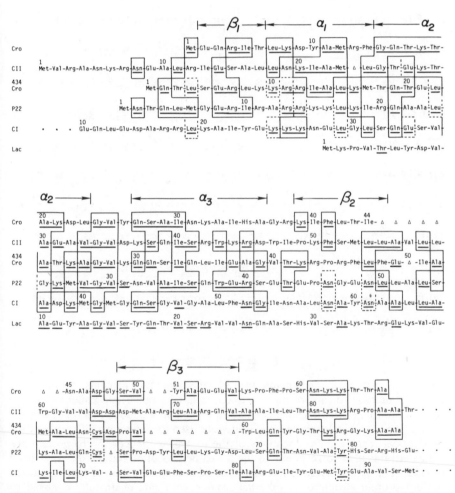

Figure 14. The amino acid sequences of cro, cII, cro 434, P22, cI and lac repressor proteins aligned to give maximum agreement of both the amino acid sequences (shown) and the DNA gene sequences. Homologous amino acids in two or more sequences are underlined. α_1, α_2, and α_3 indicate the locations of the α helices in cro; β_1, β_2, and β_3 indicate the three β strands.

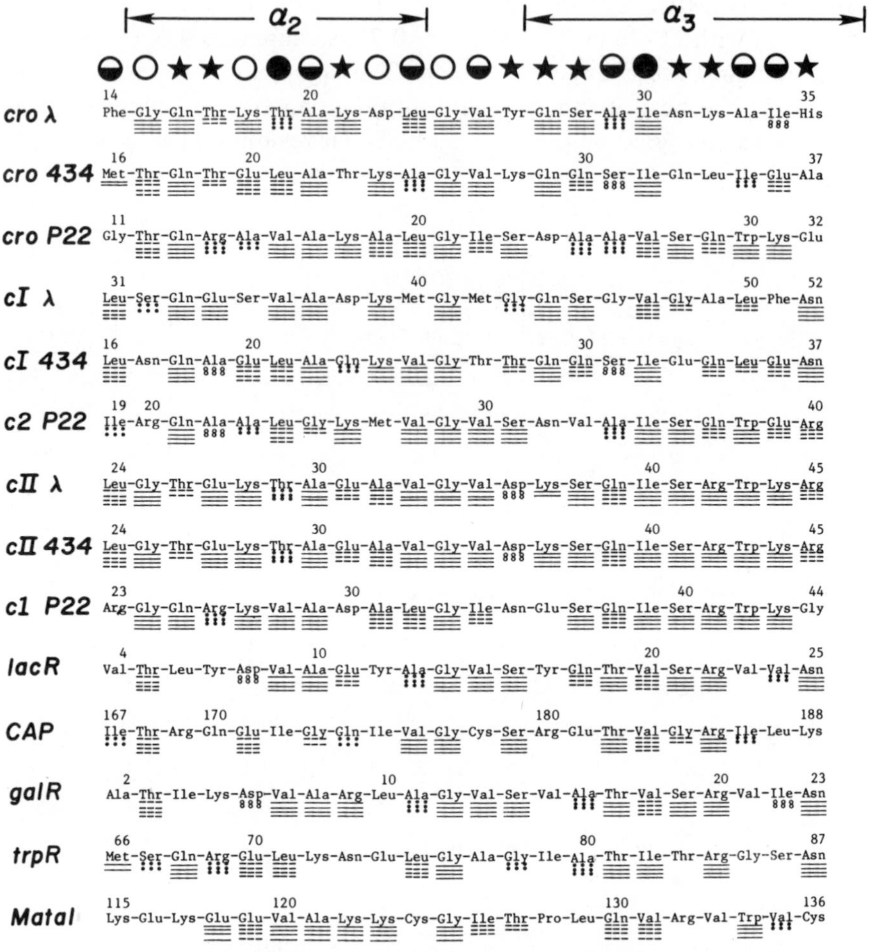

Figure 15. Amino acid segments from 14 gene regulatory proteins. Amino acids that are common to four or more sequences have a quadruple underline; those common to three or two sequences have a triple and a double underline, respectively. Solid, broken, or dotted underlines are used to distinguish different groups of identical residues. The locations of the α_2 and α_3 helices in cro protein are indicated; this also corresponds to the α_2–α_3 helices of cI and the α_E–α_F helices of CAP. The symbols above the first row indicate the location of that residue in the cro protein structure: ○ external residue with side chain fully exposed to solvent; ◐ surface residue only partly exposed; ● internal buried residue; ★ residue presumed to interact with the DNA when the protein is bound. The proteins included in this figure are: cro λ, cro repressor protein from bacteriophage λ (Hsiang et al., 1977; Roberts et al., 1977); cro 434, cro repressor protein from

and between 14 and 33 of cro 434 (MBC = 0.90; significance 3.6 σ), similar to the homology found by Grosschedl and Schwarz (1979).

The striking result is that the above pairwise alignments allow a common alignment of the cro, cII, cro 434, c2 P22, and cI sequences (Fig. 14; Anderson et al., 1982). From this initial alignment, it is possible to find significant homologies with a number of other proteins (Matthews et al., 1982b; Ohlendorf et al., 1983). Similar homologies in the sequences of a number of DNA-binding proteins have been recognized by Sauer et al. (1982) and Weber et al. (1982).

The region of maximal homology between these proteins corresponds to the $\alpha_2-\alpha_3$ helices of cro. Figure 15 shows the correspondence of the sequences of 14 DNA-binding proteins within this region. According to the proposed model for cro bound to DNA (Anderson et al., 1981; Ohlendorf et al., 1982), helix α_3 lies in the major groove of B-DNA and forms most of the specific interactions with the DNA. The α_2 helix is proposed to contribute a number of sequence-independent hydrogen bonds. In the case of the *lac* repressor protein (hereafter lacR), there is striking correspondence between the residues that are predicted to interact with DNA from these sequence comparisons and those predicted from genetic studies (Matthews et al., 1982b).

Because the gene-regulatory proteins presented in Figure 15 do not recognize the same DNA sequences, differences in the residues that specifically bind to DNA are to be expected. This variation can be seen in Figure 15. It would be expected, however, that the residues playing a role

---

bacteriophage 434 (Grosschedl and Schwarz, 1979); cro P22, cro repressor protein from *Salmonella* phage P22 (R. T. Sauer and A. R. Poteete, personal communication); cI λ, cI repressor protein from bacteriophage λ (Sauer, 1978; Sauer and Anderegg, 1978); cI 434, cI repressor protein from bacteriophage 434 (R. R. Yocum, W. F. Anderson, and R. T. Sauer, personal communication); c2 P22, gene c2 repressor protein from *Salmonella* phage P22 (Sauer et al., 1981); cII λ, product of *cII* gene from bacteriophage λ (Schwarz et al., 1978); cII 434, product of *cII* gene from bacteriophage 434 (Grosschedl and Schwarz, 1979); cI P22, product of *cI* gene from *Salmonella* phage P22 (Sauer and Poteete, personal communication); lacR, *lac* repressor protein from *E. coli* (Beyreuther et al., 1977; Farabaugh, 1978); CAP, catabolite gene activator protein from *E. coli* (Aiba et al., 1982); galR, *gal* repressor protein from *E. coli* (von Wilcken-Bergmann and Müller-Hill, 1982); trpR, *trp* repressor protein from *E. coli* (Gunsalus and Yanofsky, 1980); and Matal, mating-type protein from *Saccharomyces cerevisiae* (Astell et al., 1981).

in forming or maintaining the $\alpha_2-\alpha_3$ bihelical unit should be conserved. The following requirements are consistent with this expectation.

1. No proline should occur in positions 17–23 and 29–35, to be consistent with these residues being within α helices.

2. Gly 24 is strongly conserved. In cro the polypeptide backbone has dihedral angles corresponding to a left-handed α helix. Residues in this conformation are usually glycine (Matthews, 1977).

3. Residues 19, 23, 25, and 30 are expected to be hydrophobic and never charged, consistent with these residues being buried as in cro.

4. Residue 20 is nearly always a glycine or an alanine. In cro, the side chain of this residue is wedged in a surface crevice between the α_2 and α_3 helices, thus making β-branched side chain impossible at this location. However there is just enough room to accommodate a linear aliphatic side chain such as lysine.

7. STRUCTURAL HOMOLOGY IN GENE-REGULATORY PROTEINS

Prior to the knowledge of the amino acid sequence of CAP, a comparison was made between its structure and that of cro (Steitz et al., 1982). It was shown that the α_2 and α_3 helices in cro are structurally homologous with the E and F helices in CAP.

We found that the 24 consecutive α-carbon atoms of cro correspond within 1.1 Å to residues 166–189 of CAP. To estimate the error due to imprecision of coordinates, α-carbons 166–189 of one domain of CAP were superimposed on the corresponding α-carbons of the other domain; the root-mean-square difference in atomic coordinates was 0.7 Å in this case. For cro, the corresponding discrepancy is 0.4 Å. Therefore, the remarkable structural correspondence between CAP and cro, shown in Figure 16, approaches the experimental error of the coordinates.

An estimate of the significance of the observed agreement between the two α helices in CAP and cro was obtained in two ways. First, the empirical structure agreement probability plot of Remington and Matthews (1980), showing an agreement of 1.1 Å between 24 contiguous α-carbon atoms, is significant at the level of about 3.5 σ and is, therefore, quite

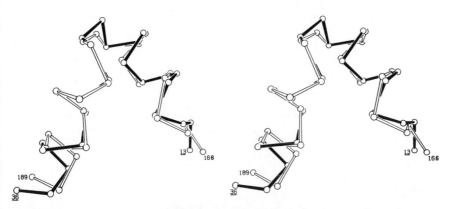

Figure 16. Stereo drawing showing the close structural correspondence between the E–F helices of CAP (open bonds) and the α_2–α_3 helices of cro (solid bonds). Reprinted, with permission, from Steitz et al. (1982).

unusual. Second, as a further test of the significance of the structural agreement between CAP and cro, we carried out a systematic search through all the proteins listed in the Brookhaven Protein Data Bank (Bernstein et al., 1977) in order to see if backbone segments of other proteins agreed with the 24-residue segment of cro as well as or better than had been observed for CAP. This search involved 21,540 comparisons of the two-helical cro segment with all possible 24-residue segments from 134 coordinate files. The best structural correspondence obtained from this search was for a part of the backbone of hen egg-white lysozyme and had rms differences of 2.8 Å for the 24 contiguous α-carbon atoms (Steitz et al., 1982). The average agreement of 6.9 Å and standard deviation of 1.7 Å are as expected for a 24-residue comparison (Remington and Matthews, 1980). It is striking to find that no other comparison even approaches the value of 1.1 Å for cro and CAP (see Fig. 17).

Subsequent to the structural comparison of cro and CAP, the amino acid sequence of CAP has been determined (Aiba et al., 1982; Cossart and Gicquel-Sanzey, 1982), and it has been found to be homologous with other DNA-binding proteins in the region of the α_2–α_3 structural unit (Weber et al., 1982; Ohlendorf et al., 1983). The structural and sequence homologies, taken together, suggest that a similar motif of α helices will be found in many of the proteins that bind specifically to double-stranded DNA. One common component of the specific recognition of the DNA

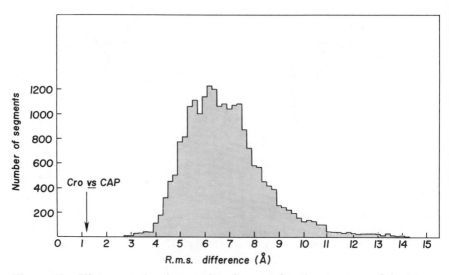

Figure 17. Histogram showing results of comparing the structure of the 24-residue segment comprising two α_2 and α_3 helices of cro with all 24 residue segments of all protein structures in the Brookhaven Data Base (Bernstein et al., 1977). The best root-mean-squared difference found is 2.8 Å. Reprinted, with permission, from Matthews et al. (1980b).

sequence is likely to be provided by the amino acid side chains of an α helix that protrudes from the surface of the protein and fits into the major groove of B-DNA. It is to be anticipated that the structures of many other proteins that specifically recognize double-stranded DNA sequences include this two-helix motif (see Figure 18).

Sequence homology between cro protein from phage λ and the cro proteins in phages 434 and P22 suggests that all of these proteins have similar structures (Grosschedl and Schwartz, 1979; Anderson et al., 1982; Sauer et al., 1982; unpublished observations). The cro proteins are all small, of about 70 amino acid residues, and in this case the polypeptide folds to form a single DNA-binding domain. Association of a pair of monomers about a twofold symmetry axis then yields a dimer in which the twofold axis of the protein can align with the local twofold symmetry axis normal to the DNA, thereby doubling the area of interaction between protein and DNA (Anderson et al., 1981). In the case of the larger proteins cI, c2 P22, and lacR, the N-terminal part of the polypeptide folds to form

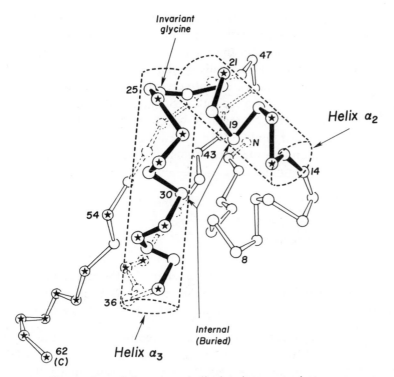

Figure 18. Illustration of the α_2–α_3 helical unit as seen in cro repressor. The connections between the 22 α-carbon atoms comprising the bihelical unit are drawn solid. Residues presumed to interact with DNA are indicated by a star (side chains are not shown). Also shown is the "invariant" glycine and the residues whose side chains are buried in cro. Reprinted, with permission, from Ohlendorf et al. (1983).

a DNA-binding "headpiece" while the carboxyl part of the molecule forms an essentially separate domain (Ptashne et al., 1980; Sauer et al., 1981; Miller and Reznikoff, 1978). CAP also has two domains although the DNA-binding region is toward the C-terminus of the molecule in this case (McKay and Steitz, 1981).

The addition of a second domain adds another level of sophistication to the function of these proteins. For example, in the case of lacR it is the C-terminal domain that binds inducers, thereby allowing the binding of the repressor to respond to changes in the level of metabolites in the cell (Miller and Reznikoff, 1978). In the case of cI repressor, the C-terminal domain is associated with cooperative binding of a pair of cI dimers to adjacent operator sites in the phage λ lytic/lysogenic control system (Ptashne et al., 1980). It is not unlikely that the first double-strand DNA-binding proteins to evolve were relatively small and had elements in common with the cro protein we see today. Subsequently, additional domains could have been added, as a result of which the basic DNA-binding function would have been modified. In the case of cI, c2 P22, and lacR, the DNA-binding region is at the N-terminus while the "modulating domain" constitutes the C-terminal part of the molecule; for CAP, the order is reversed.

8. CONCLUSIONS

Structural studies of cro repressor, together with the results of other studies, suggest the following general features for DNA–protein recognition by transcriptional repressors and activators:

1. *Symmetry.* Twofold symmetric recognition sites on the DNA are recognized by oligomeric proteins with at least one twofold axis of symmetry, the dyad axis of the protein coinciding with the dyad axis of the DNA site.

2. *DNA conformation.* The DNA maintains an essentially standard right-handed B form Watson–Crick conformation, possibly with some bending of the DNA. Neither intercalation between base pairs nor localized unwinding or conformational rearrangement is necessary for sequence-specific recognition.

3. *α-helical fold.* Many gene-regulatory proteins appear to contain a two-helical fold derived from a common evolutionary precursor.

4. *Helix in groove.* The above two-helical unit binds to DNA with the first α helix making predominantly sequence-independent interactions and the second α helix occupying, or partly occupying, the major groove and making sequence-specific interactions.

5. *Specificity.* Specificity of recognition derives in large part from multidentate hydrogen bonds between amino acid side chains and the parts of the base pairs exposed within the major groove of the DNA.

6. *Flexibility.* Local flexibility of the protein may be important in initial sequence-independent binding and in location and recognition of desired target sites.

REFERENCES

Aiba, J., Fujimoto, S., and Ozaki, N. (1982) *Nucl. Acids Res.* **10**, 1345–1361.

Anderson, W. F., Cygler, M., Vandonselaar, M., Ohlendorf, D. H., Matthews, B. W., Kim, J., and Takeda, Y. (1983). *J. Mol. Biol* **168**, 903–906.

Anderson, W. F., Ohlendorf, D. H., Takeda, Y., and Matthews, B. W. (1981). *Nature* **290**, 745–758.

Anderson, W. F., Takeda, Y., Echols, H., and Matthews, B. W. (1979) *J. Mol. Biol.* **130**, 507–510.

Anderson, W. F., Takeda, Y., Ohlendorf, D. H., and Matthews, B. W. (1982) *J. Mol. Biol.* **159**, 745–751.

Astell, C. R., Ahlstrom-Jonasson, L., Smith, M., Tatchell, K., Nasmyth, K. A., and Hall, B. D. (1981). *Cell* **27**, 15–23.

Berg, O. G., and Blomberg, C. (1976). *Biophys. Chem.* **4**, 367–381.

Berg, O. G., Winter, R. B., and von Hippel, P. H. (1981). *Biochemistry* **20**, 6929–6948.

Berg, O. G., Winter, R. B., and von Hippel, P. H. (1982). *TIBS* **7**, 52–55.

Bernstein, F. C., Koetzle, T. F., Williams, G. J. B., Meyer, E. F., Jr., Brice, M. C., Rodgers, J. R., Kennard, O., Shimanouchi, T., and Tasumi, M. (1977). *J. Mol. Biol.* **112**, 535–542.

Beyreuther, K. (1978). *Nature* **274**, 767.

Blundell, T. L., and Johnson, L. N. (1976). *Protein Crystallography.* Academic Press, London.

Bricogne, G. (1974). *Acta Crystallogr.* **A30**, 395–405.

Cossart, P., and Gicquel-Sanzey, B. (1982). *Nucl. Acid Res.* **10**, 1363–1378.

Farabaugh, P. J. (1978). *Nature* **274**, 765–769.

Flashman, S. M., (1978). *Mol. Gen. Genet.* **166**, 61–73.

Folkmanis, A., Takeda, Y., Simuth, J., Gussin, G., and Echols, H. (1976). *Proc. Natl. Acad. Sci. USA* **73**, 2249–2253.

Green, D. W., Ingram, V. M., and Perutz, M. F., (1954). *Proc. R. Soc. Ser. A* **225**, 287–307.

Grosschedl, R., and Schwarz, E. (1979). *Nucleic Acids Res.* **6**, 867–881.

Gunsalus, R. P., and Yanofsky, C. (1980). *Proc. Natl. Acad. Sci. USA* **77**, 7117–7121.

Hélène, C., and Lancelot, G. (1982). *Prog. Biophys. Mol. Biol.* **39**, 1–68.

Hsiang, M. W., Cole, R. D., Takeda, Y., and Echols, H. (1977). *Nature* **270**, 275–277.

Jack, A., and Levitt, M. (1971). *Acta Crystallogr.* **A34**, 931–935.

Johnson, A. (1980). "Mechanism of Action of the λ Cro Protein," Ph.D. Thesis, Department of Biochemistry and Molecular Biology, Harvard University, Cambridge, Massachusetts.

Johnson, A., Meyer, B. J., and Ptashne, M. (1978). *Proc. Natl. Acad. Sci. U.S.A.* **75**, 1783–1787.

Johnson, A., Meyer, B. J., and Ptashne, M. (1979). *Proc. Natl. Acad. Sci. USA* **76**, 5061–5065.

Johnson, A. D., Pabo, C. O., and Sauer, R. T. (1980). *Methods Enzymol.* **65**, 839–856.

Johnson, A. D., Poteete, A. R., Lauer, G., Sauer, R. T., Ackers, G. K., and Ptashne, M. (1981). *Nature* **294**, 217–223.

Levitt, M. (1974). *J. Mol. Biol.* **82**, 393–420.

Maniatis, T., Ptashne, M., Backman, K., Kleid, D., Flashman, S., Jeffrey, A., and Major, R. (1975). *Cell* **5**, 109–113.

Matthews, B. W. (1977) in *The Proteins*, H. Neurath and R. Hill, Eds., 3rd ed., Vol. 3. Academic Press, New York, pp. 403–590.

Matthews, B. W., Ohlendorf, D. H., Anderson, W. F., Fisher, R. G., and Takeda, Y. (1982a) *Cold Spring Harbor Symp. Quant. Biol.*, **47**, 427–433.

Matthews, B. W., Ohlendorf, D. H., Anderson, W. F., and Takeda, Y. (1982b). *Proc. Natl. Acad. Sci. USA* **79**, 1428–1432.

McKay, D. B., and Steitz, T. A. (1981). *Nature* **290**, 744–749.

Meyer, B. J., Maurer, R., and Ptashne, M. (1980). *J. Mol. Biol.* **139**, 163–194.

Miller, J. H., and Reznikoff, W. S., Eds. (1978). *The Operon*. Cold Spring Harbor Laboratories, Cold Spring Harbor, New York.

Molnar, C. E., Barry, C. D., and Rosenberger, F. U. (1976). *Tech. Memo. No. 229*. Computer Systems Laboratory, Washington University, St. Louis.

Ohlendorf, D. H., Anderson, W. F., Fisher, R. G., Takeda, Y., and Matthews, B. W. (1982). *Nature* **298**, 718–723.

Ohlendorf, D. H., Anderson, W. F., and Matthews, B. W., (1983). *J. Mol. Evol.*, **19**, 109–114.

Ohlendorf, D. H., and Matthews, B. W. (1983). *Ann. Rev. Biophys. Bioeng.*, **12**, 259–284.

Ptashne, M., Jeffrey, A., Johnson, A. D., Maurer, R., Meyer, B. J., Pabo, C. O., Roberts, T. M., and Sauer, R. T. (1980). *Cell* **19**, 1–11.

Remington, S. J., and Matthews, B. W. (1980). *J. Mol. Biol.* **140**, 77–79.

Richards, F. M. (1968). *J. Mol. Biol.* **37**, 225–230.

Richter, R. H., and Eigen, M. (1974). *Biophys. Chem.* **2**, 255–263.

Roberts, T. M., Shimatake, H., Brady, C., and Rosenburg, M. (1977). *Nature* **270**, 274–275.

Sauer, R. T., (1978). *Nature* **276**, 301–302.

Sauer, R. T., and Anderegg, R. (1978). *Biochemistry* **17**, 1092–1100.

Sauer, R. T., Pan, J., Hopper, P., Hehir, K., Brown, J., and Poteete, A. R. (1981). *Biochemistry* **20**, 3591–3598.

Sauer, R. T., Yocum, R. R., Doolittle, R. F., Lewis, M., and Pabo, C. O. (1982). *Nature* **298**, 447–451.

Schwarz, E., Scherer, G., Hobom, G., and Kossel, H. (1978). *Nature* **272**, 410–414.

Seeman, N. C., Rosenberg, J. M., and Rich, A. (1976). *Proc. Natl. Acad. Sci. USA* **73**, 804–808.

Shimatake, H., and Rosenberg, M. (1981). *Nature* **292**, 128–132.

Steitz, T. A., Ohlendorf, D. H., McKay, D. B., Anderson, W. F., and Matthews, B. W. (1982). *Proc. Natl. Acad. Sci. USA* **79**, 3097–3100.

Takeda, Y. (1979). *J. Mol. Biol.* **127**, 177–191.

Takeda, Y., Folkmanis, A., and Echols, H. (1977), *J. Biol. Chem.* **252**, 6177–6183.

von Hippel, P. H. (1979). In *Biological Regulation and Development*, R. F. Goldberger, Ed. Vol. 1. Plenum, New York, pp. 279–347.

von Hippel, P. H., and McGhee, J. F. (1972). *Ann. Rev. Biochem.* **41**, 231–300.

von Wilcken-Bergmann, B., and Müller-Hill, B. (1982). *Proc. Natl. Acad. Sci. USA* **79**, 2427–2431.

Weber, I. T., McKay, D. B., and Steitz, T. A. (1982). *Nucl. Acids Res.* **10**, 5085–5102.

Winter, R. B., Berg, O. G., and von Hippel, P. H. (1981). *Biochemistry* **20**, 6961–6977.

Winter, R. B., and von Hippel, P. H. (1981). *Biochemistry* **20**, 6948–6960.

Woodbury, C. P., Hagenbüchle, O., and von Hippel, P. H. (1980). *J. Biol. Chem.* **255**, 11, 534–11,546.

Yarus, M. (1969). *Ann. Rev. Biochem.* **38**, 841–880.

6

Structure of the Operator Binding Domain of Lambda Repressor

MITCHELL LEWIS
Smith Kline and French Laboratories
Philadelphia, Pennsylvania

JIAHUAI WANG
The Institute of Biophysics at Beijing
Academia Sinica
The Peoples Republic of China

CARL PABO
Department of Biophysics
Johns Hopkins Medical School
Baltimore, Maryland

CONTENTS

---

1. INTRODUCTION

The recognition and binding of regulatory proteins to specific sites on the DNA molecule is one of the primary mechanisms for gene regulation in prokaryotic systems. Bacteriophage λ encodes for two regulatory proteins, repressor and cro, that have antagonistic roles in phage growth (for a recent review see Ptashne et al., 1982). The repressor protein, expressed in the prophage state, binds to the DNA at specific sites and turns off all viral genes that are necessary for lytic growth. In addition, repressor acts as a positive regulator of gene expression, stimulating the transcription of its own gene. The cro protein, expressed during lytic growth, acts as a simple negative regulator that turns off the synthesis of repressor. These two proteins form a molecular switch that regulates phage growth by recognizing and binding to the same operator sites on the bacteriophage DNA. The left and right operator regions of the λ DNA are 80-base-pairs long and composed of three discrete recognition sites. Each site is 17-base-pairs long with sequences that are similar to but not identical with the sequences of the other sites. These recognition sites have an approximate twofold axis of symmetry about the central base pair. The right operator and a model for repressor binding are illustrated in Figure 1. This operator is embedded between two divergent promoters, P_R and P_{RM}. The RNA polymerase recognizes and binds either to the promoter P_R, which directs transcription of the lytic genes including cro, or to P_{RM}, which directs leftward transcription of repressor.

Lambda's repressor (monomer MW = 26,228) is a protein of 236 amino acids having two domains of approximately equal size. The N-terminal domain (residues 1–92) and the C-terminal domain (residues 132–236) are joined by a connector of some 40 residues that is extremely susceptible to proteolytic cleavage (Pabo et al., 1979). The N-terminal fragment generated by papain cleavage (residues 1–92) binds specifically to the λ operator site and makes the same operator contacts as intact repressor, although it binds less tightly (Sauer et al., 1979). Like the intact repressor, the N-terminal domain is able to mediate both positive and negative control of transcription. The C-terminal domain, residues 132–236, allows repressor to form stable dimers, which in turn increases the binding of repressor to operator DNA. To understand the structural basis for se-

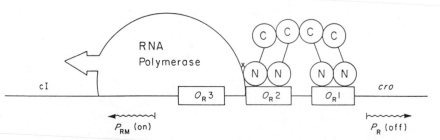

Figure 1. The configuration of proteins bound to the λ right operator. The repressor monomers are composed of two domains. The N-terminal domain binds DNA and the C-terminal domain provides important dimer contacts. Repressor dimers are bound at O_R1 and O_R2 with the RNA polymerase at P_{RM}. The repressor molecules at sites 1 and 2 sterically block polymerase from binding to the promoter P_R. The RNA polymerase bound to P_{RM} directs leftward transcription of the *cI* gene.

quence-specific recognition of DNA and the molecular basis of gene regulation, we have determined the three-dimensional structure of the N-
terminal or DNA-binding domain of repressor from bacteriophage λ.

2. THE STRUCTURE OF THE N-TERMINAL DOMAIN OF LAMBDA REPRESSOR

The operator-binding domain of λ repressor has been crystallized and the
structure solved to 3.2 Å resolution by the isomorphous replacement technique and an iterative density modification procedure (Pabo and Lewis,
1982). A preliminary electron density map was calculated from a single
heavy atom derivative with phases determined from isomorphous and
anomalous scattering information. This electron density was of sufficient
resolution to locate the three molecules in the asymmetric unit, but a
more interpretable map was obtained by exploiting the noncrystallographic symmetry. The improved electron density map was calculated
after several cycles of density modification (Bricogne, 1976). Figure 2
shows corresponding sections of electron density calculated with the isomorphous and modified phases. A complete three-dimensional model of
the protein has been built and the coordinates refined to 3.2 Å resolution
(Lewis and Pabo, to be published).

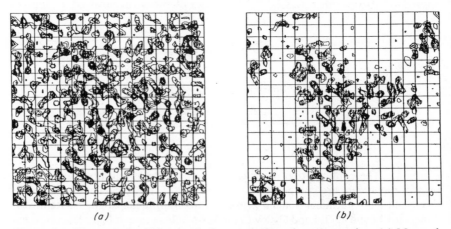

(a)   *(b)*

Figure 2.   Electron density maps before and after phase averaging. (*a*) Map calculated with SIR phases. Sections are perpendicular to the local threefold axis,
which is in the center of the map. Lines on the orthonormal grid mark at 5 Å
intervals. (*b*) Corresponding section of the final electron density map.

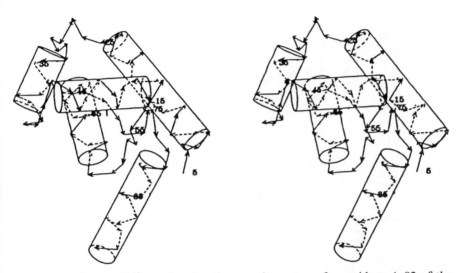

Figure 3. Stereo diagram showing the α-carbon atoms for residues 4–92 of the N-terminal domain. Superimposed on the α-carbon trace are cylinders representing the α-helical secondary structure of the molecule. In this orientation the N-terminal arm is foreshortened but the five helices are readily seen.

The structure of the N-terminal fragment is predominantly α helical, with an extended "armlike" structure (see Figure 3). The arm, projecting away from the core of the protein, is formed by the first six residues. In the crystal structure the first three residues are disordered but the position of the rest of the arm is stabilized by intermolecular contacts. The N-terminal fragment contains five helices composed of residues 9–23 (helix 1), 33–39 (helix 2), 44–52 (helix 3), 61–69 (helix 4), and 79–92 (helix 5). The first four helices and the loops connecting them form a compact domain that has a well-formed hydrophobic interior. Helix 5 is isolated from the globular domain with only a few intramolecular contacts. Biochemical experiments have shown that although the N-terminal fragment does not form a stable dimer in solution, it dimerizes with an association constant of 10^3 M and dimers of the N-terminal fragment bind to the operator (Pabo et al., 1979).

In the protein crystal, dimers of the N-terminal fragment are mediated through the C-terminal helix. The dimer contact is made by a hydrophobic patch on the surface of helix 5. A mutation in this helix that changes Ile 84 to Ser inactivates the molecule (Nelson et al., 1982). Presumably this destroys the hydrophobic dimer contact and the mutation provides further

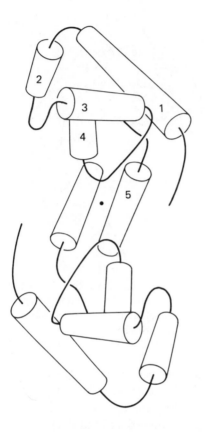

Figure 4. A sketch of the fragment dimer as seen in the crystal structure. The α-helices of the protein are represented as cylinders. The two fragments are related by a twofold axis of symmetry perpendicular to the plane of the page at a dot next to helix 5.

evidence for the physiological significance of the observed dimer. Figure 4 is a schematic representation of the dimer. In the crystal, three dimers cluster around the noncrystallographic symmetry axis to form a hexamer, which provides the basis for the crystal packing arrangement. The intermolecular contacts stabilizing the trimer of dimers are less extensive than those stabilizing the dimer, and there is no evidence that either hexamers or trimers have any biological significance.

3. BASIC FEATURES OF THE LAMBDA REPRESSOR–OPERATOR INTERACTION

A model for the repressor–operator complex was produced using the crystal structure of the protein dimer and coordinates of right-handed B type of DNA (Pabo and Lewis, 1982). Assuming the dimer observed in the protein crystal was the dimer that binds to the operator DNA, the twofold axis of the protein dimer was superimposed on the approximate twofold axis of a λ operator site. The protein dimer was rotated and translated about the dyad axis to generate an arrangement consistent with chemical and biochemical information (Johnson, 1982). Figure 5a shows the sites on the front side of the double helix that have been implicated as repressor "contacts," and Figure 5b shows how repressor may contact this side of the double helix. In the model, helix 3 of each N-terminal fragment fits into the major groove of the operator site with the N-terminal part of helix 3 closest to the double-helical axis of the DNA. Helix 2 is also near the DNA, and residues at the N-terminal end of this helix appear to be close to the sugar-phosphate backbone. Another striking feature of

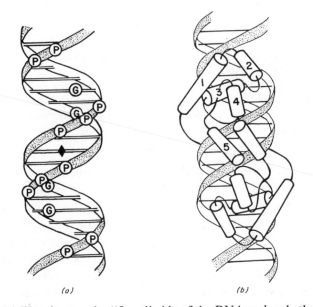

(a)   (b)

Figure 5. (a) The sites on the "front" side of the DNA molecule that have been implicated as repressor contacts. (P) represents the phosphate groups that when ethylated interfere with repressor binding. (G) represents guanine residues that are protected from methylation when repressor is bound. Details are reported by Johnson (1982). (b) Schematic representation of the protein–DNA complex. The twofold axis of the dimer is perpendicular to the plane of the page. The N-terminal arm is wrapping around the DNA molecule, and helices 2 and 3 are making contact in the major groove. Note that the protein appears to make extensive contacts in the major groove of the DNA but not the minor groove.

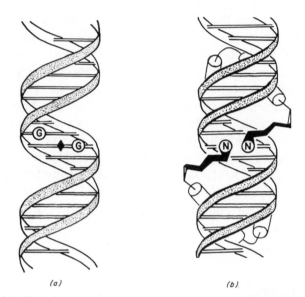

(a)   (b)

Figure 6. (*a*) The sites on the "back" of the DNA molecule that have been implicated in DNA binding. Again the (G) represents residues protected from methylation when the protein is bound. (*b*) A sketch of the fragment binding to the "back" surface of the DNA. The extended N-terminal arms follow along the major groove and could be positioned to contact the two guanine residues and protect the bases from chemical modification.

the model is the position of repressor's N-terminal arms. In the model the arms are set to the side of the double helix and are directed towards the center of the operator site. This is consistent with biochemical studies indicating that these arms wrap around the double helix and contact the major groove on the "back" of the operator site (Pabo et al., 1982). Figures 6*a* and *b* show those few sites on the "back" of the double helix that have been implicated as repressor "contacts" and how repressor's N-terminal arms may wrap around the DNA to contact these sites. Figure 7 is a space-filling representation of the repressor–operator complex.

4. SPECIFIC REPRESSOR–OPERATOR CONTACTS

For a protein to recognize and bind to a specific site on the DNA it seems that there must be specific contacts between side chains of the protein

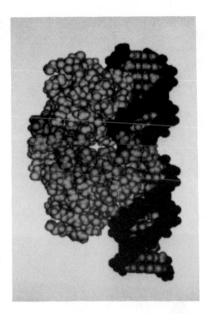

Figure 7. Space-filling model of a λ repressor dimer bound to operator DNA. Helix 3 fits into the major groove. The phosphates are shaded more darkly to help distinguish the major and minor grooves. Picture by R. Feldman.

and the bases of the DNA. In the major groove, each nucleic acid possesses several chemical groups that may act as donor or acceptor sites in hydrogen bonding (see Fig. 8); for example, the adenine base can donate a hydrogen bond from the N4 position and accept a hydrogen bond at N7, whereas guanine can accept two hydrogen bonds at the N7 and O6 positions. With respect to the protein, a large number of amino acid side chains possess chemical groups that potentially can form hydrogen bonds with the bases. For example, the hydroxyl group of serine, threonine, and tyrosine, the carboxylic or carboxylate group of glutamic and aspartic acids, and the amide group of asparagine and glutamine could be used to recognize the base sequence of the DNA. Hélène and Lancelott (1982) have reviewed many types of interactions that may occur between proteins and DNA.

To study λ repressor–operator interaction we chose the strongest repressor-binding site in the right operator. The sequence of O_R1 is shown in Figure 9. The left half of this site contains the sequence 5′ TATCACCGC 3′. This represents the "consensus" sequence that is obtained by comparing each half of the six operator sites. We have examined several plausible arrangements for sequence-specific contacts using this "consensus" operator sequence (Lewis et al., 1982). Our model-building

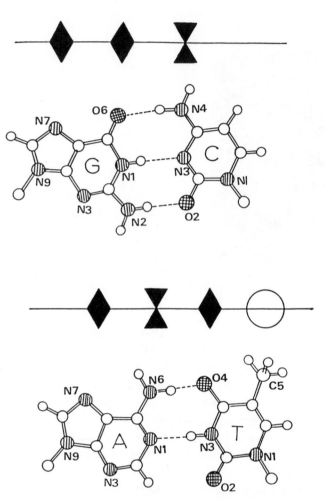

Figure 8. Line drawing of the Watson–Crick base pairing as observed in B form DNA. ⨉ hydrogen bond donor, ◆ hydrogen bond acceptor, ○ methyl group of thymine.

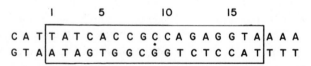

Figure 9. The nucleotide sequence for O_R1. The DNA binding site is 17 base pairs with an approximate twofold axis of symmetry through the central base pair.

274

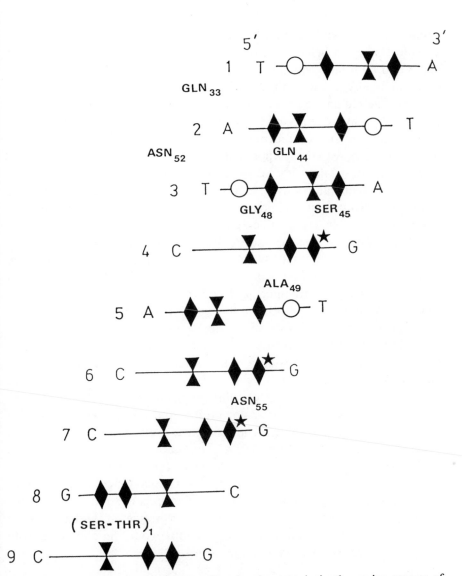

Figure 10. Proposed hydrogen bonding for the protein in the major groove of the "consensus" DNA operator site. The direction of view is directly into the major groove of the DNA with the base pairs seen edge-on. ★ guanine N7, is protected from methylation when repressor is bound.

studies suggest that sequence-specific contacts are made almost entirely by side chains on helix 3. At the N-terminus of this helix Gln 44 could make two hydrogen bonds to adenine $+2$ (numbering the bases as indicated in Figure 9 where negative numbers correspond to the opposite strand). The glutamine could accept a hydrogen bond from the N6 group and donate a hydrogen bond to the N7 group as anticipated by Seeman, et al. (1976). In our current model, Ser 45 serves as a hydrogen bond donor and pairs with the N7 position of guanine -4. However, we cannot rule out the possibility that Ser 45 could form two hydrogen bonds with adenine -3 (accepting a hydrogen bond from the N6 group and donating to the N7 position), or it is possible that Ser 45 might accept a hydrogen bond from the N6 group of cytosine $+4$. In addition to the possible hydrogen bonds, it appears that hydrophobic contacts are important in determining specificity. Gly 48 and Ala 49, which are in the middle of the binding helix, are exposed and appear to make nonbonded contacts with the methyl groups of thymine $+3$ and thymine -5. Just after helix 3, Asn 55 may donate to either the O6 or N7 positions of guanine -6. A list of specific contacts is illustrated in Figure 10.

Additional repressor–operator contacts that could help to determine specificity are made by residues to the N-terminal arm. This arm, which has the sequence Ser^1-Thr^2-Lys^3-Lys^4-Lys^5-Pro^6, has many groups that could hydrogen bond to the DNA. The lysines may form salt bridges with the phosphates spanning the major groove and position the N-terminal serine and threonine residues. These contacts are more difficult to predict because the arm has additional degrees of freedom and there is no reason to assume that the conformation observed in the protein crystal will be the same as the conformation in the repressor–operator complex. It seems likely, however, that one or more of the N-terminal residues could make specific contacts with the bases in the major groove on the back side of the operator. Furthermore, such interactions would be consistent with the biochemical studies (Pabo et al., 1982).

5. GENETIC ANALYSIS OF REPRESSOR BINDING

Several repressor mutants have been obtained that provide independent verification of the proposed complex (Nelson et al., 1982, Hecht et al., 1983). Approximately 50 different single amino acid substitutions in λ repressor's N-terminal domain have been characterized and their DNA binding properties accessed (Hecht et al., 1983). Substitutions, which decrease repressor–operator interactions, may arise from either destabilizing the tertiary structure of the whole protein or by altering the binding site itself. Figure 11 graphically illustrates the position of the mutations with respect to the protein sequence and secondary structure. Mutations, shown above the line, represent those amino acids whose side chains are exposed and on the surface of the protein. Those below the line are mutants affecting residues that are either interior or partly buried. As can be seen, the mutations that are classified as being interior are well distributed throughout the N-terminal domain. These mutations most likely affect operator binding by destabilizing the three-dimensional structure of the protein. The mutations affecting the exterior residues, those above the line in Figure 11, are not uniformly distributed but cluster in the region of helices 2 and 3. Figure 12 is a space-filling representation of the protein

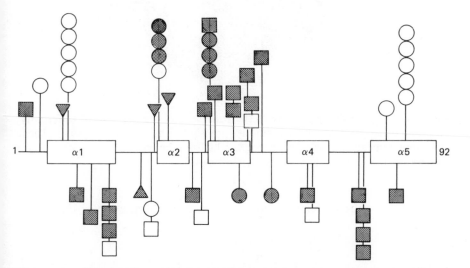

Figure 11. The positions of amino acid substitution with respect to the primary and secondary structure of λ repressor's N-terminal domain. Positions of the α helices are indicated by the boxed regions. △ Proline substitutions, □ missense substitutions, and ○ suppressed nonsense substitutions. Open symbols indicate weak or neutral substitutions and closed symbols indicate strong mutations. Symbols above the line are solvent-exposed and those below the line are completely or partly buried.

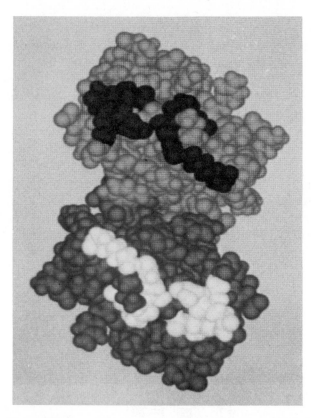

Figure 12. Space-filling models of the repressor dimer are shaded differently. Mutations that effect repressor binding are shown by contrast shading. These mutations appear to cluster on helix 3, which we believe to be responsible for binding operator DNA. Picture by R. Feldman.

dimer. The contrast shading shows a spatial representation of the strong operator mutants.

6. NONSPECIFIC BINDING OF LAMBDA REPRESSOR

Kinetic experiments have shown that λ repressor binds to specific sites on DNA at rates approaching those expected from a diffusion-controlled process (Johnson et al., 1980). Models to describe how regulatory proteins recognize specific target sites have been developed (Richter and Eigen,

1974, Berg et al., 1982). These models suggest that the recognition occurs in two steps. Initially, a nonspecific binding occurs between the protein and any part of the DNA molecule. Subsequently, the protein may slide along the DNA molecule or be transferred to another part of the DNA by an interdomain transfer. Adam and Delbruck (1968) have shown that such a mechanism in effect reduces the dimensionality or the volume of the search and therefore increases the reaction rate for the diffusion-controlled process. This nonspecific binding is thought to be largely electrostatic, involving charge–charge interactions between DNA phosphates and the basic residues on the protein. The dependence of repressor's binding constant upon ionic strength and pH clearly supports this general suggestion (Barcley and Bourgeois, 1978; Chadwick et al., 1970). To explore the electrostatic features of the protein–DNA interaction, we have examined the charge distribution of the repressor molecule (Wang and Lewis, 1982). The results are graphically illustrated in Figure 13, where the arrow represents the vector of the molecular polarity with the positive center at the arrow's head and the negative center at its tail. In this model the pseudo dyad of the operator DNA is coincident with the dyad of the protein dimer. The combined polarity vector of the two protein molecules is perpendicular to and pointing in the direction of the DNA. It is possible that before the protein dimer encounters the DNA, it may preorient itself, because the repressor dimer possesses a characteristic polarity. This would further reduce the diffusion dimensionality and the nonspecific binding would be accelerated.

Calculations of the electrostatic potential have shown that DNA, which is shielded by counterions, has a negative potential along the minor groove whereas the exterior of the major groove shows positive potential (Pullman and Pullman, 1981). In B-DNA the anionic oxygens of the phosphates are pointed toward the major groove as are the binding counterions, consequently the exterior of the major groove becomes positive. One may speculate that the polarity of the protein dimer may accelerate the nonspecific binding by aligning the polarity vector along the minor groove of the DNA. This aligns the protein so that it can search for its recognition site by sliding along the minor groove of the DNA. Once the recognition site is found, the protein dimer (and/or the DNA) could be rotated or bent around the common dyad, keeping the polarity vector of the protein dimer perpendicular to the DNA. The dimer could be readjusted until the specific complex is formed. This requires that the proposed

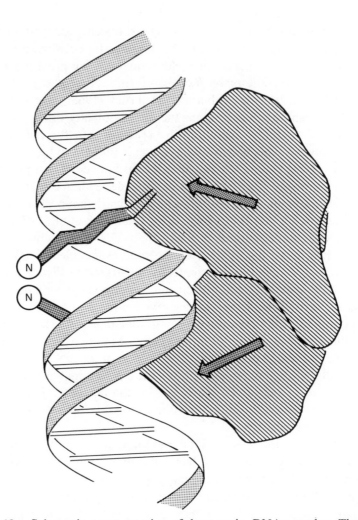

Figure 13. Schematic representation of the protein–DNA complex. The charge distribution of the protein is represented by the arrows. The arrow on each molecule represents the direction of the molecular polarity. The centroid of positive charge is at the head of the arrow and the centroid of negative charge at the arrow's end. The combined polarity vector lies along the dyad and points perpendicularly towards the DNA.

complex aligns the polarity vector of the protein dimer to point into the minor groove (see Figure 13). Once the complex has formed, additional nonspecific contacts could form. Glu 33, at the N-terminal end of helix 2, may contact the phosphate between bases 0 and 1 or between 1 and 2. Asp 52 may interact with the phosphate between bases 1 and 2 or between 2 and 3. Tyr 60, Asn 61, and Asn 58 may provide additional nonspecific interactions by contacting the phosphates between bases -5 and -6, -6 and -7, and -7 and -8, respectively.

7. IMPLICATIONS FOR GENE REGULATION

As mentioned above, the isolated N-terminal fragment, like repressor, can mediate both positive and negative control of transcription. Negative control can be explained by simple steric arguments. The bound repressor can block transcription by covering part of the promoter and preventing access to polymerase. Positive control, however, requires stimulating transcription. It has been proposed that repressor, when bound to the site O_R2, interacts with the RNA polymerase, which is bound to the promoter P_{RM}. This protein–protein contact stimulates transcription of the repressor gene. "Chemical-probe" experiments had indicated that one phosphate (in the site O_R2) is "contacted" by both repressor and RNA polymerase, and thus implied that repressor, when bound to O_R2, is very close to polymerase at P_{RM}. This observation led to the suggestion that

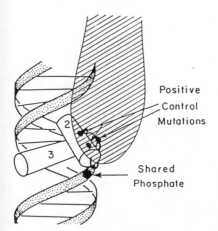

Positive
Control
Mutations

Shared
Phosphate

Figure 14. Schematic diagram showing the relationship between helices 2 and 3 of λ repressor at O_R2 and polymerase at P_{RM}. The "shared" phosphate is thought to be contacted by both bound repressor and polymerase (see Hochschild et al., 1983).

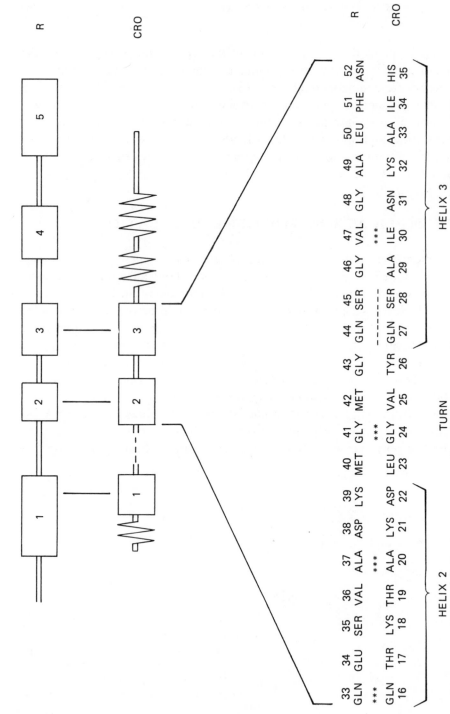

282

repressor might stimulate transcription by contacting polymerase. To further test the idea that a protein–protein contact is responsible for stimulation, mutant repressors have been isolated and sequenced (Guarente et al., 1982; Hochschild et al., 1983). These mutant repressors appear to bind the operator sites normally, but they are unable to stimulate transcription. Mutations affecting positive control have been found which change Gly 43 to Arg, Glu 34 to Lys, and Asp 38 to Asn. Examination of the model for the fragment–operator complex as sketched in Figure 14 shows that these three residues are on the surface of the N-terminal fragment and are very close to the phosphate, which is contacted by both repressor and polymerase. The model for the complex is consistent with the idea that repressor may stimulate transcription by contacting polymerase and with the assumption that the mutants which are defective in positive control have altered the residues that normally contact polymerase.

8.   COMPARISON OF LAMBDA REPRESSOR AND CRO PROTEINS

The structure of the cro protein from bacteriophage λ has been determined (Anderson et al., 1981). The λ cro protein has three α helices and three β sheets, and at first glance its structure appears rather different from the structure of the λ repressor. In fact, repressor and cro have a very limited sequence homology, but comparison of these proteins with a number of other repressor and cro proteins clearly shows that they are related and allows the λ repressor and cro to be aligned (Matthews et al., 1982; Sauer et al., 1982; Anderson et al., 1982). More recently a detailed structural comparison of λ repressor and cro revealed considerable structural homology (Ohlendorf et al., 1983). This information allowed us to revise the alignment to incorporate structural information. As indicated in Figure

---

Figure 15.   (*Opposite*) Sequence alignment of lambda's cro and N-terminal domain, illustrating the positions of the α helices and β sheets. The amino acid sequence corresponding to helix 2-turn-helix 3 is shown at the bottom of the diagram. Residues marked with asterisks are those highly conserved among several related DNA-binding proteins. The dashed line indicates those residues that would contact the same or similar bases.

15, helices 1, 2, and 3 of repressor correspond to helices 1, 2, and 3 of cro, but still the best structural correspondence involves helices 2 and 3.

Independent model-building studies for the repressor and cro suggest that these two helices form the operator binding site. In the protein–DNA complex both repressor and cro position helix 3 into the major groove with its N-terminal end closest to the double-helical axis and towards the outer end of the operator site (Pabo and Lewis, 1982; Ohlendorf et al., 1982). Helix 2, in both models, is just above the major groove, and the N-terminal region of this helix may contact the sugar-phosphate backbone of the DNA. The basic arrangement of these helices is similar, but the precise contacts made by the repressor and cro helices are quite different. In fact the helices appear to have somewhat different orientations relative to the major groove. This may reflect that only two of the nine residues in helix 3 are conserved. Both helices start with a Gln-Ser sequence, but helix 3 of repressor, unlike helix 3 of cro, has no basic residues and few hydrophilic residues.

This structural comparison pointed to another fundamental difference in the two structures. In the repressor structure, residues 25–30 form a short hairpin loop extending away from the surface of the molecule connecting helices 1 and 2. The revised sequence alignment (based on the structural correspondence) suggests that this loop is missing in the cro structure. Again these results are consistent with other chemical protection studies (Johnson, 1982). In the models of repressor and cro complexed to the DNA, this loop region between helices 1 and 2 is positioned at the ends of the operator site. The protection pattern for repressor and cro differ in that cro makes only a subset of repressor contacts. It appears that this connecting loop, which contains 3 lysine residues, could contact the phosphate backbone of the DNA and make contacts that represent the difference between the phosphate ethylation of repressor and cro.

9. CONCLUSIONS AND FUTURE DIRECTIONS

Based on the structural evidence and the sequence homologies, it appears that this two-helix motif is a basic secondary structure utilized by many duplex DNA binding proteins. The helix-turn-helix arrangement has been found in another duplex DNA-binding protein, the *E. coli* CAP protein (McKay and Steitz, 1981). A comparison of α-carbon coordinates revealed that helix E and F of CAP are extremely similar to helices 2 and 3 of cro (Steitz et al., 1982). The structural studies of cro, repressor, and CAP show that α helices are used in all known protein–duplex DNA interactions. This is consistent with the model-building studies suggesting that α helices could be used in recognition (Zubay and Doty, 1959; Sung and Dixon, 1970; Adler et al., 1972; Warrant and Kim, 1978). In fact, comparison of the protein structures and of the sequences of many duplex DNA-binding proteins suggest that a conserved supersecondary structure, which contains an α helix, a turn, and a second α helix, forms the duplex DNA-binding site of many different proteins (Steitz et al., 1982; Sauer et al., 1982; Anderson et al., 1982). Our studies of the λ repressor suggest that an extended polypeptide chain can be used to gain some additional binding energy, but it is not clear how general this type of interaction will be.

Structural studies of these duplex DNA-binding proteins have not yet given us a clear understanding of the specific interactions that are responsible for recognition of their binding sites. The fact that these detailed contacts are harder to predict, and the observation that repressor and cro must have significantly different contacts (because the amino acid sequences are so different), suggest that recognition cannot be explained on the basis of any simple "code." Thus it appears that there is no simple "rule" for recognition and that the detailed geometry of each protein will be crucial to the recognition process.

Our studies of the λ repressor–operator interaction are proceeding in several directions. Current model-building studies of the complex suggest that the model coordinates should be refined so as to fix the path of the backbone more precisely and thus provide a more rigorous constraint on the atomic positions. Energy refinement of the proposed complexes will be useful to determine the energetic feasibility of a particular complex.

We have begun the structure determination of crystals that contain the N-terminal domain of repressor and an 11-base-pair operator sequence. These crystals, although seriously disordered, will, we hope, provide a low-resolution image of the relative orientation of the protein with respect to the DNA. This structure determination will provide a direct test of our current model for the repressor–operator interaction and should provide a direct means for determining whether any conformational changes occur as repressor binds to the operator.

Acknowledgments. This research was funded by National Institutes of Health grants No. GM 22526 and No. GM 29109 to Mark Ptashne. We thank Mark Ptashne and Steve Harrison for their support. M. Lewis was supported by a fellowship from the Charles A. King Trust.

REFERENCES

Adam, G., and Delbruck, M. (1968). In *Structural Chemistry and Molecular Biology*, A. Rich and N. Davidson, Eds. Freeman, San Francisco, 198.

Adler, K., Beyreuther, K., Fanning, E., Geisler, N., Gronnenborn, B., Klem, A., Müller, B., Phfgal, M., and Schmitz, A. (1972). *Nature*, **237**, 322.

Anderson, W. F., Ohlendorf, D. H., Takeda, Y., and Matthews, B. W. (1981). *Nature* **290**, 754.

Anderson, W. F., Takeda, Y., Ohlendorf, D. H., and Matthews, B. W. (1982). *J. Mol. Biol.*, **159**, 745.

Barcley, M. D., and Bourgeois, S. (1978). *The Operon*. Cold Spring Harbor, p. 177.

Berg, O. G., Winter, R. B., and von Hippel, P. H. (1982). *Trends Biochem. Sci.* **7**, 52.

Bricogne, G. (1976). *Acta Crystallogr.* **A32**, 832.

Chadwick, P., Hopkins, N., Pirrotta, V., Steinberg, R., and Ptashne, M. (1970). *Cold Spring Harbor Symp. Quant. Biol.* **35**, 283.

Guarente, L., Nye, J. S., Hochschild, A., and Ptashne, M. (1982). *Proc. Natl. Acad. Sci. USA* **79**, 2236.

Hecht, M. H., Nelson, H. C. M., and Sauer, R. T. (1983). Submitted to *Proc. Natl. Acad. Sci. USA*.

Hélène C. and Lancelott, G. (1982). *Progr. Biophys. Mol. Biol.* **19**, 1.

Hochschild, A., Irwin, N., and Ptashne, M. (1983). *Cell* **32**, 319.

Johnson, A. D., Pabo, C. O., and Sauer, R. T. (1980). *Methods Enzymol.* **65**, 839.

Johnson, A. D. (1980). "Mechanism of Action of the λ Cro Protein," Ph.D. Thesis, Department of Biochemistry and Molecular Biology, Harvard University, Cambridge, Massachusetts.

Lewis, M., Jeffery, A., Ladner, R., Ptashne, M., Wang, J., and Pabo, C. O. (1982). *Cold Spring Harbor Symp. Quant. Biol.* **47**, 435–440.

Matthews, B. W., Ohlendorf, D. H., Anderson, W. F., Fisher, R. G., and Takeda, Y. (1982). *Cold Spring Harbor Symp. Quant. Biol.*, **47**, 427–433.

McKay, D. B., and Steitz, T. A. (1981). *Nature* **290**, 744.

Nelson, H. C. M., Hecht, M. H., and Sauer, R. T. (1982). *Cold Spring Harbor Symp. Quant. Biol.*, **47**, 441.

Ohlendorf, D. H., Anderson, W. F., Fisher, R. G., Takeda, Y., and Matthews, B. W. (1982). *Nature* **298**, 719.

Ohlendorf, D. H., Anderson, W. F., Lewis, M., Pabo, C. O., and Matthews, B. W. (1983). *J. Mol. Biol.*, in press.

Pabo, C. O., Krovatin, W., Jeffrey, A., and Sauer, R. T. (1982). *Nature* **298,** 441.

Pabo, C. O., and Lewis, M. (1982). *Nature* **298,** 443.

Pabo, C. O., Sauer, R.T., Sturtevant, J. M., and Ptashne, M. (1979). *Proc. Natl. Acad. Sci. USA* **76,** 1608.

Ptashne, M., Johnson, A. D., Pabo, C. O. (1982). *Sci. Am.* **245,** 128.

Pullman, A., and Pullman, B. (1981). *Quart. Rev. Biophys.* **14,** 289.

Richter, P. H., and Eigen, M. (1974). *Biophys. Chem* **2,** 255.

Sauer, R. T., Pabo, C. O., Meyer, B. J., Ptashne, M., and Backman, K. C. (1979). *Nature* **279,** 396.

Sauer, R. T., Yocum, R. R., Doolittle, R. F., Lewis, M., and Pabo, C. O. (1982). *Nature* **298,** 447.

Steitz, T. A., Ohlendorf, D. H., McKay, D. B., Anderson, W. F., and Matthews, B. W. (1982). *Proc. Natl. Acad. Sci. USA* **79,** 3097.

Sung, M. T., and Dixon, G. H. (1970). *Proc. Natl. Acad. Sci. USA* **67,** 1616.

Wang, J-H., and Lewis, M. (1982). Submitted to *Sci. Sin.*

Warrant, R. W., and Kim, S. H. (1978). *Nature* **271,** 130.

Zubay, G., and Doty, P. (1959). *J. Mol. Biol.* **1** 1.

7

Structure of Catabolite Gene Activator Protein

THOMAS A. STEITZ
IRENE T. WEBER
Yale University
New Haven, Connecticut

CONTENTS

1. INTRODUCTION

The catabolite gene activator protein (CAP), also known as the cyclic
AMP receptor protein, is an allosteric protein that functions both in the
positive and negative regulation of many operons in *E. coli*, including the
catabolite operons (Zubay et al., 1970; Epstein et al., 1975; de Crombrugghe and Pastan, 1978). In the presence of cyclic AMP (cAMP), CAP
binds to specific sites near promotors and switches on transcription at
some operons and off at others (Aiba, 1983). CAP binds only nonspecifically to DNA in the absence of cAMP. An excellent general review of
genetic, biochemical, and structural studies of CAP has been written by
de Crombrugghe et al. (1983). In this review we emphasize the structural

studies of CAP, and we describe other studies only as they are required to help understand the relationship between the CAP structure and its function.

The structural studies of CAP are primarily intended to address three questions. First, what is the nature of the cAMP-induced allosteric transition that changes CAP into a protein binding more tightly to a specific nucleotide sequence than to a nonspecific sequence of DNA? Second, how does the CAP complex with cAMP recognize a specific nucleotide sequence? Third, in those situations in which CAP acts as activator, how does its binding to a specific DNA binding site facilitate transcription by RNA polymerase? Although it is now possible to phrase these questions in more detailed molecular and structural terms, the final answers have not yet been obtained. However, some insight into possible mechanisms are suggested from the structural studies.

CAP is a dimer of identical 22,500 dalton subunits whose amino acid sequence is known from the DNA sequence of the gene (Aiba, et al., 1982; Cossart and Sanzey, 1982). Binding studies on the products of proteolytic digestion indicate that each subunit consists of two structurally and functionally distinct domains (Eilen et al., 1978; Aiba and Krakow, 1981) as is the case for *lac* repressor (Platt et al., 1973; Files and Weber, 1976; Geisler and Weber, 1976), and lambda (λ) repressor (Pabo et al., 1979). Digestion of native CAP with chymotrypsin results in a dimer of N-terminal domains that binds cAMP but no longer binds specific DNA in a cAMP-dependent manner (Eilen et al., 1978). A 12,500 dalton C-terminal fragment binds to DNA but does not bind cAMP (Aiba and Krakow, 1981). Thus, in solution the N-terminal domain binds cAMP and forms the dimer and the C-terminal domain binds DNA. This is precisely the picture that emerges from the crystal structure of CAP.

Table 1 shows the sequence of eight functional CAP binding sites that have been defined by DNase footprinting studies. The binding of CAP to these sites protects about 20 to 30 base pairs. The sequences have been aligned to maximize the homology on the left half of the site using either the transcribed or the nontranscribed strand in the 5' to 3' direction. The "consensus" sequence obtained is given below; capital letters indicate six or more identities at that position and lower-case letters indicate four or five identities at that position:

aApyTGTGAc-T---tCA---a

Table 1. DNA Sequences that Bind CAP

Operon	Sequence
lac 1	5′ C G C A A T T A A T G T G A G T T A G C T C A C T C A T T A G G C A 3′
gal	A T G C G A A A A G T G T G A C A T G G A A T A A A T T A G T G G
ara C	C A T A G C A A A G T G T G A C G C C G T G C A A A T A A T C A A T G
cat	C G A A A A T G A G A C G T T G A T C G G G C A C G T A A A G A
pBR322	A T A T G C G G G T G T G A A A T A C C G C A C A G A T G C G T A A A G
deo	C C T T A A T T G T G A T G T G T A T C G A A G T G T G T T G C
omp A	A C T T A C A A G T G T G A A C T C C G T C A G G C A T A T G A
crp	G C A C G G T A A T G T G A C G T C C T T T G C A T A C A T G C A

Position numbers marked along sequences: *lac* 1 (−70, −61–60, −50); *gal* (−30, −40–42, −50); *ara* C (−70, −80, −90, −93, −100, −110, −60, −50, −44, −40, −30); pBR322 (−30, −50, −40, −50, −60); *deo* (−50, −40, −30); *omp* A (−20, −30, −43, −40, −50); *crp* (50, 42, 40, 30).

	Sequence
Consensus sequence	− a A g T G T G A c g T − g − t C a − a − t − g −
Number of agreements	5 7 4 8 8 8 7 8 8 8 4 4 7 4 5 6 5 5 5 4 5

(minor entries below consensus: a 3, py; g 3, py; a 2, py)

Groove: minor — major — minor — minor — major — minor — major — minor

Twofold axis

a The sequences of eight sites known to be specifically recognized by CAP. The sites are *lac* (Simpson, 1980), *gal* (Taniguchi et al., 1979), ara C (Ogden et al., 1980; Lee et al., 1981), cat (Le Grice and Matzura, 1981), pBR322 (Queen and Rosenberg, 1981), deo (Valentin-Hansen et al., 1982; Valentin-Hansen, 1982), omp A (Movva et al., 1981) and crp (Aiba, 1983). The sequences have been aligned to show the similarities in the conserved region. They are numbered with respect to the transcription start point. Highly conserved bases are underlined. The "consensus" sequence is listed below, together with the number of times each base occurs in this alignment of sequences. Also, the position of the major and minor grooves of B-DNA is marked around an approximate twofold axis of symmetry.

This is similar to the "consensus" sequences derived previously: 5' AA-TGTGA--T----CA 3', (Ebright, 1983); TGTG · N8 · CACA, (Queen and Rosenberg, 1981); and AAAGTGTGACA, (Taniguchi et al., 1979). The site in the *lac* operon has a high degree of symmetry and many of the sites show substantial homology to the symmetrical sequence TGTGA.N6.TCACA. A similar "consensus" can be formed by aligning the eight sites in the same direction relative to the transcription start position; this necessarily results in a sequence with less consensus and more symmetry:

aagtGTGA · N7 · CA-at-A

The centers of the CAP site in different operons are all at different distances from the transcription start site. In five of these operons, CAP binds in the -42 region, presumably on the opposite side of the DNA from RNA polymerase, and in three operons CAP binds outside the promoter region. In the *lac* operon, CAP binds in the -55 region and on the same side of the DNA as the polymerase (Schmitz and Galas, 1979). The difference in the relative positions of CAP and RNA polymerase in the seven operons at which CAP functions as an activator makes a unique and direct interaction between CAP and polymerase uncertain as a general mechanism for activation of transcription. A direct interaction between CAP and polymerase would seem to require substantial flexibility in polymerase so that the portion of polymerase interacting with DNA in the -35 region and with CAP could be a variable distance from the transcription start site. A CAP-induced change in the structure of the DNA or a nonspecific interaction between CAP and polymerase also remain as possible activation mechanisms.

2.   STRUCTURE OF cAMP-CAP

The three-dimensional structure of CAP complexed with cAMP was determined from a 2.9 Å resolution electron density map that was phased by multiple heavy atom isomorphous replacement and to which the amino acid sequence was fit (McKay and Steitz, 1981; McKay et al., 1982). The crystals were grown under conditions in which CAP binds specifically to DNA: 0.5 m*M* cAMP, 50 m*M* phosphate, pH = 8.0, and about 0.3 to 0.5

mM CAP subunit (McKay and Fried, 1980). The protein coordinates have been partially refined at 2.5 Å resolution using Konnert–Hendrickson restrained least-squares refinement. Cycles of least-squares refinement have been alternated with manual interpretation of $(2 \times F_o - F_c)\alpha$ Fouriers to yield an R factor of 0.25 (I. T. Weber, unpublished).

Each subunit of the CAP dimer is folded into two distinct domains (Fig. 1). The larger N-terminal domain contains a completely buried cAMP molecule bound between the interior of a β roll structure and a very long α helix. All of the interactions between the two subunits are provided by the larger, cAMP-binding domain (Fig. 2). The smaller C-terminal domain contains three α helices and a small amount of β sheet structure. For reasons given above it is assumed that the N-terminal domain is the regulatory domain that binds the cAMP while the C-terminal domains are involved in the specific binding of CAP to DNA.

The CAP dimer is asymmetric, as a consequence of different relative

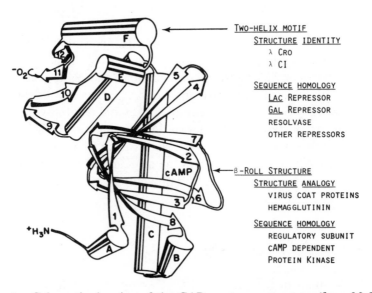

Figure 1. Schematic drawing of the CAP monomer structure (from McKay et al., 1982). The numbered arrows represent β sheets; the lettered cylinders, α helices. The binding site for cAMP in the β roll is labeled. The subunit has a modular structure consisting of two domains: the smaller C-terminal domain binds to DNA and the larger N-terminal domain binds the allosteric regulator molecule, cAMP.

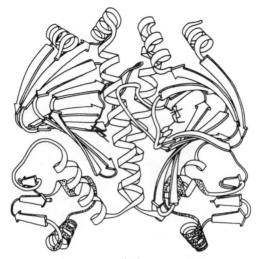

(a)

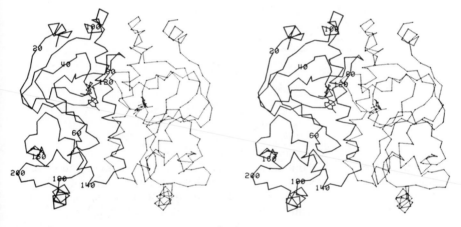

(b)

Figure 2. (a) Schematic drawing of the CAP dimer (modified from a drawing by Jane Richardson). All of the interactions between the two subunits are provided by the large N-terminal domain. The major interactions are provided by the two long C helices that lie together in the center of the dimer. The cAMP molecules are shown within the β roll structure, and the conserved, DNA-binding two-α-helix motif is shaded. (b) Stereo view of the α-carbon backbone of CAP viewed along the two F helices. One subunit is represented in bonds with solid lines and the other is in bonds with dotted lines. The closed subunit is numbered every 20 residues. Contact regions between the two subunits are particularly evident in this view. The cAMP molecules are shown. Note the proximity of the adenine to the two long C helices.

Figure 3. The α-carbon backbone of the two CAP subunits superimposed to show the difference in the relative orientations of the large and small domains. The large domains were superimposed by least-square minimization of differences between corresponding α-carbon positions. A rotation of 28° is required to superimpose one small domain onto the other (McKay and Steitz, 1981).

orientations of the large and small domains in the two subunits (Fig. 3) (McKay and Steitz, 1981). One subunit is in an "open" conformation with a large cleft between domains, while the other subunit is "closed" and without the cleft. Quantitatively, if the large domains from the two subunits are superimposed by least-squares minimization of differences in corresponding α-carbon positions, a subsequent rotation of the small domain of one subunit by 28° is required to bring it into coincidence with the small domain of the second subunit (Fig. 3). The two large domains are related by a precise twofold axis and the two small domains are related by an approximate (186.7° ± 3.7°) twofold axis. Due to the difference in subunit structures these two approximate dyad axes relating the domains deviate by 12.9° from being colinear.

All of the intersubunit contacts in the CAP dimer are between the two large domains. A major source of intersubunit contact is provided by the two 24-residue-long C helices that extend the full length of the large domains. Further intersubunit contacts are provided by a part of the β roll of one subunit interacting with the C helix of the opposite subunit.

The large and small domains are connected by a single covalent stretch of polypeptide that consists of residues 133–138. In addition to the covalent connection between the domains, numerous noncovalent interactions serve to stabilize the relative orientations of the two domains. The noncovalent interactions and the conformation of residues 133–138, which form the covalent link or possible hinge between the domains, are different in the two subunits. Whether or not there are a large number of different relative domain orientations in rapid equilibrium in solution, or only these two, is not known at this time.

3.   CYCLIC AMP BINDING IN THE LARGE DOMAIN

There is one cAMP of apparent full occupancy in each subunit of the dimer. Each cAMP molecule is completely buried within the interior of the CAP dimer and is bound between the deep pocket formed by the β roll and the C helix of the large domain in such a way that it interacts with amino acid side chains from both subunits of the dimer (Figs. 4 and 5). The cAMP binding site differs in the two subunits. The subunit in the "open" conformation provides more hydrogen bond interactions to the ribose and phosphate O atoms than are seen in the "closed" subunit. In the "closed" subunit Ser 83 and Ala 84 do not form hydrogen bonds to cAMP. The ribose and cyclic phosphate are interacting with side chains from the β roll. The buried phosphate, for example, is interacting with the guanidinium group of a buried Arg 82. The adenine is interacting with the two long C helices in the center of the dimer. Of particular interest are the hydrogen bonds made by the N6 group of the adenine with Thr 127 of the C helix in which the cAMP is bound and the side chain of Ser 128 in the C helix from the other subunit of the dimer (Figs. 4 and 5).

Because opposite sides of the C helices face different subunits, the side chains from each helix alternate between the two adenine rings from

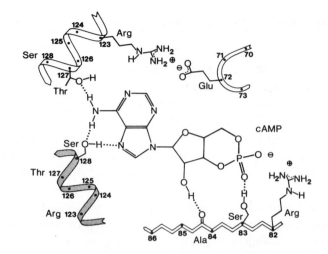

Figure 4. Schematic drawing of cAMP and the CAP residues with which it appears to be interacting. The shaded helix is from the other subunit. Some of the hydrogen bonding interactions are shown. During the course of refinement, in one subunit the Ser 83 side chain has moved away from the hydrogen bonding position shown here. However, other hydrogen bond interactions are now evident, as shown in Figure 5.

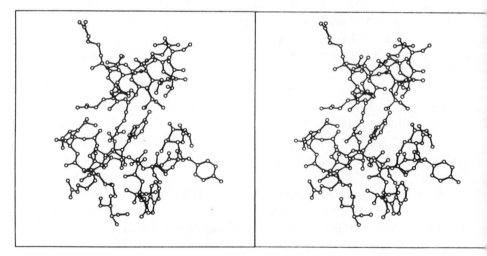

Figure 5. Stereo drawing of cAMP bound in the "open" subunit. The protein coordinates have been partially refined at 2.5 Å resolution to a crystallographic R factor of 0.25.

298

opposite subunits: Arg 123, Val 126, and Thr 127 are close to the cAMP in one subunit, while Leu 124 and Ser 128 from the same helix interact with the cAMP molecule of the second subunit. This involvement of residues from both subunits in binding each cAMP is of interest in view of the evidence for cooperative interactions in the binding of the two cAMP molecules/dimer (Takahashi et al., 1980) and suggests that subunit interactions may play an important role in the allosteric effect of cAMP binding.

4.   ALLOSTERIC MECHANISMS OF CAP ACTIVATION BY CYCLIC AMP

The transition of CAP from its DNA recognition state to a state in which it retains only a nonspecific DNA binding activity is induced by the removal of cAMP. Lacking a structure of CAP in the absence of cAMP, it is not possible to describe the structural basis for the allosteric transition. Because the effector itself makes no contact with the small domains, a mechanism in which simultaneous direct interaction of the effector with both domains induces a conformational change is ruled out. However, at least two possible allosteric mechanisms are apparent from the current structural work (McKay et al., 1982). The first possibility is that removal of cAMP may alter the relative orientation of the two domains even though cAMP is binding directly to only the large domain via the C helices that terminate in the hinge region between the domains. The second possible mechanism is suggested by the interaction of cAMP with both subunits: the binding of cAMP may alter the relative orientation of the two subunits. If the DNA binding site does indeed span both small domains, then a change in either the relative subunit orientation or the relative domain orientation or both will alter the specific DNA binding site.

The inability of cGMP and cIMP to stimulate specific DNA binding by CAP implies an important role for the N6 group of adenine in the allosteric activation. In cGMP and cIMP, the amino group at position 6 of adenine is replaced by a keto oxygen which cannot hydrogen bond with Ser 128 and Thr 127 in exactly the same way as shown in Figure 5. In the cGMP complex, the OH of Ser 128 cannot simultaneously be a hydrogen bond donor to both the N7 (as it is with cAMP) and the O6 of guanine. Therefore, the interactions of the adenine N6 group with Thr 127 of the subunit in which the cAMP is bound and, most importantly, with Ser 128 of the other subunit, are probably central for the action of CAP. These small differences in the specific interactions of cIMP or GMP with the surrounding protein (e.g., Ser 128) result in failure to induce the necessary allosteric transition for transcription activation.

5.   HOMOLOGY WITH cAMP–DEPENDENT PROTEIN KINASE

The amino acid sequence of the N-terminal domain of CAP (Aiba et al., 1982; Cossart and Gicquel-Sanzey, 1982) is highly homologous with the amino acid sequence (Takio et al., 1982) of the regulatory subunit (R_{II}) of the cAMP-dependent protein kinase from bovine heart muscle (Fig. 6) (Weber et al., 1982b). The R_{II} sequence contains two adjacent and internally homologous regions, both of which have high resemblance to the cAMP-binding domain of CAP. This homology extends over most of the N-terminal domain of CAP and is particularly good for the region of the β roll structure. For example, there is an arginine in both R_{II} domains that is homologous with Arg 82 in CAP that binds the phosphate of cAMP. This suggests that the protein kinase regulatory subunit contains two cAMP binding domains in the C-terminal region, each having a β roll structure similar to that in CAP (Fig. 7). Thus, the cAMP molecule is expected to bind to R_{II} within a pocket formed by residues from the β roll, as is the case with CAP. It is less clear from the sequence comparison whether there is an α helix corresponding to the C helix in CAP interacting with the adenine of cAMP.

   In both prokaryotes and eukaryotes, changes in the levels of cAMP act as a second messenger to signal changes in glucose concentration. When glucose levels drop, adenyl cyclase is stimulated and cAMP levels rise. In bacteria a drop in glucose levels results in switching on of the appropriate catabolic gene, (arabinose, lactose, galactose, etc.) operons, depending on what sugars are in the medium. In contrast, a drop in sugar levels in higher eukaryotes results in a breakdown of glycogen. These different responses to the same problem are required in the two cases because bacteria do not use glycogen as a reserve energy store to the same extent as mammals, and cows are not swimming in solutions of arabinose.

CAP / RIIA / RIIB sequence alignment

αA β1 β2

	10		20		30

CAP: Thr Leu Glu Trp Phe Leu Ser His Cys His Ile His Lys Tyr Pro Ser Lys Ser Thr Leu Ile His Gln Gly Glu
143 — 150 — 160

RIIA: Gln Leu Ser Gln Val Leu Asp Ala Met Phe Glu Arg Thr Val Lys Val Asp His Val Ile Asp Gln Gly Asp
265 — 270 — 280

RIIB: Glu Arg Met Lys Ile Val Asp Val Ile Gly Glu Lys Val Tyr Lys Asp Gly Glu Arg Ile Ile Thr Gln Gly Glu

β3 β4

40 · 50

CAP: Lys Ala Glu Thr Leu Tyr Tyr Ile Val Lys Gly Ser Val Ala Val Leu Ile Lys Asp Glu Glu Gly Lys Glu Met
170 — 180 — 190

RIIA: Asp Gly Asp Asn Phe Tyr Val Ile Glu Arg Gly Thr Tyr Asp Ile Leu Val Thr Lys Asp Asn Gln Thr Arg Ser
290 — 300 — 307 315 — 320

RIIB: Lys Ala Asp Ser Phe Tyr Ile Ile Glu Ser Gly Glu Val Ser Ile Leu Ile Lys Asp Gly Glu Asn Gln Glu Val

β5 β6

60 · · · · 80 · · · ·

CAP: Ile Leu Ser Tyr Leu Asn Gln Gly Asp Phe Ile Gly Glu Leu Gly Leu Phe Glu Glu Gly Gln Glu Arg Ser Ala
200 — 210

RIIA: Val Gly Gln Tyr Asp Asn His Gly Ser Phe — Gly Glu Leu Ala Leu Met Tyr Asn — Thr Pro Arg Ala Ala
330 — 340

RIIB: Glu Ile Ala Arg Cys His Lys Gly Gln Tyr Phe Gly Glu Leu Ala Leu Val Thr Asn — Lys Pro Arg Ala Ala

β7 β8 αB

· · 90 100 240

CAP: Trp Val Arg Ala Lys Thr Ala Cys Glu Val Ala Glu Ile Ser Tyr Lys Lys Phe Arg Gln Leu Ile Gln Val Asn
220 — 230 — 240

RIIA: Thr Ile Val Ala Thr Ser Glu Gly Ser Leu Trp Gly Leu Asp Arg Val Thr Phe Arg Arg Ile Ile Val Lys Asn
350 — 360 — 370

RIIB: Ser Ala Tyr Ala Val Gly Asp Val Lys Cys Leu Val Met Asp Val Gln Ala Phe Glu Arg Leu Leu Gly Pro Cys

αC

110 120 · · · · · · 130 134

CAP: Pro Asp Ile Leu Met Arg Leu Ser Ala Gln Met Ala Arg Arg Leu Gln Val Thr Ser Glu Lys Val Gly Asn Leu
110 — 120 — 250 — 260 — 265

RIIA: Asn Ala Lys Lys Arg Lys Met Phe Glu Ser Phe Ile Glu Ser Val Pro Leu Leu Lys Ser Leu Glu Val Ser Glu
— 380 — 390 — 395

RIIB: Met Asp Ile Met Lys Arg Asn Ile Ser His Tyr Glu Glu Gln Leu Val Lys Met Phe Gly Ser Ser Met Asp Leu

Figure 6. Alignment of two adjacent regions of the R_II sequence with the CAP sequence. R_IIA and R_IIB are homologous regions of R_II sequence. Solid underlining indicates amino acids that are identical to one or more residues in the other sequences; dashed underlining indicates closely similar amino acids. The positions of the α helices and β strands of the CAP molecule are indicated. Asterisks indicate amino acids that are close to cAMP in CAP (from Weber et al., 1982b).

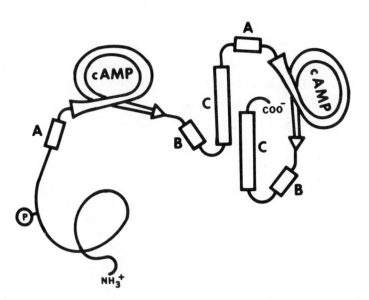

Figure 7. Schematic drawing of the regulatory subunit of cAMP-dependent protein kinase (R$_{II}$) showing that it consists of two cAMP binding domains homologous with that of CAP and a third domain involved in binding to the catalytic subunits.

It appears that a common ancestral precursor protein capable of binding cAMP has evolved into the appropriate receptor or transducer of cAMP levels in both bacteria and mammals (Weber et al., 1982b). In prokaryotes, the cAMP-binding domain is attached to a DNA-binding domain that shares structural and sequence homologies with the DNA-binding portion of bacterial and viral repressors (Steitz et al., 1982a; Weber et al., 1982a; Steitz et al., 1982b). In mammals, the cAMP-binding domain is part of a subunit that regulates, in a cAMP-dependent manner, the activity of protein kinase, the first enzyme in a cascade resulting in activation of the phosphorylase-catalyzed breakdown of glycogen. Thus, the modular "design" of proteins allows homologous cAMP-binding domains to switch on transcription of catabolic genes in bacteria and activate protein kinase in mammals when glucose levels fall and cAMP levels rise.

Recently, Shaw et al. (1983) have found that the amino acid sequence of the protein specified by the *fnr* gene is homologous with the sequence of CAP. The *fnr* gene is essential for the expression of anaerobic respiratory metabolism in *E. coli* and is probably a protein that regulates the transcription of genes encoding anaerobic respiratory functions. Extensive sequence homologies with CAP exist in both the cAMP binding domain and the DNA binding domain. Because the residues that interact specifically with cAMP in CAP are different in *fnr* (e.g., Arg 82 in CAP is Ser in *fnr*), it is assumed that *fnr* binds some regulatory small molecule other than cAMP. Thus, there may be several regulatory proteins with structures similar to CAP which evolved from the same ancestral protein. These different regulatory proteins have evolved DNA binding sites specific for different sequences and binding pockets that bind different effector molecules.

6.   THE DNA BINDING DOMAIN: THE TWO-HELIX MOTIF

Comparison of the structures of cro and CAP (Fig. 8) has revealed several striking similarities (Steitz et al., 1982a). In both molecules an α helix protrudes from the surface of each of the two subunits of the dimer and is separated across the molecular twofold axis by 34 Å from the same helix of the other subunit. In the case of cro, it was clear that these two protruding α helices, one from each subunit, would just fit into successive major grooves of right handed B-DNA (Anderson et al., 1981). Furthermore, there appears to be an identical supersecondary structure consisting of a two-α-helix motif in both proteins. The 24 α-carbon atoms of the E and F helices of CAP superimpose on the 24 α-carbon atoms of the α_2 and α_3 helices of cro with a root-mean-square (rms) difference of 1.1 Å. A search of the Brookhaven data bank showed that no other proteins on file contain the same two-α-helix motif. The best comparison gives an rms difference of 2.7 Å. Thus, Steitz et al. (1982a) concluded that there is an identical two-α-helix structural motif occurring in cro and in the DNA-binding domain of CAP, and that this two-helix motif is not found in any known protein structure that does not bind duplex DNA. This structural comparison was the first indication of the existence of a common, conserved two-helix motif for recognition of specific DNA sequences.

   Inspection of the structure of the λcI repressor fragment shows that it also contains the same two-helix motif (Fig. 8). Furthermore, as in the case of CAP and cro, the second of these two helices protrudes from the

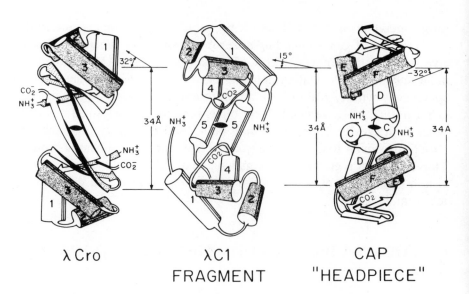

λ Cro

λ C1 FRAGMENT

CAP "HEADPIECE"

Figure 8. Comparison of schematic drawings of cro repressor, the amino-terminal fragment of cI repressor from λ phage, and the DNA-binding domains of CAP. The view is down the twofold axis of each dimer. Regions of α helix are represented as tubes and the common two-helix motif is shaded. The second helix of the two-helix motif protrudes from the surface and is separated by 34 Å from the dimer-related helix in each case. However, the tilt of these protruding helices relative to the line connecting their centers differs in each protein.

surface of the protein and is separated from the same helix in the other subunit by 34 Å across the dimer twofold axis (Pabo and Lewis, 1982). This suggests that the two-helix structural motif is important for the sequence specific interactions of all three of these proteins with DNA.

The two-helix motif probably occurs in many regulatory proteins from both phage and bacteria (Anderson et al., 1982; Matthews et al., 1982; Weber et al., 1982a; Sauer et al., 1982). Anderson et al. (1982) first showed the existence of sequence homologies between cro and several other phage repressors and activators. The list of homologous sequences was found to include the *lac* repressor of *E. coli*, and a detailed model for the repressor–operator interaction was proposed (Matthews et al., 1982). Weber et al. (1982a) found that the amino acid sequence of the two-helix region in CAP is significantly homologous with sequences in *lac* and *gal* repressors and exhibits weaker homologies with the phage repressors, showing that gene activators as well as repressors share this homologous

Table 2. Sequences Homologous with CAP

α Helix E

				(167)										
				(4)										
				(2)										
				(159)										
CAP	Gln	Ile	Lys	Ile	Thr	Arg	Gln	Glu	Ile	Gly	Gln	Ile	Val	Gly
Lac R	Met	Lys	Pro	Val	Thr	Leu	Tyr	Asp	Val	Ala	Glu	Tyr	Ala	Gly
Gal R			Met	Ala	Thr	Ile	Lys	Asp	Val	Ala	Arg	Leu	Ala	Gly
Tn3	Gln	Lys	Gly	Thr	Gly	Ala	Thr	Glu	Ile	Ala	His	Gln	Leu	Ser
Cro	Ala	Met	Arg	Phe	Gly	Gln	Thr	Lys	Thr	Ala	Lys	Asp	Leu	Gly

α Helix F

	1	2	3	4	5	6	7	8	9	10	11	12	13	14	15
CAP	Cys	Ser	Arg	Glu	Thr	Val	Gly	Arg	Ile	Leu	Lys	Met	Leu	Glu	Asp
Lac R	Val	Ser	Tyr	Gln	Thr	Val	Ser	Arg	Val	Val	Asn	Gln	Ala	Ser	His
Gal R	Val	Ser	Val	Ala	Thr	Val	Ser	Arg	Val	Ile	Asn	Asn	Ser	Pro	
Tn3	Ile	Ala	Arg	Ser	Thr	Val	Tyr	Lys	Ile	Leu	Gln	Asp	Glu	Arg	Ala
Cro	Val	Tyr	Gln	Ser	Ala	Ile	Asn	Lys	Ala	Ile	His	Ala	Gly	Arg	Lys

[a] Amino acid sequence of CAP, the *E. coli lac* and *gal* repressors, and Tn3 resolvase and cro repressor in the region corresponding to the E and F helices in CAP. *Lac* repressor contains six residues that are identical with and six similar to the 21 residues of the E and F helices of CAP. Tn3 resolvase has seven residues identical with and five similar to the 21 residues in CAP. The solid underline indicates two or more identical amino acid residues and the dashed underline indicates similar residues. Tn3 resolvase has a Ser residue at the position of the "invariant" Gly, but the rest of the sequence homology is strong, particular for those residues involved in formation of the two-helix structure.

sequence (Table 2). CAP also shows significant sequence homology (seven of 21 residues identical) to resolvase (Table 2), a site-specific recombination enzyme from transposable elements Tn3 and γδ (I. T. Weber and T. A. Steitz, unpublished). Sequence homologies among the phage repressors were also observed and the list extended by Sauer et al. (1982).

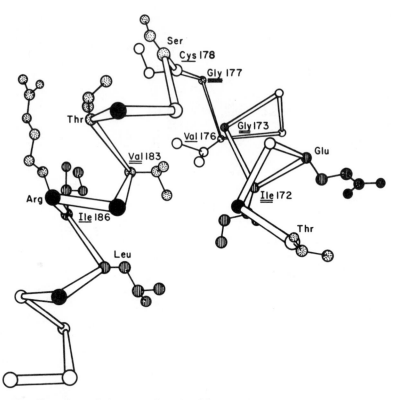

Figure 9. Drawing of the α-carbon backbone of the two-helix motif and the homologous side chains. Stippled atoms are side chains that are identical in CAP, *lac*, and *gal*. Striped atoms are closely similar in CAP and *lac* repressor and include side chains that seem to be important in interactions between the two helices. The triply underlined glycine is absolutely essential, the doubly underlined residues are either of two types and the singly underlined residues appear to be one of several hydrophobic amino acids. The completely filled α-carbon atoms indicate some side chains that may interact with DNA, derived from model building with CAP and DNA. Some additional interactions with DNA are also possible.

Thus it appears that the two-helix motif is likely to exist in many of the regulatory proteins that bind specifically to double-stranded DNA and presumably is important for their specific interaction with DNA.

The amino acid residues that are conserved or semiconserved among the various sequences are those involved in the formation of the two-helix structure, either the bend or the interaction between the two helices (Weber et al., 1982b; Sauer et al., 1982). These conserved residues and their location in the two-helix motif are shown in Figure 9. The side chains of the majority of residues are facing solvent so that their extreme variability among the repressors and activators is not surprising. Of course, the residues that actually interact with the DNA vary from protein to protein because the sequences that these proteins specifically recognize also vary.

Genetic studies also support the idea that the two-helix motif is directly involved in recognition of DNA sequences. Most, though not all, of the mutations that reduce the affinity of *lac* repressor for operator occur in the sequence that is homologous with the two-helix motif of cro (Matthews et al., 1982) and CAP (Weber et al., 1982a). These mutations are in such positions that they would either affect the ability to form the two-helix motif or would alter the nature of specific *lac* repressor interactions with DNA. Further, mutations of λ repressor that are deficient in binding the operator DNA also map in the N-terminal domain. Virtually all of these mutations that are in residues whose side chains lie on the surface of the protein occur in the two-helix motif (F. Hutchinson, private communication, 1982; Hecht et al., 1983). Finally, mutations in λ repressor that affect its ability to serve as an activator but not its ability to act as a repressor occur within this two-helix motif (Guarente et al., 1982). These data taken together strongly suggest that the two-helix motif is essential for duplex DNA interaction and, in the case of λ repressor, for the ability of the protein to act as an activator.

7. MODEL BUILDING A CAP–DNA COMPLEX

The only way to unambiguously establish how CAP binds to a specific DNA sequence is to cocrystallize CAP complexed with an appropriate double-stranded oligonucleotide. With this objective, we are attempting to cocrystallize CAP with a 16-base-pair fragment of double-stranded

DNA whose sequence corresponds to the CAP binding site in the *lac* operon. This DNA fragment was synthesized by Ikuta and Itakura. We are also examining sequences corresponding to half the DNA binding site. As these experiments have not been successfully completed, we have turned to model building to see whether it is possible to construct a reasonable complex.

In model building a protein–DNA complex for CAP (and also with cro and λc1 protein fragment) it has been assumed that: (1) the structure of the protein will remain essentially unchanged except for the conformation of specific side chains, (2) the approximate twofold axis of the DNA will be coincident with the twofold axis relating the two subunits of the protein, (3) there will be a large degree of shape, charge, and hydrogen bonding complementarity between the protein and the DNA extending over 16 to 20 base pairs of DNA, (4) models of the complex must account for data on chemical and enzymatic protection of the DNA binding site by specifically bound protein.

The assumption that the structure of CAP in the DNA complex is approximately the same as in this crystal form seems plausible, because crystals are grown under conditions similar to those under which CAP binds specifically to DNA in solution. Further, dissolved crystals of CAP are 100% active in DNA binding whereas the CAP solution from which they are grown has a lower activity (M. Fried and D. Crothers, private communication). It certainly could be true however, that the relative orientations of the DNA binding domains change upon complex formation.

CAP has only an approximate twofold axis (186° rotation) relating the DNA binding domains, which may be related to the asymmetry in most of the DNA sequences to which CAP binds specifically. Nevertheless, there is twofold symmetry in the sugar-phosphate backbone and approximate symmetry in the base sequence, suggesting that the approximate DNA and protein twofold axes should be aligned. This leaves two ends of CAP where the DNA could in principle be binding. We have excluded models using the large domain because the proteolytically-derived large domain does not bind specifically to DNA (Aiba and Krakow, 1981). Making this assumption, there are only two parameters that can be varied in the overall relationship between the DNA and the protein: the DNA can be rotated about the mutual twofold axis, and the DNA-to-protein distance can be adjusted.

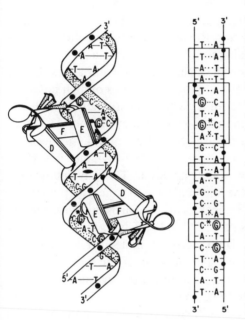

5'   3'

-T···A-
-T···A-
-A···T-
-A···T-
-T···A-
-Ⓖ···C-
-T···A-
-Ⓖ·*·C-
-A·ˣ·T-
-G···C-
-T···A-
-T···A-
-A···T-
-G···C-
-C···G-
-T·ˣ·A-
-C·*·Ⓖ-
-A···T-
-C···Ⓖ-
-T···A-
-C···G-
-A···T-
-T···A-

3'   5'

Figure 10. Schematic drawing showing one way that CAP might interact with B-DNA (redrawn from Steitz et al., 1982b). On the right is a summary of results from chemical and enzymatic protection by CAP binding to the CAP site in the *lac* operon. This indicates that CAP is interacting with 18 to 20 base pairs. Dots mark phosphates whose ethylation prevents CAP binding; circled G's are protected from methylation when CAP binds; mutations in asterisked bases reduce CAP affinity. The boxed sequences are more than 75% conserved in eight sequenced CAP sites; the X indicates a T that can be cross-linked to CAP (Simpson, 1980). The model is illustrated on the left. The small domains of CAP are shown with the F helices penetrating into the major grooves of right-handed B-DNA. Side chains from helix F can interact with about four base pairs in the major groove. Other interactions seem possible only with the sugar-phosphate backbone.

An additional important constraint on any model of CAP interacting with DNA is provided by data (Simpson, 1980; Majors, 1979) on CAP protection of DNA from chemical and enzymatic modification (Fig. 10) and by the common sequences shared by the several known CAP binding sites (Table I). These data suggest that CAP is interacting with a span of DNA that is 18 to 20 base pairs long. Of particular interest is the comparison of some eight sequences of CAP binding sites that have been examined (Table 1). The consensus sequence shows a great deal of conservation for the sequence 5'AA-TGTGA 3'. This implies that one subunit is capable of recognizing specifically seven of eight successive DNA base pairs. An additional consideration in building a model of a complex between CAP and DNA is the degree of structural complementarity between these two macromolecules. We expect that the shape and nature of the binding site for DNA on CAP should allow extensive contact between the two macromolecules over a surface that is 65 to 70 Å by 20 Å.

The electrostatic complementarity between CAP and DNA has been evaluated by calculating the electrostatic potential surfaces around both CAP and DNA (Steitz et al., 1982b; Weber and Steitz, 1984). The algorithm for calculating the electrostatic potential at and near the molecular surface is described by Matthew and Richards (1982). The method uses a modified Debye–Huckel screening term which assigns a local ion concentration to charge sites based on their calculated solvent accessibility. Counter-ion "specific binding sites" are located at sites of high potential energy and the potential at the site for a given ionic strength dictates the binding constant and therefore the partial occupancy. Figure 11 shows the positive and negative electrostatic potential energy, ϵ, surfaces for CAP contoured at a level of 2kT (i.e., $\epsilon = 2kT$). The asymmetric surface charge distribution of CAP is immediately apparent. The only net positive electrostatic charge density on the molecule lies on the C-terminal domains just outside the two F α helices. This is consistent with the assumption that the small domains of CAP bind to DNA. Furthermore, it appears reasonable to place the DNA across a molecular twofold axis interacting with both small domains. The position of the positive charge density at the extremities of the two small domains, that is, away from the twofold axis, would appear to have an orienting effect on CAP when it binds to DNA. Figure 11 also shows the negative electrostatic potential map of CAP contoured at a level of 2kT. The net negative charge density appears to be concentrated around the center of the large domains. It is striking that the positive electrostatic potential is concentrated away from the molecular symmetry axis and more on the sides of the DNA binding domains rather than on the top, as viewed in Figure 11a. The same calculation for cro (Ohlendorf et al., 1983) shows the positive electrostatic potential to be concentrated in the area between the two protruding helices and on the twofold axis. It may be that CAP bends the DNA and cro does not (see below).

The electrostatic complementarity between CAP and DNA suggests a model for the interaction of CAP with a nonspecific sequence of DNA (I. T. Weber and T. A. Steitz, unpublished results). A complex of CAP and B-DNA in which four ion pairs are formed between basic side chains of the F helices and the phosphate backbone of DNA appears to be elec-

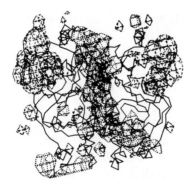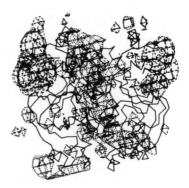

(a)

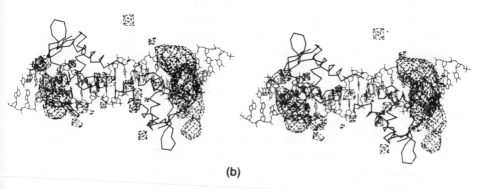

(b)

Figure 11. (*a*) A stereo drawing of the positive and negative electrostatic potential energy surfaces superimposed on the α-carbon backbone of CAP. The surfaces are contoured at a level of 2kT. Smooth lines represent positive and dashed lines negative potential. (*b*) A stereo drawing of the positive electrostatic potential energy surfaces superimposed on the α-carbon backbone of the small domains and viewed down the molecular twofold axis relating them. This view is approximately orthogonal to that in (*a*). The DNA is positioned as in the model complex.

trostatically stable, because the calculated electrostatic contribution to the binding energy of the nonspecific CAP–DNA complex is $\Delta G \sim -8$ kcal/mole for low ionic strengths. For comparison, the experimental binding energy is -9 kcal/mole (M. Fried, 1982). The electrostatic field also favors the interaction between CAP and DNA even when the two molecules are separated by about 12 Å. Therefore, the interaction of the electrostatic fields of CAP and DNA tends to orient the two molecules and is effective even at a distance.

8.   MODELS OF SPECIFIC CAP–DNA COMPLEX

A model for CAP–DNA complex precisely analogous to a model simultaneously proposed for cro (Anderson et al., 1981) was suggested by McKay and Steitz (1981), who noted that the overall shape of CAP is complementary to left-handed B-type DNA with the two F helices of the dimer fitting into successive major grooves of left-handed rather than right-handed B-DNA. In this model, as with the cro model, the helices lie parallel to the major groove. Specific interactions can be made between side chains of CAP and the exposed edges of base pairs in the major groove of DNA as well as with the backbone, covering a DNA site approximately 20 base pairs long, exactly as suggested for cro. However, this model requires that the specific binding of CAP to DNA converts two turns of right-handed DNA to a left-handed B-DNA, which would change the superhelix density of closed circular DNA (McKay and Steitz, 1981). To test this prediction, Kolb and Buc (1982) measured the change in linking number produced by binding CAP to closed circular PBR322 DNA, both with and without an added CAP binding site. They found that the binding of CAP, presumably to a specific site on the DNA, failed to change the linking number significantly. The most straightforward conclusion from these experiments is that CAP does not bind to left-handed DNA.

Various models for CAP bound to right-handed B-DNA have been also suggested (McKay and Steitz, 1981; Steitz et al., 1982a; Salemme, 1982; Ebright and Wong, 1981; Pabo and Lewis, 1982). The model of Salemme (1982) is of interest in explaining many of the features of the nonspecific complex that high concentrations of CAP form with DNA in the absence of cAMP: these fibrous complexes have been observed by electron microscopy (Chang et al., 1981). In this model the two subunits of the CAP dimer interact with adjacent coils of a tightly supercoiled B-DNA helix. It is possible in this way to have the F helices lying in the major grooves of right-handed DNA. This may be valid for the cooperative binding of CAP to nonspecific DNA sequences (Saxe and Revzin, 1979; Takahashi et al., 1979). However, this model does not apear to explain data pertinent to specific CAP-binding to DNA. Most important, it is not consistent with data showing that the CAP dimer must interact with 18–20 consecutive base pairs of DNA. It is not possible for one DNA binding domain to protect this length of DNA from nuclease digestion or to interact in such a way as to explain the effect of ethylation of phosphates on CAP binding (Figure 10).

The model published by Ebright and Wong (1981) is the only one that the CAP crystal structure rigorously excludes. Based on their reported conclusion that molecules like indole acetic acid are competitive with cAMP and the fact that their model requires the carboxyl of indole acetic acid to be protonated at neutral pH and a hydrophobic indole moiety to bind in the same place as the hydrophilic ribose-phosphate, they proposed that the adenine of cAMP is exposed and inserts into double-stranded DNA. This proposal is inconsistent with the crystal structure because the cAMP is observed to be completely buried in the interior of CAP (McKay and Steitz, 1981; McKay et al., 1982).

The positive electrostatic charge potential indicates a possible model for right-handed B-DNA interacting specifically with CAP (Steitz et al., 1983; Weber and Steitz, 1984). The tilt of the protruding F alpha helices in CAP is such that these helices cannot lie completely parallel to the major groove of right-handed B-DNA analogous to the model with cro (McKay and Steitz, 1981). However, if the DNA is placed diagonally across the two helices, such that it overlaps the highest positive electrostatic charge potential (Steitz et al., 1982b), then it is possible for the amino end of each of the two F helices to interact in successive major grooves (Figs. 10–13, Steitz et al., 1982a). In some ways this model is analogous to the model of λ repressor fragment complexed with DNA (Pabo and Lewis, 1982). As with cro, side chains emanating from the end of the F helix can be built to make specific hydrogen bonding interactions with the edges of four base pairs exposed in the major groove (Figs. 14 and 15). Arg 180, Glu 181, Arg 185, and Lys 188 interact with the conserved GTGA base pairs. However, it does not appear possible to make specific hydrogen bond interactions directly with the bases for more than eight base pairs, in a region spanning 14 base pairs, which is shorter than the expected size of the DNA binding site to which CAP binds (Steitz et al., 1982a). This model cannot account for the consensus sequence 5′AA-TGTGA by direct interactions between CAP side chains and the exposed edges of the bases. The fact that a sequence this long is substantially

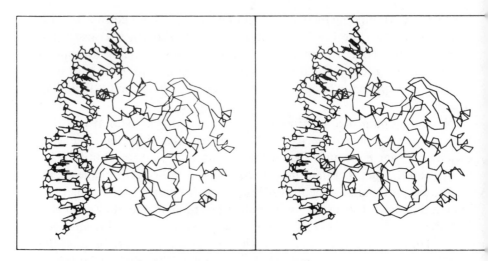

Figure 12. An α-carbon backbone stereo drawing of the CAP dimer viewed along the two F α helices interacting with 24 base pairs of DNA in the B form, whose sequence corresponds to that of the *lac* operon binding site.

conserved among the eight known DNA binding sites implies that CAP is recognizing some feature of the DNA structure within this eight base-pair sequence. This recognition would have to be accomplished by one subunit, and in this model the protein is interacting directly with the edges of the bases in only four of the seven conserved base pairs.

One way to increase the number of contacts between the DNA and protein in this orientation is to bend or kink the DNA so that the outer parts of the DNA can make contact with the protein. Such bending or kinking of the DNA allows more overlap of the negative electrostatic potential of the DNA with the positive potential of CAP. The electrostatic potential calculations appear to be consistent with cro binding to straight DNA and CAP binding to bent DNA.

If the DNA is smoothly bent to a radius of curvature of 70 Å, additional hydrogen bonds and salt links can be made to the sugar-phosphate backbone, and the bases, extending the interaction site size to as much as 20 base pairs (Weber and Steitz, 1984; Steitz et al., 1983). Some of the possible interactions between CAP and its binding site in the *lac* operon are shown in Figures 14–16. If this model proves correct, it would appear that some sequence recognition by CAP arises from its recognition of nucleotide-sequence-dependent variations in the sugar-phosphate backbone.

A schematic drawing of the model of CAP bound to a specific sequence of DNA is given in Figure 16. Note that interactions with the sugar-phosphate backbone are not identical in both subunits. Support for this model comes from genetic experiments of Ebright et al., 1984, who selected a CAP revertant to the L8 and L29 mutations in the DNA binding site in the *lac* operon. They found that a change from Glu 181 to Leu, Val or Lys increased CAP affinity for *lac* L8 or L29 DNA, consistent with Glu 181 interacting with the GC base pair proposed in our model (Figs. 14–16).

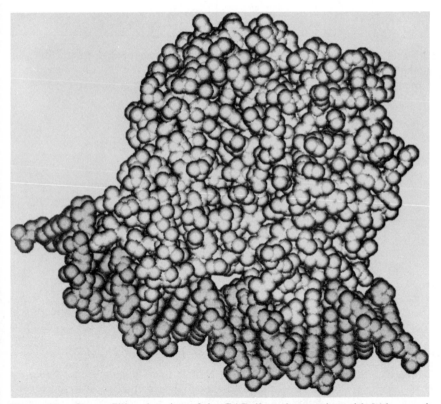

Figure 13.   Space-filling drawing of the CAP dimer interacting with 24 base pairs of right-handed B-DNA (program by R. Harrison). The DNA is bent with a 70 Å radius of curvature.

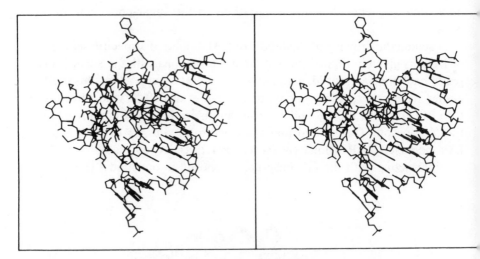

Figure 14. Stereo drawing of one DNA binding domain of CAP interacting with one-half of the CAP binding site in the *lac* operon site. The DNA is 13 base pairs of the right-handed B form bent to a radius of curvature of 70 Å.

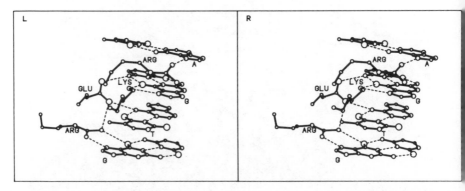

Figure 15. Stereo drawing showing the F helix interacting with four base pairs in the major groove. Four amino acid side chains interact with the base pairs. This model has been built to maximize the number of hydrogen bond interactions between the amino acid side chains of the F helix and the base pairs of DNA. The amino acids include two arginines and a lysine, so an alternative is that these basic residues interact with the phosphates of DNA.

It is possible that some residues of the extreme C-terminus of CAP make specific interactions with the DNA (McKay and Steitz, 1981). The terminal four residues are disordered in the electron density map, presumably because they are mobile. Pabo et al. (1982) have provided evidence that the N-terminal residues of λc1 fragment which are observed to be flexible in the native protein make specific contacts with operator DNA. Similarly, Ohlendorf et al. (1982) have suggested that the flexible C-terminal residues of cro interact in the minor groove of operator DNA upon complex formation. Krakow and co-workers (private communication, 1983) find that carboxypeptidase digestion of the extreme C-terminal residues of CAP results in a reduction of CAP's affinity for DNA. Because the last visible residues at the C-terminus of CAP are a significant distance from the DNA in our current model (Figs. 10 and 14), it is not clear from presently available data how the terminus might interact with the DNA.

A few words of caution are in order concerning the details of the specific CAP–DNA models presented in Figures 12–16, and these doubtless apply to the models of cro-operator and λc1 fragment-operator complexes as well. It is possible to construct detailed models of CAP complexed with left-handed as well as right-handed B-DNA. The number of hydrogen bonded interactions between CAP and DNA are similar in both cases and are about the same as in the proposed model for cro-operator (Ohlendorf et al., 1982). It seems that this reflects the ingenuity of the model builders and not necessarily the validity of the models. Many aspects of the models of cro, CAP, and λc1 fragment are probably accurate because the involvement of the two-helix motif is firmly established and the number of possible ways of interacting the protein and DNA are limited. Many specific details, however, may be incorrect. For example, the role of bound water, which is likely to be important, has thus far been ignored. Another ambiguity in the model building is posed by arginine and lysine side chains. Frequently, they can be positioned with the amino or guanadinium groups either binding to the phosphate in the backbone or interacting with the bases in the groove. Finally, it has been assumed in all three cases , CAP, cro and λc1 fragment, that the structure of the protein, except for side chains, does not change upon formation of the complex with DNA. This may not be correct. The structure of DNA in the complex also may vary locally from the B-DNA conformation. Clearly, the crystal structure of a suitable complex with DNA is essential for obtaining the correct detailed model for each of the specific DNA binding proteins.

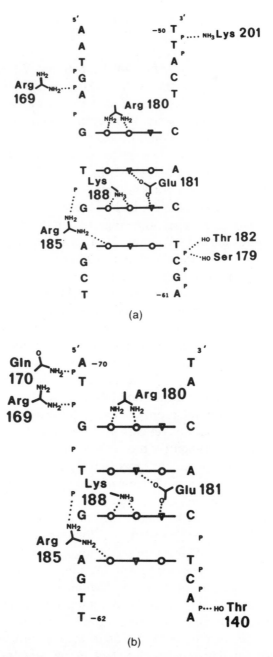

Figure 16. A schematic diagram of some of the interactions proposed between the two small domains of CAP and its DNA binding site in the *lac* operon. Hydrogen bond donors on the bases are indicated by Δ and acceptors by O. Phosphates within 5 Å of protein atoms are indicated by P. Interactions made by each subunit with half of the DNA site are shown separately.

318

9. GENERAL CONCLUSIONS ON SPECIFIC PROTEIN–DNA INTERACTIONS

Although the specific details of interactions between CAP, cro, or λc1 fragment and the duplex DNAs to which they bind should be regarded with some caution, several general conclusions can be drawn from these studies about the mechanisms of DNA sequence recognition by proteins. It is clear that much of the specificity is achieved by a two-α-helix structure that is common to these three proteins (Steitz et al., 1982a; B. W. Matthews, private communication, 1983) and probably to many sequence-specific DNA binding proteins (Anderson et al., 1982; Matthews et al., 1982; Sauer et al., 1982; Weber et al., 1982b). In each case the second helix of the two-helix structure: (1) protrudes from the surface of the protein, (2) is separated by 34 Å across a molecular diad axis from a dimer-related mate, (3) interacts in the major grooves of B-DNA. Although the angle between the protruding α helices and the major grooves is probably different in each protein, some specific sequence recognition is achieved by hydrogen bonds formed between side chains from the protruding helices and the edges of base pairs exposed in the major grooves (McKay and Steitz, 1981; Anderson et al., 1981; Ohlendorf et al., 1982; Weber et al., 1982b; Pabo and Lewis, 1982). Additional specific interactions, both with the bases and with the backbone, are likely to be made by either the extreme N- or extreme C-terminus of the protein (Sauer et al., 1982; Ohlendorf et al., 1982). The apparent need for segmental flexibility of the whole DNA binding domain relative to other domains and the use of flexible chain ends for recognition may account for the occurrence of DNA binding domains at either the N-terminus or C-terminus of a protein but not at its center.

Acknowledgments.   This research was supported by U.S. Public Health Service Grant No. GM-22778 and National Science Foundation Grant No. PCM-31-46549.

REFERENCES

Aiba, H. (1983). *Cell* **32,** 141–149.

Aiba, H., Fujimoto, S., and Ozaki, N. (1982). *Nucl. Acids. Res.* **10,** 1345–1361.

Aiba, H., and Krakow, J. S. (1981). *Biochemistry* **20,** 4774–4780.

Anderson, W. F., Ohlendorf, D. H., Takeda, Y., and Matthews, B. W. (1981). *Nature* **290,** 754–758.

Anderson, W. F., Takeda, Y., Ohlendorf, D. H., and Matthews, B. W. (1982). *J. Mol. Biol.* **159,** 745–751.

Chang, J.-J., Dubochet, J., Baudras, A., Blazy, B., and Takahashi, M. (1981). *J. Mol. Biol.* **150,** 435–439.

Cossart, P., and Gicquel-Sanzey, B. (1982). *Nucl. Acids. Res.* **10,** 1363–1378.

de Crombrugghe, B. and Pastan, I. (1978). In *The Operon,* J. H. Miller and W. S. Reznikoff, Eds. Cold Spring Harbor Laboratory, pp. 303–324.

de Crombrugghe, B., Busby, S., and Buc, H. (1983). In *Biological Regulation and Development,* Vol. 3, K. Yamamato, Ed., Plenum, New York, in press.

Ebright, R. M. (1983). In *Molecular Structure and Biological Function,* J. Griffen and W. Duax, Eds. Elsevier, North Holland, New York, pp. 91–100.

Ebright, R., Beckwith, J., Cossart, P. and Gicquel-Sanzey, B. (1984) *Nature,* in press.

Ebright, R. H., and Wong, J. R. (1981). *Proc. Natl. Acad. Sci. USA* **78,** 4011–4015.

Eilen, E., Pampeno, C., and Krakow, J. S. (1978). *Biochemistry* **17,** 2469–2473.

Epstein, W., Rothman-Denes, L. B., and Hesse, J. (1975). *Proc. Natl. Acad. Sci. USA* **72,** 2300–2304.

Files, J. G., and Weber, K. (1976). *J. Biol. Chem.* **251,** 3386–3398.

Fried, M. (1982). Ph.D. Thesis, Yale University, New Haven.

Geisler, N., and Weber, K. (1975). *Proc. Natl. Acad. Sci. USA* **73,** 3103–3106.

Guarente, L., Nye, J. S., Hochschild, A., and Ptashne, M. (1982). *Proc. Natl. Acad. Sci. USA* **79,** 2236–2239.

Hecht, M., Nelson, H. and Sauer, R. (1983). *Proc Natl. Acad. Sci. USA* **80,** 2676–2680.

Kolb, A., and Buc, H. (1982). *Nucl. Acids. Res.* **10,** 473–485.

Lee, N. L., Gielow, W. O., and Wallace, R. G. (1981). *Proc. Natl. Acad. Sci. USA* **78,** 752–756.

Le Grice, S., and Matzura, H. (1981). *J. Mol. Biol.* **150,** 185–196.

Majors, J. (1979). Ph.D. Thesis, Harvard University.

Matthew, J. B., and Richards, F. M. (1982). *Biochemistry* **21,** 4989–4999.

Matthews, B. W., Ohlendorf, D. H., Anderson, W. F., and Takeda, Y. (1982). *Proc. Natl. Acad. Sci. USA* **79,** 1428–1432.

McKay, D. B., and Fried, M. G. (1980). *J. Mol. Biol.* **139,** 95–96.

McKay, D. B., and Steitz, T. A. (1981). *Nature* **290,** 744–749.

McKay, D. B., Weber, I. T., and Steitz, T. A. (1981). *J. Biol. Chem.* **257,** 9518–9524.

Movva, R. N., Green, P., Nakamura, K., and Inouye, M. (1981). *FEBS Lett.* **128,** 186–190.

Ogden, S., Haggerty, D., Stoner, C. M., Kolodrubetz, D., and Schleif, R. (1980). *Proc. Natl. Acad. Sci. USA* **77,** 3346–3350.

Ohlendorf, D. H., Anderson, W. F., Fisher, R. G., Takeda, Y., and Matthews, B. W. (1982). *Nature* **298**, 718–723.

Ohlendorf, D. H., Anderson, W. F., Takeda, Y., and Matthews, B. W. (1983). *J. of Biomolecular Structure and Dynamics* **1**, 553–563.

Pabo, C. O., Krovatin, W., Jeffrey, A., and Sauer, R. T. (1982). *Nature* **298**, 441–443.

Pabo, C. O., and Lewis, M. (1982). *Nature* **298**, 443–447.

Pabo, C. O., Sauer, R. T., Sturtevant, J. N., and Ptashne, M. (1979). *Proc. Natl.. Acad. Sci. USA* **76**, 1608–1612.

Platt, T., Files, J. G., and Weber, K. (1973). *J. Biol. Chem.* **248**, 110–121.

Queen, C., and Rosenberg, M. (1981). *Nucl. Acids Res.* **9**, 3365–3377.

Salemme, F. R. (1982). *Proc. Natl. Acad. Sci. USA* **79**, 5263–5267.

Sauer, R. T., Yocum, R. R., Doolittle, D. F., Lewis, M., and Pabo, C. O. (1982). *Nature* **298**, 447–451.

Saxe, S. A., and Revzin, A. (1979). *Biochemistry* **18**, 255–263.

Schmitz, A., and Galas, D. J. (1979). *Nucl. Acids Res.* **6**, 111–137.

Shaw, D. J., Rice, D. W., and Guest, J. R. (1983). *J. Mol. Biol.* **166**, 241–247.

Simpson, R. B. (1980). *Nucl. Acids Res.* **8**, 759–766.

Steitz, T. A., Ohlendorf, D. H., McKay, D. B., Anderson, W. F., and Matthews, B. W. (1982a). *Proc. Natl. Acad. Sci. USA* **79**, 3097–3100.

Steitz, T. A., Weber, I. T., and Matthew, J. B. (1982b). *Cold Spring Harbor Symp. Quant. Biol.* **47**, 419–426.

Steitz, T. A., Weber, I. T., Ollis, D., and Brick, P. (1983). *J. of Biomolecular Structure and Dynamics* **1**, 1023–1037.

Takahashi, M., Blazy, B., and Baudras, A. (1979). *Nucl. Acids Res.* **7**, 1699–1712.

Takahashi, M., Blazy, B., and Baudras, A. (1980). *Biochemistry* **19**, 5124–5130.

Takio, K., Smith, S. B., Krebs, E. G., Walsh, K. A., and Titani, K. (1982). *Proc. Natl. Acad. Sci. USA* **79**, 2544–2548.

Taniguchi, T., O'Neill, M., and de Crombrugghe, B. (1979). *Proc. Natl. Acad. Sci. USA* **76**, 5090–5094.

Valentin-Hansen, P. (1982). *EMBO J.* **1**, 1049–1054.

Valentin-Hansen, P., Aiba, H., and Schumperli, D. (1982). *EMBO J.* **1**, 317–322.

Weber, I. T., McKay, D. B., and Steitz, T. A. (1982a). *Nucl. Acids Res.* **10**, 5085–5102.

Weber, I. T., and Steitz, T. A. (1984). *Proc. Natl. Acad. Sci. USA,* in press.

Weber, I. T., Takio, K., Titani, K., and Steitz, T. A. (1982b). *Proc. Natl. Acad. Sci. USA* **79**, 7679–7683.

Zubay, G., Schwartz, D., and Beckwith, J. (1970). *Proc. Natl. Acad. Sci. USA* **66**, 104–110.

8

The Gene 5 Protein and Its Molecular Complexes

ALEXANDER McPHERSON
University of California
Riverside, California

GARY D. BRAYER
University of British Columbia
Vancouver, British Columbia
Canada

CONTENTS

1. INTRODUCTION

Bacteriophage fd is a filamentous virus that infects *E. coli*. It is closely related to phages M13 and fl. The genome of this phage provides for the expression of ten genes encoded in a circular, single DNA strand consisting of 6408 nucleotides whose sequence is known (Schaller et al., 1978). The native phage particle is a long, flexible cylinder of length 8700–8900 Å and a diameter of about 60 Å. The core of the virus consists of two antiparallel strands derived from the collapsed circular DNA. These strands are complexed on the exterior by the coat protein, the product of gene 8 (Marvin and Hohn, 1969). In addition, the capsid contains five gene 3 and five gene 6 proteins that cap one end of the particle and three or four gene 7 and gene 9 protein molecules that complete the opposite end (Ray, 1977; Denhardt, 1975; Coleman and Oakley, 1980).

Following infection of the *E. coli* cell by absorption of the phage to the bacterial pili and insertion of viral DNA, the replication of fd follows the rolling circle mechanism described by Denhardt (1975). The coat protein molecules are not lost but disappear into the cell membrane, to be reutilized by daughter phage particles when assembly later occurs. This assembly process occurs at or in the host cell membrane, where the strongly hydrophobic gene 8 protein recombines with phage DNA as it is extruded from the bacterium.

The replication process occurs in three stages. Initially, the parental DNA is converted into a double-stranded replicative form (RF) composed of the viral DNA strand paired with a complement, and this includes synthesis of an RNA primer at a specific locus on the DNA circle, extension by DNA polymerases III and I, and ligation into a closed circle (Salstrom and Pratt, 1971; Mazur and Model, 1973; Mazur and Zinder, 1975). The DNA is then supercoiled by a cellular gyrase. In the second state, the RF-DNA is cleaved specifically by the product of gene 2, unwound by an *E. coli* enzyme, and replicated by the cellular DNA polymerase.

In the final stage of infection, termination of RF form DNA must be effected and the synthesis of single-stranded daughter virions completed. When the RF-DNA pool reaches a size of 100–200 copies per cell, the product of gene 5 is elaborated (Alberts et al., 1972; Pratt et al., 1974). These protein molecules bind with high affinity to the newly synthesized viral strands and preclude complementary strand formation by the polymerase. The combination of the gene 5 protein with the daughter strands is stoichiometric, and the complex is observed to be a long helical filament that somewhat resembles the mature phage. This gene 5 protein–DNA complex migrates to the host cell membrane where the gene 5 protein is displaced in a poorly characterized process by the gene 8 coat protein. The virus is capped and transported across the membrane to the exterior. The gene 5 protein returns to the cytoplasm for reuse in the formation of new nucleoprotein particles (Salstrom and Pratt, 1971; Mazur and Model, 1973; Mazur and Zinder, 1975).

A feature of the infection cycle that has provided molecular biologists, physical chemists, and X-ray crystallographers with a very useful tool is the appearance in the individual *E. coli* cells of upwards of 100,000 copies of the gene 5 protein (Alberts et al., 1972). In practical terms, this means that hundreds of milligrams of a pure DNA binding protein can be isolated from a rather modest amount of infected cells. Thus, the gene 5 protein is one of the few protein molecules of this kind available in adequate amounts for crystallographic and physical studies. Such studies allow us to directly visualize and deduce the mechanisms by which proteins recognize and bind to nucleic acids and, specifically here, single-stranded DNA. Elucidation and elaboration of the structural basis for recognition and interaction of these two classes of macromolecules are essential to our understanding at the molecular level of control and expression of genetic information, DNA replication, the processing of nucleic acid transcripts, and the interplay of the molecular components engaged in protein synthesis.

2. STRUCTURE AND PROPERTIES OF THE GENE 5 BINDING PROTEIN

The amino acid sequence of the gene 5 DNA binding protein (G5BP) has been determined (Cuyper et al., 1974; Nakashima et al., 1974). It contains 87 amino acids in a single polypeptide chain, giving a molecular weight of 9700 daltons. No disulfide bridges are present. This relatively small protein exists predominantly as a dimer in solution, although at elevated concentrations higher order aggregates are observed (Pretorius et al., 1975; Cavalieri et al., 1976). Both circular dichroism (CD) experiments (Day, 1973) and sequence predictions (Anderson et al., 1975a) indicate that this protein is composed for the most part of β-pleated sheet structure with little or no helical component.

The binding affinity of G5BP for bacteriophage fd DNA has been estimated to be of the order of 10^9 M (Coleman and Oakley, 1980). Chemical modification studies have shown three of five tyrosines to be involved in DNA binding (Pretorius et al., 1975; Anderson et al., 1975a). At least one phenylalanine is also involved in the binding process (Coleman et al., 1976; Garssen et al., 1980; Lica and Ray, 1977). Also, lysine and arginine residues have been shown to participate in liganding DNA (Anderson et al., 1975a; Alma-Zeestraten, 1982).

The stoichiometry of G5BP-DNA binding has also been investigated. The number of nucleotides bound per protein monomer has been variously reported to be five (Pratt et al., 1974; Gray et al., 1982b; Cavalieri et al., 1976; Torbet et al., 1981), four (Cavalieri et al., 1976; Day, 1973; Alberts et al., 1972; Oey and Knippers, 1972), and three (Anderson et al., 1975a; Gray et al., 1982a). It is noteworthy that this protein will tightly bind single-stranded DNA of any base sequence or composition. A large cooperative effect is associated with complexation of longer polynucleotides. That is, once one G5BP protein dimer is bound to a DNA strand, binding of the next dimer immediately adjacent to the first is enhanced by an estimated 60-fold (Cavalieri et al., 1976; Alberts et al., 1972). It is also known that bound DNA is completely unstacked and in an extended conformation regardless of its original state (Day, 1973; Alberts et al., 1972).

Although the primary physiological role of G5BP is to switch on synthesis of single-stranded DNA daughter virions and then stabilize as well as protect these strands, it has also been found to be a strong helix-destabilizing protein. That is, the melting temperatures (Tm) of a variety of double-stranded DNA's are lowered by approximately 40°C in the presence of G5BP, irrespective of nucleotide composition (Alberts et al., 1972). This melting or unwinding capacity appears to result from a strong preference for single-stranded DNA over its double-stranded counterpart. In this way, G5BP disturbs the natural equilibrium between native and single-stranded regions of duplex DNA leading to unwinding of the nucleic acid. Thus, this protein belongs to a class of helix-destabilizing proteins that includes the calf thymus unwinding protein (Herrick and Alberts, 1976), the gene 32 protein from T_4 (Alberts et al., 1968), and bovine pancreatic ribonuclease (Felsenfeld et al., 1963).

3.   DETERMINATION OF THE STRUCTURE OF THE GENE 5 BINDING PROTEIN

3.1.   Isomorphous Replacement Solution

Crystals of quality suitable for high resolution X-ray diffraction study of the G5BP were grown from polyethylene glycol using the vapor diffusion method (McPherson et al., 1976; McPherson, 1976). The crystals, shown in Figure 1, are of space group C2 with cell dimensions $a = 76.08$ Å, $b = 27.78$ Å, $c = 42.00$ Å, and $\beta = 103°$. The diffraction pattern, a sample of which is shown in Figure 2, extends to at least 1.5 Å resolution and shows little deterioration after more than 100 hr of X-ray exposure. The crystals have as their asymmetric unit a single molecule of the G5BP. Because the protein exists in solution as a dimer, this requires that the dimer possess an exact dyad axis relating the two subunits.

The structure of the gene 5 protein crystals was analyzed using the conventional isomorphous replacement method. A total of six heavy atom derivatives were used in the investigation, none of which were of partic-

Figure 1. Monoclinic crystals of the gene 5 DNA unwinding protein (G5BP) from bacteriophage fd grown by vapor diffusion from 10% w/w polyethylene glycol 4000.

ularly good quality, as it turned out. In addition, the distribution of the heavy atoms in the cell was poor in that all were clustered about the crystallographic dyad axis. As a result, the initial interpretation of the electron density map (McPherson et al., 1980) contained a number of errors. A reevaluation and analysis of the data, however, produced what has proved to be the correct structure by the refinement techniques described below and in detail by Brayer and McPherson (1983a).

The mode of binding of some of the heavy atoms are of some interest in itself. All of the platinum heavy atom compounds, K_2PtCl_4, $PtBr_2(NH_3)_2$, and numerous others, consistently bound to the sulfur of Met 1, ignoring the single histidine, the cysteine, and the second Met 76. The Pt atom in most cases appeared to occupy one of two discrete positions with respect to the sulfur and had apparently two modes of binding. The ReO_4^{2-} ion, which was quite isomorphous, bound exactly on the

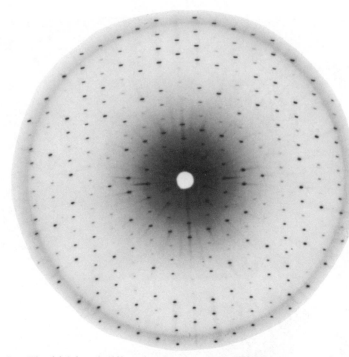

Figure 2. The hk0 level diffraction pattern from G5BP crystals. The precession angle was 16° and the exposure time was approximately 18 hr on an Elliott rotating anode generator operated at 40 kV and 40 mA with a focal spot size of 200 μm².

crystallographic twofold axis. Thus it had to be shared equally between two molecules of G5BP. The ReO_4^{2-} ion appears to be bound by lysine 7 provided as a pair by two symmetrically-related molecules. A derivative useful only at low resolution was produced by iodination of the protein in the crystal. The principal site of substitution was found to be almost equally close to both tyrosine 26 and tyrosine 56 with the former the most likely site. The large degree of nonisomorphism that accompanied this modification is interesting and may have functional implications with regard to the features described in Section 8.2.

From the interpretation of the isomorphous replacement results, an initial model was obtained at 2.3 Å resolution that lacked many of the details of a complete structure. As described by Brayer and McPherson (1983a), details were both deduced and refined, using more or less conventional refinement procedures.

3.2. Structure Refinement

The progress of refinement of the complete structure of G5BP using re-strained parameter least-squares is shown in Figures 3 and 4. The version of the refinement program, used throughout, applied restraints to bond distances, interbond angle distances, planar groups, peptide bond planarity, nonbonded contacts distances, chiral volumes of asymmetric carbon atoms, and to the thermal parameters of bonded atoms (Konnert, 1976; Hendrickson and Konnert, 1980). Variable thermal parameters on bound solvent molecules were also utilized as was the Lagrange constraint on the mean shift in the y coordinate. The structure factor scale was obtained from a Wilson plot analysis (Thiessen and Levy, 1973). The overall B estimated from this procedure was 24.8 A^2.

The starting model used for the refinement of G5BP was measured by plumb line and was not subjected to an idealization program. As shown in Figure 3, data from 6.0–3.8 Å resolution was initially used, followed by increments to 3.2, 2.8, and 2.3 Å. Only structure factor data greater than three sigma were used, for a total of 3529 reflections to 2.3 Å resolution. Individual atomic isotropic temperature factor refinement was started after 65 cycles of refinement. Prior to this, all atoms had been assigned the overall isotropic B value estimated for the entire protein molecule. The first part of refinement was deemed complete after 84 cy-

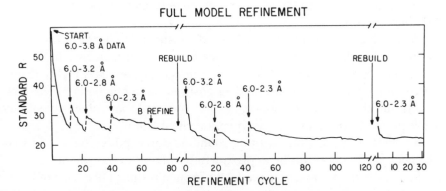

Figure 3. The progress of restrained parameter least-squares refinement for all atoms of the G5BP structure. Points of resolution extension and manual intervention are noted.

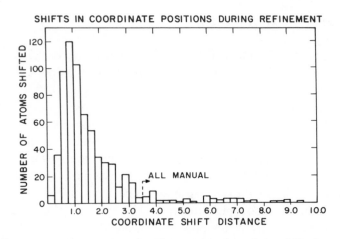

Figure 4. A histogram representing the magnitudes of the shifts in atomic positions incurred by all atoms of the G5BP structure during the refinement procedure.

cles with a standard R of 24.5%, as atomic shifts at this point had become minimal. The refined protein model was then closely examined using $2F_O-F_C$ and F_O-F_C Fourier maps. Further fragment maps covering the entire course of the polypeptide chain were computed, where all atoms except those in deleted regions were used as the phasing model. Significant structural revision was required for residues 4–8, 23–29, and 64–67, and the orientations of a number of surface side chains were altered. These regions had been the most difficult to determine in earlier structure-building attempts.

Five solvent molecules were located for inclusion in the refinement. As might be expected from the rather high overall isotropic B for this crystalline form, solvent about the protein surface appears highly disordered and few actual solvent binding positions could be determined. The criteria used to select solvent molecules required that solvent positions be removed from regions of polypeptide chain still in question; that they form suitable interactions with surrounding polypeptide chain, and that the occupancy observed be approximately that of an oxygen atom.

Following the first protein model revision, a further 119 cycles of least-squares refinement were completed, leading to a standard R of 21.2%. During this period all atoms and solvent molecules were refined with individual isotropic B values. Occupancies for solvent molecules were held fixed at 1.0. The conclusion of this part of refinement was followed by a second examination of the resultant G5BP model using $2F_O–F_C$, $F_O–F_C$, and fragment Fourier maps. Revisions at this point were very minor, in general involving side chains in those regions modified in the first model rebuild. A further seven solvent molecules were also located and added to the refinement model.

The revised protein model then was subjected to a further 30 cycles of refinement, at which point it was clear that further cycles would not substantially improve the model of G5BP. The final standard crystallographic R at this point was 21.7%. A further round of $2F_O–F_C$, $F_O–F_C$ and fragment maps were computed, and upon examination no further adjustments to the protein model were indicated.

At the end of refinement the root-mean-square differences between ideal and the refined model coordinates were 0.04 for bond distances, 0.043 for planar groupings, and 0.523 for chiral centers. The final root-mean-square shift for atomic coordinates on the last cycle was 0.001 Å

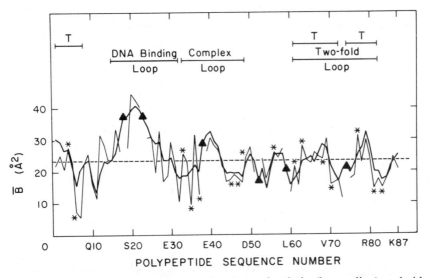

Figure 5. Thermal factor parameters for the main chain (heavy line) and side-chain (light line) atoms of G5BP as a function of position along the polypeptide chain. Also specifically denoted are glycine residues (▲) and those residue side chains resident in hydrophobic regions (*). The overall average isotropic thermal factor for all nonhydrogen protein atoms is indicated by the dashed line. Above the main graph, residues forming part of the three major β loops of G5BP are indicated. Those residues which are part of the intermolecular β barrel of the G5BP dimer are designated by a "T".

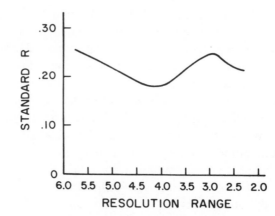

Figure 6. The variation of standard crystallographic R with resolution as observed for the final cycle of full structure least-squares restrained parameter refinement.

and 0.02 $Å^2$ for individual isotropic B values. The overall scale, which throughout refinement had been held to that determined from the Wilson plot analysis, had an indicated shift of -0.008. Figure 5 illustrates the range of individual isotropic B values over the polypeptide chain obtained at the end of refinement.

The standard crystallographic R as a function of resolution is shown in Figure 6. The initial average $\Delta[F_O-F_C]$ discrepancy at the start of refinement was 176.53, and was subsequently reduced to 39.02 at the end. The overall rms coordinate shift for all atoms during refinement was 2.265 Å. Root-mean-square shifts along axial directions were: x (1.087 Å); y (1.551 Å); z (1.239 Å). In the optical comparator the y axis direction was perpendicular to the mirror. All atom shifts of more than 3.5 Å were the result of manual intervention.

4. MOLECULAR CONFORMATION OF THE GENE 5 BINDING PROTEIN

4.1. Monomer Structure

A schematic drawing illustrating the overall polypeptide chain folding of G5BP is shown in Figure 7. As predicted from CD-ORD studies (Day, 1973) and structure-sequence prediction rules (Anderson et al., 1975a),

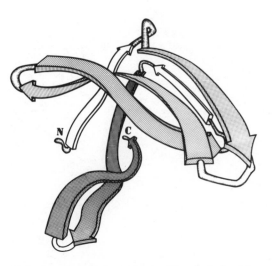

Figure 7. Schematic drawing of the polypeptide chain backbone of G5BP illustrating the three major β loops present. Dotted strands are the "DNA binding loop," striped loop is the "complex loop," and checkered loop the "dyad loop." The N- and C-termini are also indicated.

the molecule is composed entirely of β structure. The overall dimensions are approximately 35 × 30 × 47 Å. These are rather large for a protein composed of only 87 amino acids, and as one might expect the G5BP monomer structure is quite open.

Figure 7 shows that from the N-terminus, the first 10 residues form a short β strand. Then, following a β turn (residues 10–13, Type I; Venkatachalam, 1968), the polypeptide chain completes the first major loop of the molecule. It is composed of residues 15 through 32 and we have termed it the "DNA binding loop." This loop rises far above the bulk of the protein mass of G5BP in the view of Figure 7. The return strand of polypeptide chain then goes on to create a second major β loop composed of residues 33–49. We have termed this the "complex loop." The polypeptide chain then descends to the back of the molecule to form a short eight-residue strand (residues 52–59) before reemerging to generate a third large β loop. This final β loop is composed of residues 61–82 and is termed the "dyad loop." In this orientation, the C-terminus lies near the N-terminus, towards the back of the molecule.

A stereo drawing of the refined α-carbon polypeptide chain backbone conformation of G5BP is shown in Figure 8a. Immediately below in Figure 8b, a stereo drawing of all the atoms of the G5BP structure is presented.

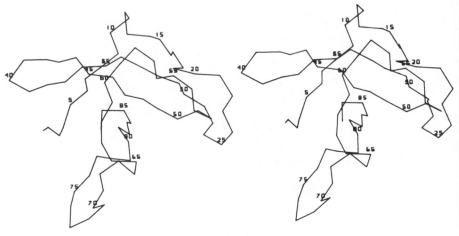

(a)

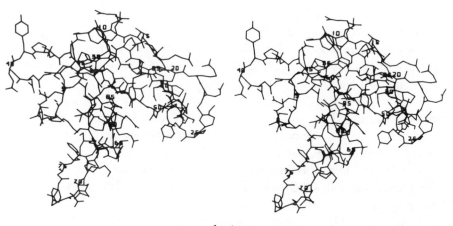

(b)

Figure 8. (*a*) Stereo drawing of the α-carbon backbone of G5BP following completion of 2.3 Å resolution refinement. Orientation is that of Figure 11, for ease of correlating the structural features illustrated in both. (*b*) Stereo drawing of all the nonhydrogen atoms of the G5BP structure. The view presented here is the same as that of the α-carbon drawing above. In both drawings the α-carbon of every fifth amino acid is labeled as to its position in the amino acid sequence.

Both drawings are in the same orientation as the stylized drawing of Figure 7. Reference to Figures 8*a* and 8*b* shows that the G5BP molecule is composed essentially of three major β loops which project outwards from a common hydrophobic core. These loops are arranged such that a three-stranded β sheet is formed across the midportion of the molecule. This sheet consists of portions of the DNA binding loop and the complex loop, and in conjunction with the short eight-residue strand to the back of the molecule, forms a partial β barrel that encloses the hydrophobic core of G5BP. This hydrophobic core is made up of the side chains of Ile 6, Cys 33, Val 35, Ile 47, Leu 49, Tyr 56, Tyr 61, Leu 81, and Leu 83.

4.2. Dimer Structure

The major species of G5BP found in solution is the dimer unit (Pretorius et al., 1975; Rasched and Pohl, 1974). It has been proposed that the forces involved in dimer association are mainly hydrophobic, and it is known this specie is stable under conditions of high ionic strength, pH extremes, dilution, and elevated temperatures (Cavalieri et al., 1976). Other observations indicate that dimer units do not self-associate into larger complexes unless single-stranded nucleic acid material is present. One major physiological role G5BP plays in the bacteriophage fd life cycle is the protection of the newly formed, covalently closed circular, single-stranded virus strands. This is accomplished by coalescing with the viral strand to form a helical, linear, nucleoprotein rod. No such self-assembling rods are observed in the absence of viral DNA.

It is also the dimer form of G5BP that is found in the crystalline state. That is, two G5BP molecules are closely associated about a perfect dyad axis, which is coincident with a twofold symmetry element of the crystallographic unit cell. Examination of the dimer unit reveals a striking feature of the association—the extraordinary degree of intermolecular bonding and complementarity achieved between the dyad-related subunits.

In Figure 9, a schematic drawing of the G5BP dimer unit is presented in the same orientation as that of the G5BP monomer of Figure 7. The twofold axis relating monomers runs directly into the plane of this illustration, and its position is evident upon comparison of common monomer features. The β loop coding of Figure 7 has also been preserved in Figure 9. The overall dimensions of the dimer are approximately 55 × 45 × 36

Figure 9. Stylized drawing of the polypeptide chain backbone of the G5BP dimer in a similar orientation to that shown for a G5BP monomer in Figure 10. The three major β loops of each monomer are also illustrated similarly. Dotted strands, DNA binding loops (residues 15–32); striped strands, complex loops (residues 33–49); checkered strands, dyad loops (residues 61–82). The N- and C-termini of each monomer are also indicated.

Å. This is considerably smaller than one might expect from the dimensions of a G5BP monomer alone. Indeed, by comparing Figures 7 and 9 the structure can be seen to be a much more compact and globular structure than an isolated G5BP monomer. Examination of the monomer–monomer interface shows that each monomer unit actually encroaches by over 10 Å into the structure of the related monomer by extension around their common twofold axis. Interactions between monomer units are found to be so exact and cohesive that solvent molecules are completely excluded from the interface region and no significant interstices are present.

It seems apparent that dimer association must be a prerequisite for stabilizing the rather open structure of isolated G5BP molecules. As seen in Figure 9, this union occurs with the imposition of extended β loops into the open channels that cross each individual monomer. Playing a principal role in this aggregation phenomena are the complex (residues 33–49) and dyad (residues 61–82) β loops. Only the region immediately adjacent, the DNA binding loop (residues 15–32), is not obscured by dimer formation.

The stereo drawing of Figure 10*a* shows the course of the refined α-carbon backbone of the G5BP dimer. As in Figure 8, the twofold axis relating monomers is perpendicular to the plane of the illustration. A stereo drawing of all the atoms of a dimer G5BP unit is presented in Figure 10*b*. This drawing most clearly illustrates the precise fit of G5BP molecules about the common dyad.

4.3. The Hydrophobic Core

A major structural element is created in the G5BP dimer by the quaternary interactions between symmetry-related monomers. A compact and well-delineated six-stranded antiparallel β cage or barrel is formed from the two extended dyad β loops in conjunction with the amino-terminal strands of both monomers. The intermolecular twofold axis is perpendicular to and bisects this barrel structure (see Figs. 10*a* and *b*). This joint structural unit is internally fortified by a well defined and extensive bonding network involving a number of different kinds of side-chain and main-chain interactions. Figure 11*a* shows the atomic details of the β strands and side chains that form the intermolecular β barrel of the G5BP dimer. Particularly noteworthy is the accumulation of hydrophobic side chains that have coalesced within the cage structure (Fig. 11*b*). This is further seen in Figure 11*c*, which provides a view looking down the internal core of the intermolecular β barrel.

The interactions responsible for maintaining the composite β barrel are hydrophobic. Bearing in mind the symmetrical contribution made by each monomer, the number of hydrophobic side chains involved is at least 20 (Fig. 11*b*). These include (from each monomer) Val 4, 45, 63, 70; Leu 37 and 81; Ile 6 and 47; Phe 68 and Met 77. Two of these symmetrical pairs of interactions are particularly interesting. Both Phe 68 and Met 77 lie very near, and in the first case parallel to, the intermolecular dyad axis (see Fig. 11). As a result, the two phenylalanines are in close contact and essentially coplanar. Similarly, the two sulfur atoms of the symmetry-related methionines are in close contact across the twofold axis.

It is also of interest that the intermolecular hydrophobic core so formed

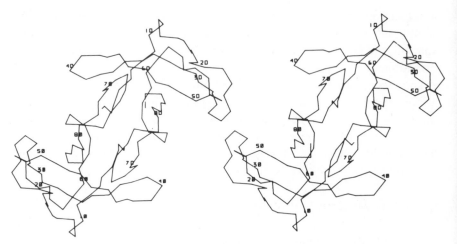

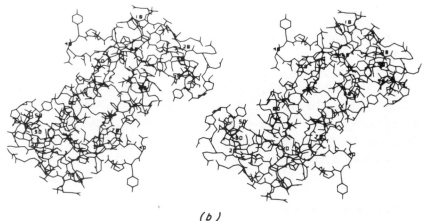

Figure 10. Stereo drawings of (*a*) the α-carbon backbone, (*b*) all the atoms of the G5BP dimer unit. The twofold axis relating G5BP monomers runs directly into the plane of this illustration and its location is immediately obvious by comparison of common structural elements from each monomer. In addition, every fifth amino acid is numbered to allow comparisons with the amino acid sequence.

is contiguous with the individual hydrophobic regions of each monomer, which in turn contribute support to their respective three-stranded β sheets. The result is a continuous, distinctly hydrophobic spine to the entire dimer unit. This very likely explains the predominance and stability of the dimer species in solution. Given the stabilizing effects of dimer association, it is not clear that an isolated monomer unit could even retain the conformation that it is observed to have in the dimer state.

In addition to the extensive hydrophobic linkages, there are also (again bearing in mind the pairwise occurrence) at least 10 electrostatic interactions joining monomers as well. The amide of residue 74 on one monomer hydrogen bonds with the carbonyl oxygen of residue 46'. Similarly, the amide of 46 interacts with the carbonyl oxygen of 73', and the charged amine group of Lys 46 is also near the carbonyl oxygen of 73'. The most interesting constellation, however, is formed by the lone histidine on each monomer, which lies close to the center of a cluster of polar groups extending from its respective twofold-related monomer. The cluster include the carboxyl group of Glu 5 (3.7 Å), the side chain of Asn 39 (4.2 Å), and the carboxyl group of Glu 40 (4.5 Å). A number of solvent molecules are also found in the neighborhood of these residues. The atomic level detail of the histidine clusters is illustrated in the stereo drawing of Figure 12. Shown in this figure are the interactions about one histidine, and an identical set of interactions is present about its twofold-related mate. It is intriguing from a mechanistic standpoint that this histidyl interaction occurs with residues quite near Tyr 41, an amino acid that has been strongly implicated in binding DNA to the surface of G5BP (Anderson et al., 1975a).

4.4. The DNA Binding Loop

It is clear from a comparison of Figures 8 and 10 that dimer formation has substantially altered the outer surface of G5BP from that exposed by the isolated monomer structure. However, the outward appearance of one portion of polypeptide chain remains unperturbed in dimer association. Those residues are 20 through 30, forming the extremity of the DNA binding loop. Due to the considerable flexibility and the large thermal parameters associated with atoms of this loop, they proved the most difficult to position during the structure refinement process. This loop and the manner in which it dramatically projects away from the bulk of the molecule is evident in Figure 13, which shows the profile of a dimer unit. In this view, the dimer twofold axis is aligned vertically and in the plane of the illustration rather than perpendicular to it as in Figure 10. The only likely interaction of this loop with the remainder of the molecule that might secure its position is a rather long salt bridge between Arg 80 and Glu 30. Aside from this tie, there is little to restrain the mobility of this

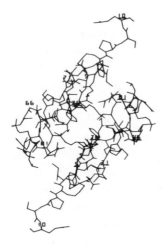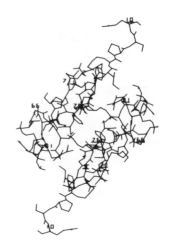

(a)

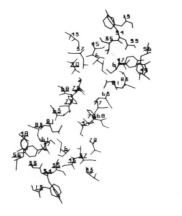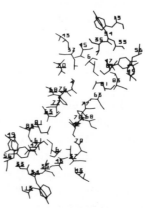

(b)

Figure 11. Stereo drawings of the structural elements of the two G5BP monomers forming the dimer intermolecular β barrel structure. (*a*) View of the two "dyad β loops" and N-terminal strands forming the skeleton of the intermolecular barrel in an orientation the same as that of Figure 10. (*b*) A view in the same orientation but now showing only those amino acids whose hydrophobic side chains contribute to the intermolecular barrel. (*c*) View down the center of the hydrophobic core linking dyad-related monomers.

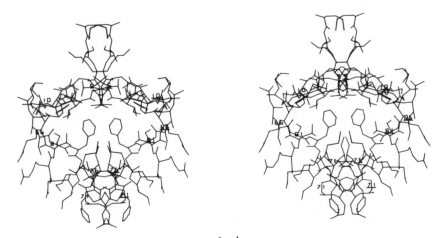

(c)

Figure 11.   (Continued)

extended β loop, which, incidently, carries Tyr 26 on its tip. This residue has also been strongly implicated in direct interaction with bound DNA (Anderson et al., 1975; O'Conner and Coleman, 1983).

Figure 14 is a detailed drawing of the conformation of this loop and the three-stranded β sheet of which it is a part. Note that the hydrogen bonding in the extended portion of the DNA binding loop is irregular, consistent with its high mobility. In contrast, the hydrogen bonding in the central portion of the compact three-stranded β sheet of each monomer is much more consistent. All three tyrosines implicated in DNA binding are resident on this β sheet. Two are located on β loop extremities (Tyr 26, Tyr 41) and one in a central position (Tyr 34). The only cysteine (Cys 33) in the sequence is also centrally located, and it too has been shown by photocrosslinking and chemical modification experiments to be near bound DNA in the liganded state (Anderson et al., 1975b; Lica and Ray, 1977; Paradiso and Konigsberg, 1982). It may be significant that the two tyrosines found at the apices of extended β loops have a neighboring proline residue. This may be to confer some measure of structural rigidity to the tips of those loops and somehow facilitate DNA binding.

It can also be seen from Figure 13 that a smaller projection of atoms created by the dyad loops of each monomer (downwards in this view) occur at the twofold interface, and these are centrally located between

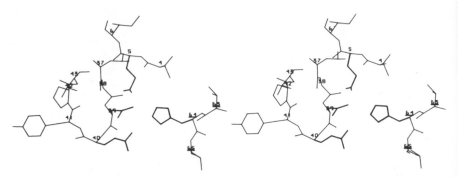

Figure 12. Stereo drawing of the interactions formed between each histidine residue and side chains projecting from the twofold-related monomer. The most interesting bonds involve Glu 5, Asn 39, and Glu 40. Tyr 41 is also in the vicinity and has been implicated in DNA binding.

the two extended β loops. These projections, in conjunction with the extended β loops, form two shallow channels oriented almost perpendicular to the plane of Figure 13. In the view of Figure 10, these extended clefts traverse the face of the G5BP dimer. These two depressions are, we feel, almost certain to be binding sites for single-stranded DNA. All

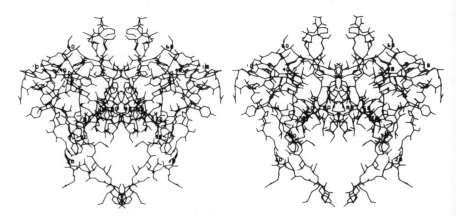

Figure 13. Stereo drawing of the G5BP dimer, in a view perpendicular to that of Figure 10. The dramatically extended β loop in this drawing is composed of residues 20 through 30 and comprises the outer extremity of the DNA binding loop.

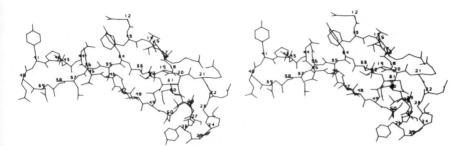

Figure 14. Stereo drawing of the G5BP extended DNA binding and complex loops, both of which form a part of the central three-stranded β sheet that crosses the midportion of each G5BP monomer. The three tyrosines (26, 34, 41) and Cys 33 are all resident on this β sheet and have been implicated directly or indirectly in DNA binding.

of the amino acid side chains implicated in DNA binding by other physical-chemical means are located here. Each DNA binding channel is approximately 10 Å in width and 35 Å in length, dimensions which one would expect necessary to accommodate a fully extended nucleic acid chain of four or five nucleotides in length.

At this point, it is useful to examine the gradation and range of amino acid mobilities along the course of the G5BP polypeptide. Figure 5 is a graph of thermal parameters for main-chain and side-chain atoms as a function of position along the polypeptide. The average overall isotropic thermal parameter for all nonhydrogen protein atoms is 23.5 Å². Evident from this plot is the considerable variance in mobility as a function of position along the polypeptide chain. There are three major segments which demonstrate greater than average flexibility. Not coincidentally, we believe, these occur at the extremes of the three β loops that together constitute a substantial portion of the G5BP molecule. Residues comprising these β loops are indicated above the thermal parameter graph. Of note is the high degree of motion found for the extended DNA binding loop (residues 15–32). This is almost predictable, given the manner in which this loop projects from the surface of the G5BP dimer (Fig. 13). Indeed, values that approach twice the average thermal factor are observed for the most distal residues. We note that, in the crystal, this loop is fixed only by virtue of its contact with another independent molecule in the crystal in the adjacent unit cell. We are inclined to believe that were the gene 5 dimers removed to an aqueous environment and unfettered by lattice constraints, this loop would be rather dynamic in its properties. This would not be inconsistent with a DNA interactive function.

Those portions of polypeptide chain forming the hydrophobic core of each monomer or dimer intermolecular β barrel, in general, have considerably less mobility. Intermolecular β barrel polypeptide strands are indicated in Figure 5 by the uppermost bars, and residues with side chains forming a part of hydrophobic regions are denoted by asterisks. This is consistent with the more rigid nature of these structural elements.

5.   SPECIFIC AMINO ACID INTERACTIONS

With the investigation of the structure and function of G5BP by biochemists and physical chemists, a number of specific amino acid residues have come under particular scrutiny. Many of these groups have been subjected to a variety of chemical modifications or examined by several different spectroscopic techniques. Their positions, orientations, neighbors, and possible tertiary interactions are, therefore, of some general interest. We briefly review some features of the amino acids of most interest or those that form otherwise noteworthy interactions. The present discussion is limited to those residues in a single G5BP polypeptide chain. Amino acid residues in the twofold-related monomer of the G5BP dimer have exactly the same environments.

5.1.   Phenylalanine

There are three phenylalanine residues in the polypeptide sequence of G5BP. Phe 13 is, for the most part, internal and positioned directly behind Tyr 34 in the view of Figure 15. The planes of the two aromatic side chains of these residues are roughly perpendicular to each other. Phe 13, therefore, has little or no access to the putative DNA binding channels traversing the dimer face. Phe 68 is near the dyad axis of the G5BP dimer and is part of the hydrophobic core of the intermolecular β barrel formed between monomer units. As such, the two Phe 68 residues of a dimer unit are positioned close together (see Figure 11c). The main interaction is ring-edge to ring-edge across the dimer twofold axis that relates them, as the two phenyl rings are essentially coplanar.

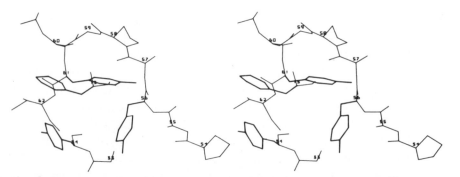

Figure 15. Stereo drawing of the molecular conformation of the unusual G5BP polypeptide sequence Pro(54)-Ala-Tyr-Ala-Pro(58), about Tyr 56 and the orientation of Try 61 lying nearby. Also shown are the positions of two additional aromatic residues, Tyr 34 and Phe 13. Although not shown, Cys 33 is also near these four aromatic residues.

Phe 73 has been implicated as having a role in binding single-stranded DNA to G5BP (Lica and Ray, 1977; Coleman and Armitage, 1978; Hilbers et al., 1978; Garssen et al., 1980). This residue is positioned at the tip of the β-bend turn of the "dyad loop," which is the major component of the G5BP dimer intermolecular β barrel. However, this residue is not a part of that structure; instead it is found on the surface, positioned in the DNA binding channel of the twofold-related monomer. This is remarkable given the large distance between channels (approximately 35 Å) and the relatively small size of each G5BP monomer molecule. This observation implies that a single nucleic acid binding site is of a composite nature, requiring residues from both monomers within the dimer, and it further supports the contention that the G5BP dimer is the essential binding species in solution.

5.2. Lysine and Arginine

There are six lysines in each monomer, and evidence indicates that some of these form electrostatic interactions with bound DNA (Anderson et al., 1975a; Alma-Zeestraten, 1982). Three of these, Lys 3, 7, and 87 are not, we feel, likely to be involved. All are on the exterior of the protein, removed from the binding channels and fully exposed to solvent. None of them form obvious interactions, although Lys 3 could bind the C-terminal of the same monomer upon a slight rearrangement. Lys 46, on the other hand, lies exactly in the DNA binding cleft next to Tyr 34. Towards the opposite end in each binding channel is the side chain of Lys 69. Lys 24 is somewhat ambiguous as it lies on the outside of the extended "DNA binding loop" that also carries the side chain of Tyr 26. Slight rearrangement of this loop could cause this residue to be thrust into the DNA binding channel area. Thus, our analysis of the native G5BP structure leads us to conclude that three lysines may be involved in DNA binding while three others likely are not.

The four arginines of each G5BP monomer are on the protein surface and may also be considered possibilities for DNA–phosphate interactions. One of these, Arg 82, is unlikely to be involved because it is removed from the putative binding channel and could not be readily rearranged to be appreciably nearer. The side chain of Arg 82 does, however, form a salt bridge to Asp 50. Arginines 16 and 21 are, like Lys 24, poised on the extended DNA binding loop and could be brought into the binding cleft region with a slight rearrangement of that loop. In the native, uncomplexed state, these two arginines are found simply extended into nearby solvent channels. The two side chains of Arg 80 and 80' in the G5BP dimer contribute to the bottom surface of the two binding channels of the dimer. This residue would appear to be a prime candidate for complexation with DNA backbone phosphates.

5.3. Tyrosine

All studies of DNA binding to G5BP have clearly indicated the essential involvement of tyrosine residues (Day, 1973; Anderson et al., 1975a; Hilbers et al., 1978; Alma et al., 1981). From the refined structure it can be said with some confidence that two of the five G5BP tyrosines are clearly not involved. Both Tyr 56 and Tyr 61 are part of the hydrophobic core of each monomer and on the side of the molecule opposite from the DNA binding channel (Fig. 10). Interestingly, Tyr 56 is part of an unusual amino acid sequence, Pro-Ala-Tyr-Ala-Pro and rather near Tyr 61. The molecular conformation of this peculiar segment of polypeptide chain and the association of the two tyrosyl rings are shown in Figure 15. A distinct kink in the course of the polypeptide chain within this segment orients the tyrosyl rings very near to one another. Both residues are essentially internal, with Tyr 61 being the most inaccessible and only the edge of the tyrosyl ring of residue 56 exposed to solvent. Figure 15 also illustrates an interesting grouping of aromatic side chains (residues 13, 34, 56, 61) occurring in the structure of G5BP. Note that the side chain of Cys 33 is also in this vicinity, to the right and at an intermediate position between the two pairs of this planar grouping.

The side chain of Tyr 34 is positioned directly in the DNA binding cleft, exposed to solvent and accessible for modification. Directly behind this tyrosyl ring are Phe 13 and Leu 43 which serve to severely limit the mobility of this side chain. The two aromatic groups of Tyr 34 and Phe 13 are near enough to stack but instead take up an edge on conformation (see Fig. 10).

Tyrosine 26 is perhaps the most intriguing residue in the G5BP structure. It is found adjacent to a proline residue, at the very tip of the extended DNA binding loop that projects far above the core of both the G5BP monomer and dimer structure (Fig. 13). In the native molecule, Tyr 26 is totally accessible to solvent. This residue at the end of the structure refinement had a relatively large thermal factor (see Fig. 5) and would be expected to be highly mobile in solution. By a simple rotation about the C-α to C-β bond, the aromatic ring could swing without difficulty into the binding cleft. If there should occur a minor conformational change at the tip of the extended β loop containing Tyr 26, this residue would be even more profoundly driven into the DNA binding channel. This same motion, incidently, would also move Lys 24, Arg 16, and Arg 21 into position to interact with bound DNA, as described in more detail in Section 8.2.

The last tyrosine, residue 41, is at the far end of the three-stranded β sheet of G5BP from Tyr 26 (see Figures 10 and 14). As such, it is at the opposite extreme of the DNA binding cleft. This residue, like Tyr 26, is adjacent to a proline and fully extended into solvent, making no contacts with other parts of the molecule.

5.4. Cysteine and Methionine

The only cysteine in the amino acid sequence, residue 33, is located exactly in the center of the binding cleft but turned inside and completely buried in the hydrophobic core beneath the central three-stranded β sheet. Its closest neighbors include Tyr 56 and 61. Its position is entirely consistent with the observation that it cannot be reacted with Ellman's re-

agent and titration with mercury acetate causes disruption of native conformation and loss of DNA binding capacity (Anderson et al., 1975b).

Methionine 77 is in a fully extended conformation in the G5BP monomer and, as described above, lies very near the dyad axis forming a part
of the intermolecular β barrel of the dimer. It is closely associated with
the symmetry equivalent Met-77' of the other monomer (Figures 11a and
11c). Methionine 1 is also fully extended and in this case exposed to
solvent. It also falls close to the intermolecular dyad and therefore not
far removed from its symmetry mate. The molecular conformations of
residues at both the N- and C-termini of the G5BP dimer are shown in
Figure 16. Also evident is the conformation of Met 1 and two lysine residues (3, 87). This methionine was the primary locus of binding for all the
platinum heavy atom derivatives used in the structural analysis.

5.5. Proline

All of the five proline residues of G5BP are on the surface of the protein
and accessible to solvent. Two of these, Pro 25 and 42, are adjacent to
tyrosines, appearing at the tips of β loops, where they participate in a
dramatic alteration in chain direction (Figs. 8 and 10). Prolines 54 and
58 occur in the unusual polypeptide Pro-Ala-Tyr-Ala-Pro sequence. The
last, Pro, 85, occurs very near the C-terminus. It is also fully exposed to
solvent and probably free to move about judging by the lack of restraining
interactions.

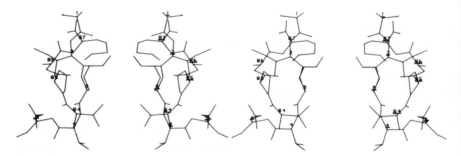

Figure 16.   Stereo drawing of both the C- and N-termini of the G5BP dimer. Of
particular interest are the conformations of Met 1, Lys 3, and Lys 87.

5.6. Histidine

The single histidine, as mentioned earlier, forms an interesting constellation of charged and polar residues that consolidates the association between the two monomers in the G5BP dimer (Fig. 12). Although this residue is involved in several interactions, it would also appear to be accessible to solvent. However, it appears to be too far removed from the G5BP binding channels to have any direct role in binding DNA.

6. THE MECHANISM OF DNA BINDING TO THE GENE 5 PROTEIN

At the outset, it seems clear from the three-dimensional structure of G5BP and other physical-chemical techniques (Cavalieri et al., 1976; Pretorius et al., 1975; Rasched and Pohl, 1974; Coleman et al., 1976), that the basic DNA binding unit is the G5BP dimer. Indeed, the extensive interbonding and overlap between dyad-related monomers, results in a structure as compact as if it had been created from a single polypeptide chain. The formation of a mutual intermolecular hydrophobic β barrel and its continuity with the hydrophobic regions found nucleating each G5BP molecule, strongly suggests that dimer formation is an essential element in the maintenance of the overall stability of each monomer. In addition, the presence of twofold-related monomer polypeptide chains in each of the two opposing DNA binding clefts further implies that the G5BP dimer is the active DNA binding species.

Given the extensive degree of monomer–monomer overlap and the apparent rigidity of the hydrophobic β barrel so formed (Figs. 10 and 11), it seems unlikely that subsequent DNA binding events would substantially alter the observed dimer protein conformation. Probable exceptions to this, however, are the extended wing or "DNA binding loop" (residues 20–30) and perhaps the tip of the "complex loop" (residues 39–43), both of which are unrestrained by dimer formation (Figs. 9, 10, and 13). It is important that the two rather well-defined binding channels crossing the dimer face include these more mobile loops as structural elements. Figure 17 shows schematically the positions of these channels and their relationship to other features of the dimer unit. Essentially, they traverse

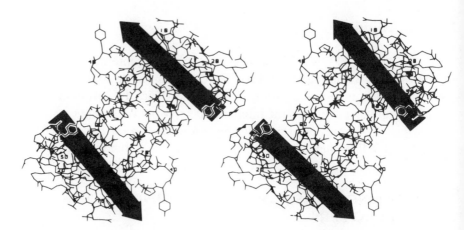

Figure 17. Stereo drawing of all the atoms of the G5BP dimer unit in an orientation such that the twofold axis relating monomers is running directly into the plane of the illustration. Also shown schematically are the positions and polarity of the two DNA binding channels crossing the dimer face. Every tenth amino acid residue has been numbered to allow comparison with the amino acid sequence.

paths on either side of the molecular twofold axis and are sandwiched between the outer extended DNA binding loops and the ridge of atoms formed at the dyad interface (Fig. 13). Note that the environment of each binding cleft is identical as a result of the twofold axis, but the polarity or direction of each is reversed from the other by the same symmetry element.

Attention has focused on the interactions between G5BP and various nucleic acids and nucleic acid fragments. Several investigations have identified a number of amino acid residues essential to the binding of nucleic acids and have provided insight into the conformational state of DNA in such complexes. We have attempted to produce below a description of that region of the G5BP molecule that evidence indicates is directly in contact with nucleic acid, and an analysis of the atomic contacts likely to be formed. In addition to the physical and chemical data, we have relied on the refined structure of G5BP at high resolution and the results of several difference Fourier experiments involving the diffusion of oligonucleotides into native G5BP crystals.

6.1. General Approaches

Three approaches were taken towards elucidating the details of DNA binding to G5BP. Two of these attempted the direct visualization of bound DNA through crystallographic techniques. The first relied on the introduction of various deoxyoligonucleotides directly into the mother liquor of native G5BP crystals of the type originally used in the structure determination (McPherson et al., 1976). Variables examined included oligonucleotide length, sequence, concentration, and soaking time. Diffraction data from four of the most promising experiments [d(pA)$_4$, d(CTTC), d(CG)$_3$, d(CCG)] were collected to 2.8 Å resolution and processed as described by Brayer and McPherson (1984). Difference Fourier methods were employed to localize bound nucleotides using the phases calculated at the end of native G5BP structure refinement.

A second approach towards visualizing the bound DNA was cocrystallizing deoxyoligonucleotides with G5BP. Unfortunately, all such attempts have resulted either in native G5BP crystals or crystal forms having a probable twelve G5BP molecules per asymmetric unit (McPherson et al., 1980b). Each of these multiple-copy G5BP crystals represents a difficult crystallographic problem. Although a structure determination using molecular replacement is in progress on one such form, even with the knowledge of the G5BP dimer structure, considerable additional time and effort will be required before binding information from this source becomes available.

In view of the problems associated with direct crystallographic visualization of the G5BP–nucleic acid complexes, we also attempted to model this interaction. The direction and constraints for this approach were grounded in the extensive physical-chemical literature that has accumulated concerning complexation, and the structural basis was provided by the refined coordinates for native G5BP.

6.2. Modeling Procedures

Modeling involved the use of a Kendrew model (2 cm/Å) constructed according to the refined coordinates of G5BP obtained as described above. In addition, a portion of the dyad-related G5BP monomer of the dimer unit was constructed. This requirement arose because amino acid side groups from both G5BP molecules of the dimer are present in each of the two DNA binding clefts traversing the dimer face. A model of five covalently-linked DNA bases was also assembled from Kendrew parts and then systematically fitted into one of the G5BP binding clefts.

Following nucleic acid positioning, atomic coordinates were measured, using the plumb-line method. These coordinates were then geometrically idealized. Subsequent comparison of idealized coordinates against measured values indicated no significant readjustments had occurred during this procedure. Similarly, adjustments in polypeptide chain conformation to accommodate the positioning of bound DNA and to optimize favorable interactions were modeled and measured. Stereochemical idealization of measured polypeptide coordinates was then accomplished using the method of Hendrickson and Konnert, 1980. Inspection of the idealized nucleic acid and protein models indicated that these remained spatially compatible.

6.3. Difference Fourier Methods

Difference Fourier maps of four putative nucleic acid complexes formed by diffusing short [d(pA)₄,d(pCTTC),d(pCG)₃,d(pCCG)] oligonucleotides into native G5BP crystals were examined. In each case the resultant difference electron density maps were disappointing in that they could not be interpreted on the basis of the oligomer structure diffused into each crystal. However, it was of some support to the DNA binding model proposed below that the major positive electron density peaks of the maps did appear in two locations within each DNA binding channel. These occurred at the ends of the two arrows in Figure 17. Specifically, the probable binding occurs near Tyr 34 and Tyr 26, that is, respectively, near the upper and lower ends of binding clefts. Both of these residues have been implicated as playing essential roles in G5BP–DNA complexation by other physical-chemical techniques (Anderson et al., 1975a; Hilbers et al., 1978; Garssen et al., 1980; Alma et al., 1981; O'Connor and Coleman, 1983).

Examination of molecular packing down the y axis of the native crystal indicates the probable cause for our inability to bind deoxyoligomers to G5BP in an ordered way. Two G5BP dimers as they are positioned along this axis are shown in Figures 18a and b. In these illustrations, the crystallographic y axis is coincident with the twofold axis of the G5BP dimers.

Comparison with Figure 10 shows that the DNA binding channels in Figure 18 run almost directly into the plane of the illustration. They are positioned just inside the two extended β loops and separated by the ridge of atoms at the molecular twofold axis. In Figure 18, the adjacent dimer molecule packed along the crystallographic y axis is positioned so that its N- and C-termini fit directly into the two DNA binding channels of the origin molecule. Thus, the greater part of both binding channels are

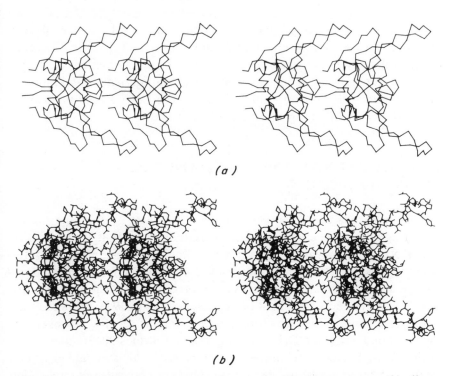

(a)

(b)

Figure 18. Stereo drawings of (a) only α-carbon backbone atoms, (b) all non-hydrogen atoms, of two G5BP dimers as packed along the y axis in the crystal. The crystallographic twofold symmetry element is coincident with the intermolecular dyad bisecting each dimer unit. The primary lattice interactions along the y direction involve the N- and C-termini of one dimer being positioned directly in the DNA binding channels of the adjacent G5BP dimer. The salt bridges and hydrogen bonding network thus formed apparently preclude the specific binding of oligonucleotides in native G5BP crystals. Binding is possible at the extremities of each binding channel as indicated by difference Fourier electron density maps.

occluded, making it physically impossible to bind contiguous strands of oligonucleotides across these regions. Further inspection shows that the termini of the translationally adjacent dimer form salt bridges to charged residues in the binding channels that otherwise might bind the phosphate backbone of DNA. Hydrogen bonding between other residues further restricts access to these channels. This leaves only portions of the outer extremities of each binding channel accessible and it is here that there is evidence of nucleotide binding in our difference Fouriers.

It is our feeling that the difference electron density at either end of the DNA binding clefts in fact represents disordered oligonucleotides that are weakly or nonspecifically bound so that the resultant difference electron density is diffuse. We believe that in spite of the fact that the difference electron density is not observed to inhabit the entire binding cleft, due to lattice interactions, it does at least provide a rudimentary definition of the end points of a bound strand. Because the DNA strand is linear, it marks the course that the nucleic acid must take.

7. STRUCTURAL AND MECHANISTIC CONSIDERATIONS

Given the extensive degree of monomer–monomer overlap and the apparent rigidity of the intermolecular hydrophobic barrel, it seems unlikely that DNA binding would substantially alter the dimer protein conformation. This is supported by the absence of evidence for conformational change upon complex formation as monitored by CD-ORD spectroscopy, NMR, and chemical modification techniques (Day, 1973; Hilbers et al., 1978; Coleman et al., 1976). Probable exceptions to this, however, are the extended wing or "DNA binding loop" (residues 20–30) and the tip of the "complex loop" (residues 39–43), both of which are unconstrained by dimer interactions (Figs. 18 and 19). This does not rule out movements of interacting side chains or minor readjustments of short segments of polypeptide chain. It does, however, support our contention that a DNA chain can be fitted to the crystallographically-determined G5BP structure without distortion or substantial change in conformation, and with a reasonable expectation that the interactions suggested do, in fact, occur.

It might be anticipated simply from our knowledge of the binding sites of enzymes that interact with extended linear substrates that the DNA

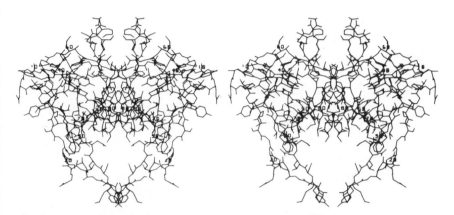

Figure 19. Stereo drawing of all the nonhydrogen atoms of the G5BP dimer with the view perpendicular to that of Figure 10. The twofold axis of the dimer unit runs in the plane of the illustration as is evident from comparison of the two extended β loops. These extended β loops consist of residues 20 through 30 and comprise the outer extremity of the DNA binding clefts. The protrusion of atoms at the monomer–monomer interface results from the juncture of dyad β loops. The two channels formed between the extended DNA binding loops and the juncture of dyad loops run approximately perpendicular to the twofold dimer axis and are the binding sites for single-stranded DNA.

binding site of G5BP would be a groove, cleft, or extended depression in the surface of the protein. As described above, in the G5BP dimer there are two such well-defined channels capable of simultaneously binding two single strands of DNA. For the gene 5 protein, however, there are additional constraints on the characteristics of the DNA binding clefts. The gene 5 protein exists as a dimer and each subunit binds to opposite strands of duplex DNA or the opposing sides of a covalently closed single-stranded circular DNA. The polarity in space of the two DNA binding sites, therefore, must be opposite, binding a DNA strand running 3′ to 5′ on one monomer and 5′ to 3′ on the other. This reversal of binding sense is accomplished by the intermolecular twofold axis of symmetry and is, in fact, the only means by which this can be achieved and still maintain the two binding sites, otherwise identical. This requires, however, that the binding paths for the DNA lie more or less in a plane perpendicular to the direction of the intermolecular twofold axis, otherwise they could not bind contiguously along DNA single strands without producing severe distortions at the interfaces between dimers. This further implies that the two DNA binding clefts in the dimer must be essentially parallel across the dimer face and not cross over into the region of the twofold-related binding cleft.

We would further expect that the dyad-related, and therefore parallel, binding sites would separate the two single strands of DNA by a distance greater than the width of the DNA double helix, that is, by 20 Å, because the protein is known to produce complete physical separation of the two polynucleotide chains when complexation occurs. At the same time, it cannot separate the two strands, and therefore the binding sites cannot in general be further apart, than the maximal distance between strands anticipated from electron micrographs of the helical gene 5-DNA complex, that is, not greater than 45 Å. The gene 5 protein binding site must also be relatively straight and without marked deviations from linearity because the polynucleotide chain has been shown to bind with little or no distortion from its natural extended conformation (Day, 1973; Anderson et al., 1975a; Coleman et al., 1976). This is apparent also from the gross structure of the gene 5 DNA helical complex as characterized by electron microscopy (Gray et al., 1982b).

Given the extended nature of bound DNA, one would expect nucleotide binding sites along the length of each binding channel at approximately 7.0 Å intervals. Furthermore, because each channel could bind as many as five deoxynucleotides (Pratt et al., 1974; Gray et al., 1982b; Cavalieri et al., 1976, and Torbet et al., 1981), the expected length of such a cleft would be of the order of 35 Å. Each binding channel must be accessible to a DNA strand over its entire length, because the nucleic acid bound physiologically by G5BP (bacteriophage fd DNA) is continuous and could not thread its way through a constricted opening.

The considerations discussed above were made at the level of the gross structure of the protein and the overall conformation of bound DNA. Reference to Figure 17 shows that at this level the binding channels of the G5BP dimer satisfy all the necessary requirements to allow complexation. Both, for example, are nearly linear and are freely accessible to solvent. The two are separated by 30 Å (center to center) and therefore the two bound DNA chains would not interact with each other. Each channel is sufficiently wide (10 Å) and long (35 Å) to accommodate extended DNA of up to five bases. Also, the two channels are essentially parallel and lie in a plane nearly perpendicular to the dyad, allowing contiguous DNA to span the entire face of the G5BP dimer. Finally, no grooves or channels on the G5BP dimer other than the ones we propose could accommodate linear DNA.

Given the compatibility of the G5BP binding channel structure to the conformational requirements of bound DNA, it was relatively straight-foreward to meld the two macromolecular elements into a model protein–nucleic acid complex. From a relatively coarse initial fitting, specific interactions consistent with the available data were brought into play to achieve a better fit. We chose to use guanine bases throughout because

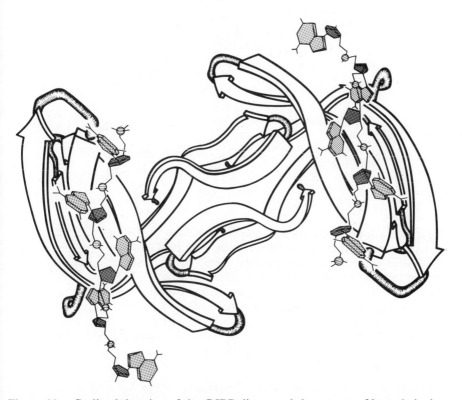

Figure 20. Stylized drawing of the G5BP dimer and the course of bound single-stranded DNA. Features of the bound chains are coded as follows: Striped circles, phosphate groups; dark gray rings, sugar moieties; light gray rings, bases. Differentiation of polypeptide chains and the location of specific amino acids can be made by comparing with the stereo drawings of Figures 7 and 9, all of which view the G5BP dimer from a similar orientation. A total of 10 nucleotides are illustrated, five in each of the two bound DNA strands.

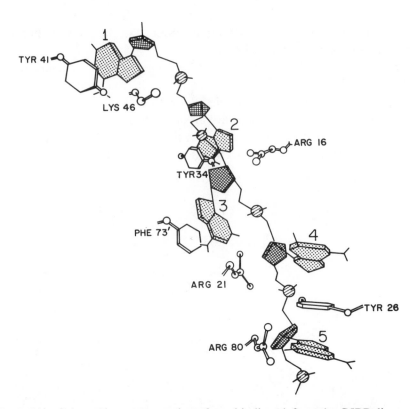

Figure 21. Schematic representation of one binding cleft on the G5BP dimer and the major interactions formed with DNA. The twofold-related binding channel exhibits an identical interactive environment. Two types of chemical interactions appear to be responsible for complexation. Aromatic side chains of the protein stack upon base rings of the DNA, and the phosphate backbone is neutralized by a combination of lysyl and arginyl side chains. The distribution of nucleotide binding sites is such that the DNA must be in a nearly fully extended state. To facilitate discussion in the accompanying text, each nucleotide has been arbitrarily numbered starting at the top of the drawing.

this purine moiety would impose the greatest number of spatial constraints during model building. Upon maximization of mutually compatible contacts, the protein model was then examined to determine if small conformational changes could enhance binding without significantly distorting the native structure of the protein. Minor polypeptide rearrangements were reconstructed in two regions leading to the model for the G5BP–DNA complex shown in Figures 20 and 21.

8. THE PROTEIN-NUCLEIC ACID COMPLEX

The orientation of the single-stranded DNA with respect to the surface of the gene 5 DNA binding protein is shown schematically in Figure 20. The general course of the nucleic acid chain is that suggested in the representation of Figure 17. That is, the two DNA strands are positioned just inside the extended DNA binding loops and held apart by a smaller protrusion of atoms at the molecular twofold axis. Strand separation is approximately 30 Å and the two DNA chains run essentially antiparallel to one another.

For the most part, each strand is positioned across the central three-stranded β sheet which forms the core of the G5BP molecule. Major interactions are also made with the "complex" and "DNA binding" loops. As Figure 20 shows, the β bend tips of the "dyad" loops are also in the vicinity of the DNA strands. Note, however, that this latter interaction occurs within the twofold-related G5BP monomer binding channel as opposed to the monomer unit of which the dyad loop is a part. These dyad loop interactions are essential components in DNA complexation. Crosslinking binding channel interactions of this sort suggest that the G5BP dimer may be the smallest specie capable of binding DNA.

It is apparent from Figure 20 that the phosphate backbone is bound closely to the surface of the protein throughout its course, whereas the base rings project somewhat further from the surface. The DNA conformation is almost fully extended and bases unstacked. Even in this extended state it was necessary to utilize five nucleotides to traverse the full length of each binding channel. The average phosphate-to-phosphate distance in the model is very nearly that expected for a fully extended polynucleotide. Thus the structure is most consistent with those investigations that have found five bases bound per G5BP monomer (Pratt et al., 1974; Gray et al., 1982b; Cavalieri et al., 1976; Torbet et al., 1981).

8.1. Interactions

The most prominent interactions between bound DNA and the G5BP are detailed in Figure 21. Only one binding channel is illustrated, because its twofold-related counterpart has exactly the same chemical and physical environment. The base of nucleotide 1 stacks coplanar with the aromatic side chain of Tyr 41 and the side chain of Tyr 34 stacks upon base 2. Lys

46 is positioned such that its free amino group is approximately equidistant from the phosphates of bases 1 and 2. Nucleotide 3 lies flat against the exposed side of the phenylalanyl side chain of residue 73. Note that this residue is a part of the polypeptide chain of the twofold-related G5BP monomer. The phosphate groups of bases 2 and 3 are located on either side of the side chain of Arg 16. Tyrosine 26, at the tip of the DNA binding loop, intercalates between base rings 4 and 5. The guanidinium group of Arg 21 lies near phosphate 3, and the same group of Arg 80 is positioned between phosphates 4 and 5. Besides these specific interactions, we observe that there are groupings of hydrophobic side chains clustered on the sides of base rings not stacked against aromatic groups and additional polar residues in the vicinities of bound ribose rings and phosphate groups. Two general mechanisms appear responsible for complexation. Base rings stack on protein aromatic side chains, and the bound phosphate backbone is fixed by appropriately positioned lysyl and arginyl side chains. A molecular model of a G5BP binding channel with bound DNA is illustrated in Figures 22 and 23.

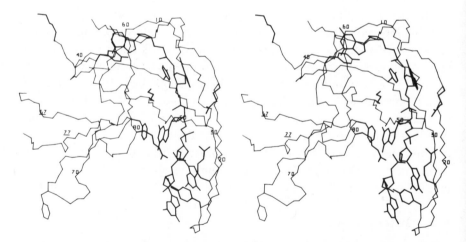

Figure 22.   Stereo drawing of a DNA binding cleft with a bound oligomer of five bases in length, also showing the polypeptide chain of the G5BP monomer, which forms most of the binding channel surface. An additional binding element, the dyad loop of a twofold related G5BP monomer (residue numbers underlined) is illustrated as well. Specific interactive groups discussed in the text are drawn with dark lines along with the bound DNA to accent their positions.

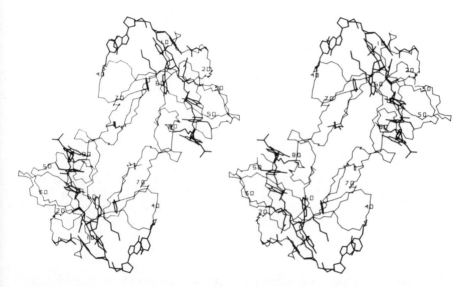

Figure 23. Stereo drawing of the polypeptide chain backbone of the G5BP dimer structure (light bonds) with two single strands of bound DNA (dark bonds). The protein conformation is that after modeling the DNA complex. For comparison, see the structure of the native G5BP dimer illustrated in Figure 10. Every tenth residue of both G5BP monomers has been numbered; side chains interacting with DNA are drawn with heavy lines.

8.2. Conformational Changes

Optimization of the binding interactions was effected by allowing minor conformational changes in the native high resolution G5BP structure. The most extensive of these involved the tip of the DNA binding loop. In particular, the side chain of Tyr 26 was reoriented so that this residue pointed more directly between the base rings of nucleotides 4 and 5. This rearrangement also served to draw the DNA binding loop over the DNA strand. In doing so, Arg 16 and 21 assumed positions adjacent to the phosphates of nucleotides 2 and 3. These proposed structural changes are found in Figure 24, which illustrates the conformation of the DNA binding loop before and after complexation.

Movement of the DNA binding loop is easily accomplished because, for the most part, it is extended away from the core of the G5BP molecule (Fig. 20 and 22) and therefore makes few contacts with the remainder of

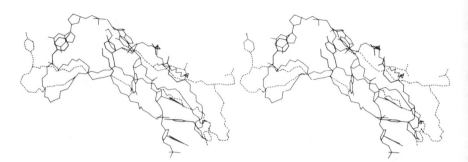

Figure 24. Stereo drawing of the G5BP binding channel in the vicinity of the DNA binding loop and in the region about the tip of the complex loop. Dashed lines indicate the native positioning of polypeptide chains, while solid lines show the conformations of these same chains upon DNA complexation. Bound DNA and interactive groups are drawn with heavier lines.

the protein. The polypeptide segment most affected involves residues 24 to 27. This β loop, and particularly residues 24 to 27, exhibits considerable mobility even in the crystalline state where it is characterized by uniquely high temperature factors of nearly twice the mean for the remainder of the molecule. Given its positioning over the DNA binding channel, it is probable that the DNA binding loop makes initial contact with free nucleic acid strands and therefore may require the observed flexibility to sense and fix the DNA. In our model of the G5BP–DNA complex, the extensive contacts formed by this loop with the DNA would cause it to coalesce into a considerably more stable conformation.

The structure of G5BP, in the absence and presence of DNA, was examined in an effort to determine how the proposed conformational change in the DNA binding loop might be induced by polynucleotide binding. We suggest that the linchpin may be Arg 80. In the uncomplexed state the guanidinium group of this residue lies midway between Asp 79 and Glu 30 (see Fig. 8 and 10). The bridge to Glu 30 represents the last, and essentially the only, interaction between the core of the G5BP molecule and the DNA binding loop before it freely extends into surrounding solution. Our model building shows that when single-stranded DNA is inserted into the binding cleft, the phosphate groups of nucleotides 4 and 5 fall immediately above and below the guanidinium group of Arg 80 where they would vigorously compete for its attention. The resultant rearrangement of this side chain to accommodate the phosphate groups would disrupt the original Asp 79 : Arg 80 : Glu 30 bridge. Thus, when DNA enters the binding cleft, the last remaining constraint that may have been imposed on the DNA binding loop is removed thereby initiating the conformational change (Fig. 24).

A modest shift in the native position of Tyr 41 was also allowed to optimize the protein–nucleic acid fit. For the most part, rearrangement was restricted to residues 40–42, which occur at the tip of the complex loop and are relatively open to solvent. These conformational changes are also detailed in Figure 24. The high resolution refinement of G5BP has shown these residues also have considerable mobility as evidenced by their above average thermal motion parameters. The side chains of Lys 46 and Phe 73 were also repositioned, as illustrated in Figures 22 and 23. We feel the conformational changes proposed are realistic, minimal, and consistent with expectations from solution studies (Day, 1973; Coleman et al., 1976; Hibers et al., 1978).

9. CORRELATION OF PROPOSED MODEL WITH DATA FROM OTHER TECHNIQUES

With regard to the extensive research of G5BP–DNA complexation using noncrystallographic methods, a number of correlations with the proposed model can be made. It was demonstrated by chemical modification and NMR experiments that three G5BP tyrosines occupy surface positions while two others are internal (Pretorius et al., 1975; Coleman et al., 1976; Alma et al., 1981). Upon complexation with DNA, there are marked changes in the CD tyrosyl bands (Day, 1973) and NMR spectra, which have been interpreted as indicating that all three surface tyrosines are involved in binding (Pretorius et al., 1975; Coleman et al., 1976; Garssen et al., 1977, 1978; Coleman and Armitage, 1978). Tyrosyl chemical shifts indicate that stacking of these aromatic groups on the DNA bases occurs (O'Connor and Coleman, 1982 and 1983). Nuclear Overhauser effects imply that nucleotide sugar protons are near the ring protons of at least two of the aromatic residues for which resonances undergo substantial shifts during complex formation (Alma et al., 1981).

Chemical modification studies show three of five G5BP tyrosine residues can be reacted with either tetranitromethane or N-acetylimidazole (Anderson et al., 1975a). DNA binding is lost upon derivatization, and conversely, the presence of DNA protects the side chains from modification. Subsequent peptide analysis has identified the modified tyrosine rings as belonging to residues 26, 41, and 56. These chemical modification results are consistent with the spectral evidence implying that three tyrosine residues play a pivotal role in DNA complexation.

Examination of the three-dimensional structure of G5BP shows, in agreement with the data discussed above, that there are three surface and two internal tyrosine side chains. These are residues 26, 34, and 41, and residues 56 and 61, respectively. Furthermore, all three surface tyrosine residues are found in close proximity to the putative DNA binding channels. The structure of G5BP presented here is not, however, entirely in agreement with the assignment of surface tyrosines derived from chemical modification experiments (Anderson et al., 1975a). In the three-dimensional structure, tyrosines 26 and 41 are fully exposed and would be expected to undergo chemical modification. Both occur at opposite extremes of the DNA binding cleft, and it is plausible that their reaction alone would be sufficient to explain the loss of DNA binding. In the G5BP structure, Tyr 56 is largely buried in the monomer hydrophobic core near Tyr 61. Its edge, however, is exposed to solvent and it is possible that it could be nitrated. It would not, however, have been our choice as the next most accessible tyrosine. Tyrosine 34, at the center of the DNA binding cleft, seems a more reasonable candidate for modification as it appears more accessible than 56, though less so than either 26 or 41. The only explanation we can offer is that there may be an orientation or environment effect that prevents nitration of Tyr 34. Indeed, Leu 45 is stacked directly against the aromatic ring on the interior side, and Phe 13 is also hovering nearby. In addition, it does not protrude overtly into the solvent but lies back flat against the monomer hydrophobic core.

In the model we present here, Tyr 26, 34, and 41 are directly involved in the binding of DNA through stacking of their respective aromatic groups. Good agreement with experimental results is thus realized both in the number of tyrosine residues and the mode of interaction involved. Judging from their central role in each DNA binding channel, modification of these residues would almost certainly lead to loss of DNA binding.

Correlation of the model with experimental data can also be made in the case of phenylalanine residues. Several lines of evidence imply that a phenylalanine side chain is intimately involved in DNA complexation. Ultraviolet irradiation of the complex of G5BP and bacteriophage fd DNA induces the formation of a covalent crosslink between protein and DNA

(Lica and Ray, 1977). Peptide analysis revealed this crosslink occurred between residues 70 and 77 in the polypeptide chain of G5BP, with Phe 73 being the most likely candidate. Reference to Figures 10 and 23 shows Phe 73 is indeed located in the DNA binding channel. Of the three phenylalanines in the amino acid sequence, Phe 73 is the only residue having direct access to DNA complexation regions.

This phenylalanine was utilized in the model building process to stack on the base ring of nucleotide 3. It originates from the twofold-related monomer and extends across the breadth of the dimer face in assuming its position. NMR evidence (Coleman and Armitage, 1978; Hilbers et al., 1978; Garssen et al., 1980; Alma et al., 1981; O'Connor and Coleman, 1983) indicates that ring protons of a phenylalanine residue show large shifts upon DNA complexation. As with the G5BP tyrosyl residues, the magnitude and direction of this effect is consistent with aromatic ring stacking on bound DNA bases.

Other irradiation studies utilizing thymidine containing oligonucleotides and G5BP have identified a crosslink formed to Cys 33 (Lica and Ray, 1977; Anderson et al., 1975b; Paradiso et al., 1982). Although inaccessible to the bulky Ellman's reagent, this residue reacts with Hg^{2+} ion and this leads to protein unfolding and loss of DNA binding (Anderson et al., 1975b). The presence of fd DNA prevents the reaction of Cys 33 with mercurials. These studies imply that Cys 33 is in or near the protein–nucleic acid interface. Reference to Figures 8 and 10 show this to be consistent with the model in that this residue is for the most part buried in the hydrophobic interior but also borders the center of the DNA binding cleft. Thus, while it is unlikely that Cys 33 plays an active role in DNA complexation, perturbation of its native state by chemical modification would have dramatic consequences on the polypeptide chain conformation of G5BP as a whole and could readily produce loss of binding affinity. Its proximity to the G5BP binding surface is also consistent with the photocrosslink that has been shown to form.

G5BP–DNA complexes are susceptible to dissociation as a function of increasing cation concentration. This suggests that positively charged amino acid side chains interact with the phosphate backbone of DNA. Indeed, the relative sensitivity of complex dissociation by various cations as measured by CD correlates with known cation affinities for nucleotide phosphates (Anderson et al., 1975a). Abolition of DNA binding upon acetylation of the seven lysine residues of G5BP also supports this likelihood.

It has further been observed that the side chains of some arginine residues undergo significant chemical shifts and line broadening upon G5BP complexation with d(pA)$_4$ or d(pA)$_8$ (Coleman et al., 1976). The immobilization of as many as two lysine side chains has also been detected using NMR (Alma-Zeestraten, 1982).

The structure of G5BP places all lysine and arginine side chains on the surface of the molecule. However, only two of these have immediate access to the DNA binding channel in the native state of the protein. These are Lys 46 and Arg 80. Nevertheless, the minor rearrangement of the DNA binding loop (Fig. 24) would position Arg 16 and 21 in this region as well. Thus, from structural considerations, at least four positively charged protein side chains would play a role in DNA complexation.

The model of G5BP–DNA complexation shows the role that Lys 46 and Arg 16, 21, and 80 would play in binding the phosphate backbone of oligonucleotides. With the possible exception of Arg 80, these charged side groups remain fairly open to the surrounding solvent. This is in keeping with the observation that DNA complexation provides no protection against the acetylation of lysine residues (Anderson et al., 1975a) and subsequent complex dissociation.

The only other side chain that has been investigated is the single histidine residue at position 64 in the amino acid sequence. The proton resonances of this histidine remain unchanged upon formation of G5BP–DNA complexes (Alma et al., 1981). This is consistent with its position towards the back of the protein, removed from the general area of the DNA binding channels.

10.   THE GENE 5 PROTEIN–DNA HELICAL COMPLEX

Knowing the structure of the gene 5 protein and the DNA to which it binds, as well as the manner by which it is bound, and given the parameters of the gross helical structure formed by the association of the macromolecules, it should be possible to describe the atomic structure of the entire gene 5–DNA complex. Much, if not all, of this information is in fact available, and we have applied it to the determination of a consistent model for the gene 5–DNA complex using a rigorously quantitative approach that, we believe, systematically samples all possible alternatives.

10.1. Features of the Assembly

The gene 5–DNA complex formed between the protein and the circular single-strand DNA of the fd bacteriophage has been investigated using an extensive variety of techniques. Table 1 summarizes most of the known properties of the helical complex. Electron microscopy studies show the assembly to be a long, flexible nucleoprotein helix (Alberts et al., 1972; Pratt et al., 1974; Oey and Knippers, 1972; Gray et al., 1982b; Torbet et al., 1981). One turn of this helix, it was judged, contains about six dimers of the gene 5 protein. The helix has a width of about 100 Å and a rise of about 90 Å per turn (Alberts et al., 1972; Gray et al., 1982a; Torbet et al., 1981). Titration studies of the phage DNA with the gene 5 protein *in vitro* have yielded associations of three to five DNA bases per gene 5 protein (Alberts et al., 1972; Pratt et al., 1974; Oey and Knippers, 1972; Cavalieri et al., 1976; Gray et al., 1982a; Torbet et al., 1981; Day, 1973; Gray et al., 1982b). Our examination of the DNA binding surface of the native protein suggests the number to be five. The structural parameters obtained for the complex by electron microscopy are in good agreement with those obtained by low-angle X-ray scattering and neutron diffraction studies (Torbet et al., 1981; Gray et al., 1982b).

In the presence of short deoxyoligonucleotides of four to six in length, the gene 5 protein appears under some circumstances to form aggregates reflecting the mode of linear polymerization directed by long strands of DNA. These aggregates, however, are self-limiting, and both crosslinking studies in solution (Rasched and Pohl, 1974) and X-ray crystallographic studies on single crystals of the protein complexed with oligonucleotides (McPherson et al., 1980b) tend to suggest that the predominant aggregate size is fixed at 12. Further implications of these analyses were that the duodecamer possessed a sixfold axis and that the inherent dyad axis of each gene 5 protein dimer was aligned perpendicular to the hexagonal axis in such a way that the entire arrangement exhibited 622 (D_6) point group symmetry (McPherson et al., 1980a, 1980b). Such a toroidal structure, or disk, was believed related to one turn of the gene 5-DNA helix by simply opening it at one point and spreading the junction to produce a "lock washer" kind of structure, as illustrated in Figures 25a and b. This would, then, be somewhat akin to the similarity between the disk and helix turn in the tobacco mosaic virus. A salient feature of this closed

Table 1. Quantitative Features of the Gene 5 fd–DNA Helical Complex

Parameter	Observed or Measured	Calculated for Model	Reference
Hand of helix		Right	
Number of DNA strands in complex	2 Antiparallel Dimer	2 Antiparallel Dimer	(b,c,g)
Protein unit of construction			(f,g)
Disposition of DNA	Unstacked-extended	Unstacked-extended	(b,i,j,m)
Total number of nucleotides in complex	6408	6408	(a)
Contour length of entire complex	8800 Å	9000 Å	(b,h,k)
Number of gene 5 molecules in entire complex	1300	1280	(c)
Outer radius of complex	100 Å	93 Å	(b,g,h)
Helical pitch	89–93 Å	90 Å	(b,g,h)
Number of turns in entire complex	96–98	100	(g,h)
Number of gene 5 protein molecules per turn	12	12.8	(b,h)
Number of nucleotides per helix turn for each strand	33	32	(g)
Number of nucleotides bound per gene 5 molecule	3–5	5	(b,c,d,e,g,h,i,j,k)
Mass per unit length	1400–1600 daltons/Å		(h,k)
Radius of DNA gyration	34.5 Å	35.1 Å	(h)
Radius of DNA in helix	18–36 Å		(h,k)
Rise per residue for each nucleotide	2.75 Å	2.83 Å	(g)
Angle of rise	23°	23°	(g)
Separation distance of opposing DNA strands		32 Å	(g)

(a) Schaller et al., 1978; (b) Alberts et al., 1972; (c) Pratt et al., 1974; (d) Oey and Knippers, 1972; (e) Cavalieri et al., 1976; (f) Brayer and McPherson, 1983a; (g) Gray et al., 1982b; (h) Torbet et al., 1981; (i) Day, 1973; (j) Anderson et al., 1975a; (k) Gray et al., 1982a; (l) Brayer and McPherson, 1983b; (m) Coleman and Armitage, 1978

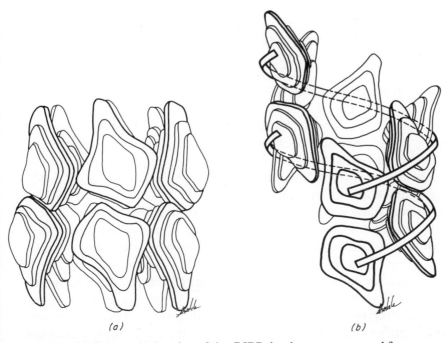

(a)   (b)

Figure 25. (a) Schematic drawing of the G5BP duodecamer proposed from crystallographic studies of the protein complexed with oligonucleotides to be the asymmetric unit. This duodecamer, having 622 point group symmetry, was thought to be related in a simple manner to one turn of the G5BP–DNA helical complex, as shown in (b).

arrangement of gene 5 dimers was that the protein–protein interactions between adjacent molecules, presumably a consequence of DNA binding, would closely resemble the homologous interactions resulting from contiguous aggregation along a linear DNA strand. The gross dimensions of the toroidal duodecamer proposed, based only on the native structure of the gene 5 protein closely packed as a disk, was a thick cylinder of height about 90 Å and width about 100–120 Å, consistent with the observed helical parameters of the gene 5–DNA complex.

10.2. Elements of the Complex

The dominant constraints imposed on any model for the gene 5–DNA helical structure arise from the structures and character of the two components, the gene 5 protein and the two antiparallel strands of the DNA, and limitations on the manner by which they interact with one another. Spectroscopic studies indicate that in the complex the bases of the DNA are fully unstacked and the polynucleotide chain more or less extended (Alberts et al., 1972; Day, 1973). There can, of course, be no base pairing between the two antiparallel strands of DNA because they are not complementary in the case of the phage DNA. Because the complex does represent an "unwound" form of DNA, one would expect the opposing sugar-phosphate backbones to be separated by a distance greater than is found in native duplex DNA. Indeed, by examining the relative locations of the two dyad-related binding sites for DNA on the gene 5 protein dimer, we estimate this distance to be 30 to 35 Å. As noted above, the gene 5 monomer must cover from 4 to 5 bases of the DNA along each strand. Any model of the complex must not require or permit any discontinuities in the two DNA strands, but the DNA must progress in a continuous linear path from one gene 5 binding site to the next without interruption. This requirement would not, it might be noted here, hold for the proposed dodecamer disk structure, which employs discontinuous oligomers.

The structure of the gene 5 protein is literally the keystone element of the complex, and any model must utilize the native structure as determined by X-ray diffraction analysis or essentially that structure, because we cannot rule out some small conformational changes occurring as a result of DNA binding. A broad range of physical and chemical studies, however, indicate that there are no gross changes in the gene 5 protein upon complexation with DNA (Pretorius et al., 1975; Day, 1973; Anderson et al., 1975a; Coleman and Armitage, 1978). Although small and subtle alterations must surely be allowed, and indeed expected as described above, we believe the assumption is valid that the protein structure in the complex is essentially the same as determined in the crystal. Our analysis of the protein structure in terms of secondary and tertiary bonding interactions and thermal or motion patterns over the molecule show that although there are some segments of the polypeptide likely to be more dynamic and subject to conformational change than others, these motion-sensitive areas did not occur where we ultimately found the protein–protein contact surfaces to be.

The details of the structure determination and subsequent refinement of the gene 5 protein structure as well as a full description of the molecule are presented above. Several features, however, should be reiterated as they are relevant to a model of the protein–nucleic acid complex. Seen in Figures 8 and 10 are illustrations of the gene 5 monomer and dimer. The two monomers within the dimer are tightly coupled in both a structural and functional sense, so much so that it may be misleading to treat the 10,000 dalton species as an independent entity. The coupling between the two about the dimer twofold axis is particularly rigid and strong. It involves the formation of an intradimer β barrel of six antiparallel strands, three contributed by polypeptide chain from one-monomer hydrogen bonded to three symmetry-related strands donated by the mate (see Fig. 11a–c). This secondary structural scheme is liberally augmented by hydrophobic contacts and several strategically placed salt bridges between the two monomers. The bonding between the two is so extensive we feel it unlikely that the gene 5 monomer commonly exists free in solution. A consequence of this interlocking of the two subunits is that the dyad axis relating the pair must be stable and, therefore, unlikely to be lost upon complex formation with the DNA. Thus we feel it a fair assumption that the dimers retain a perfect dyad, as observed in the crystal, when a part of the helical complex.

The binding site for the DNA on the surface of the protein can be identified with some measure of confidence simply by inspecting the structure of the gene 5 molecule. The single long cleft formed by the β strands of the polypeptide run over the face of the molecule for a distance of about 35 Å or more, about the distance expected if four to five nucleotides, fully extended, were to be covered. This alone, of course, is not sufficient to fix the DNA binding site, but its identity, as detailed above, is reinforced by the location in its interior of many residues shown by physical, spectroscopic and chemical modification studies to be intimately involved in DNA binding.

Using the available evidence and data from other laboratories as well as our own, we described above what we believe to be a plausible and likely mechanism for the binding of a DNA strand to the gene 5 protein. We believe the mechanism shown in Figure 21 is essentially correct, and that a model of the gene 5 DNA helical complex must be in substantial agreement with this mode of binding. The greatest degree of uncertainty with respect to the binding mechanism lies in those residues making up the Tyr 26 loop. This loop appears to be somewhat mobile, has relatively higher temperature factors and, as we described above, is likely to experience some limited conformational change upon binding of DNA. We do not, however, see this as producing any appreciable distortion in the

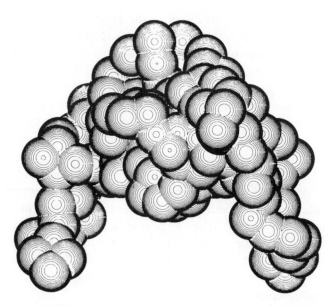

Figure 26. Space-filling model of the G5BP dimer created by placing a 3.5 Å diameter sphere around each α-carbon and omitting all other atoms. Note the wedge, or keystone, shape of the dimer, which has at the top, narrow surface the carboxyl and amino termini and at the broad, bottom surface the two symmetrical DNA binding loops.

course of the DNA strand as we have delineated it, nor in the major mass distribution of the protein itself.

One obvious feature of the gene 5 protein dimer, seen also in Figure 26, that is at once very suggestive, is the wedge or keystone shape that it demonstrates. This trapezoidal protein block has on its broad face the two DNA binding clefts of opposite polarity while the narrow face exhibits the N and C termini of both monomers, all four available for salt linkages. It requires little imagination to perceive at a gross level how these wedge shapes could be arranged to form a closed circle or the turn of a helix.

11. MODELING PROCEDURE

From the physical, chemical, and structural evidence at our disposal, we assembled a collection of stereochemical and functional properties that must be demonstrated by an acceptable model of the gene 5–DNA complex. These are listed in Table 1. Needless to say, each of the items in the table does not represent an independent variable, the observations and constraints are for the most part interdependent, and some shown in Table 2 are not even quantifiable or not easily quantifiable. Nevertheless, if a particular constraint is violated in the modeling process, it can be used to eliminate any otherwise acceptable possibility.

We began modeling the helix by making the assumption that the dyad axis observed for the dimer in the crystal was maintained in the complex, as discussed above, and that it would be perpendicular to the helix axis. This seems very probable because any other arrangement would require that there be a distribution of compensating angles for the various dimers in a turn, that is, the dimer units of the helix would be nonidentical or the two single strands of DNA in the complex would be bound differently. Neither of these possibilities would seem to conform to any observations, and there is no evidence to suggest that either is true.

All possible helices were then defined by (1) the radius of the helix, (2) the repeat distance along the helix (pitch) or the height of one turn, h, and (3) the orientation of the gene 5 dimer as determined by its rotation about the intramolecular dyad axis. For every possible helix within a reasonable range of these parameters—a number of resultant quantities were calculated. These included the number of DNA bases per turn of the helix and the number of gene 5 protein dimers bound per turn.

Table 2. Qualitative Properties of the Gene 5 fd–DNA Helical Complex

1. The structure must have the gross physical properties observed by electron microscopy and enhanced by three-dimensional reconstruction.
2. The helical complex must be quite open and solvated with free access to the interior.
3. The complex must have short range rigidity but long-range flexibility.
4. The DNA must pass continuously from one protein subunit to the next without discontinuity or dislocation at the interfaces.
5. The binding of the DNA to the protein must utilize the binding cleft and mechanism in a manner consistent with all chemical and physical studies.
6. Assembly of the complex must not require any large conformational changes in the protein structure or distortion of the DNA backbone.
7. There must be a large number of protein–protein atomic contacts to account for the observed cooperativity, self assembly, and stability.
8. There must be no prohibited atomic contacts between atoms of adjacent helix elements.

The radius of the helical complex was taken to be the mean distance of the path of the bound polyphosphate backbone. This lies, for the most part, on the exterior of the helix at essentially constant radius, but because the nucleic acid binding cleft is embedded, to some extent, in the protein it does not represent the limit perimeter of the structure.

Helices were calculated assuming that an integral number of nucleotides were bound to each monomer of protein and were taken to be three, four and five. Values of three nucleotides per gene 5 monomer gave helices which were accompanied by large numbers of prohibited contacts. In addition, the helices were extremely tight and narrow in gross appearance and their properties could not be correlated with the observed properties shown in Table 1. In particular, they would not predict the observed contour length and number of turns, helix width, or number of nucleotides and gene 5 monomers per turn. An alternative approach to obtaining a small DNA radius was to reverse the sense of the gene 5 dimer axis so that the broad face of the molecular pair was turned inward and the narrow face outward. This helix, however, was very irregular, exhibited a large number of prohibited contacts, and required considerable distortion of the DNA at the protein–protein interfaces. Furthermore, it was inconsistent with most of the observed properties of the helical complex seen in Table 1.

From our analysis of the DNA binding site and from NMR and spectroscopy studies, the DNA was taken to be in extended form and a phosphate–phosphate distance of 7.1 Å (Franklin and Gosling, 1953) was used to calculate the number of nucleotides per turn of the helix based on the particular radius. From this, the number of gene 5 dimers forming one turn of the helix at the given radius and pitch was calculated.

The search was initially conducted using fairly coarse increments of the variables but adequate to assure continuity over the range. At those values of radius, pitch, and angular rotation where possibilities were encountered, the increments were reduced appropriately and a search made over a fine grid of values.

For each model helix, using the atomic resolution structure of the native protein, the total number of close but acceptable interatomic contacts was calculated at the interfaces. Even more important, the number of prohibited contacts for all sets of parameters (1), (2), and (3) were computed. The number of allowed contacts served as a measure of the fit or degree of complementarity between successive dimer units along the DNA strands in the helix. This must be substantial as these contacts are required to explain the extensive cooperativity of protein binding that is observed as well as the structural basis underlying the formation and stability of the helix. The unacceptable contacts are the most sensitive determinant in choosing which models are plausible. Our approach was to map these two contact parameters as a function of the variables and to search for points having maximum acceptable contacts and minimum unacceptable contacts. It was then our intention to examine, using a variety of graphics devices, all reasonable possibilities emerging from this analysis. We anticipated this array of plausible and probable structures to be rather large. In fact, we found quite the contrary.

We should point out here that at no time did we incorporate in the modeling process, either implicitly or explicitly, the constraint that the DNA had to fit reasonably into the binding cleft of the protein in a linear fashion, that it be continuous in passing from one gene 5 unit to the next without abrupt change or distortion, or that the final helical structure resemble that observed in the electron microscope. These were tests held in reserve to provide a basis for choosing the correct models from the probable models presented by the atomic contact analysis.

The features on the contact map were quite negative in character, with large numbers of acceptable contacts generally accompanied by a prohibitively large number of unacceptable contacts, particularly for low radii, therefore yielding impossible structures. On the other hand, when the number of unacceptable contacts was nonprohibitive there were two few total contacts to produce a plausible model. This occurred, for example, when the radius became large and the gene 5 dimers began separating from one another. Instead of finding a broad spectrum of likely candidates for the gene 5–DNA helix, there was in fact only one distinct minimum in the search, that is, one set of parameters that provided large favorable contact surfaces between the consecutive gene 5 protein units and was accompanied by very few prohibited interactions. This occurred when $r = 35$ Å, $h = 90$ Å, and the dimer angle was $23\frac{1}{2}°$. This point in the analysis was clearly pronounced and distinct from the general noise level produced by the remaining structures.

At this point, the best candidate was examined on a computer graphics system in order to visualize the interactions between the protein molecules, to examine the orientation of the binding clefts on adjacent molecules, and to determine if a permissible mode of binding was provided for the DNA. In addition, the other characteristics not directly employed as constraints in the analysis were examined for compatibility.

12. THE HELIX MODEL

The arrangement, which we contend is likely to be the fundamental unit of the protein–DNA helical complex, is illustrated in Figures 27 through 29. The first feature to strike one's attention is that the DNA binding clefts on the individual protein units are perfectly contiguous and aligned in a linear course without any requirement for distortion of the polynucleotide chain as it passes from one DNA binding site to the next. Thus the protein arrangement is entirely compatible with a smooth flow of two single strands of DNA along the length of the helix with no local disruptions—a consequence of, not a constraint in, the contact analysis. The DNA chains formed by linking the ends of the fragments bound by each protein dimer are shown in Figure 28. The two strands of DNA in the complex, running antiparallel courses, are separated by a relatively constant distance of 32 Å and form, by themselves, an unpaired double helix having a radius of 35 Å.

Another feature of the model is that the entire structure of the helical complex does, in fact, very closely resemble in gross appearance the structures observed in the electron micrographs between the gene 5 protein and the phage DNA (Alberts et al., 1972; Gray et al., 1982b; Torbet et al., 1981; Gray et al., 1982a). This applies as well to electron micrograph images enhanced and refined by spatial filtering and three-dimensional

---

Figure 27. (*Opposite*) Stereo drawing of the gene 5 protein–DNA helical complex constructed according to the optimal parameters derived from the contact analysis. Only the backbone α-carbons of the protein are shown for clarity. Shown here is 2.5 turns of DNA and 1.5 turns of the protein. There are 6.4 gene 5 dimers per turn, or 12.8 monomers, each monomer having 5 nucleotides of DNA bound. Because of the dyad axis in the dimer, one monomer of each pair binds a DNA strand running 3′–5′ and the other a strand running 5′–3′. The DNA strands are 32 Å apart and have a radius of 35 Å. The entire complex has a width of about 93 Å and the height of one turn is 90 Å. The complex has a wide and exposed major groove; the minor groove is filled by the densely packed hydrophobic cores of the protein.

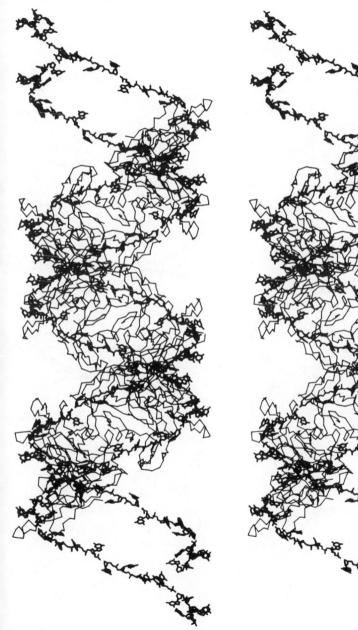
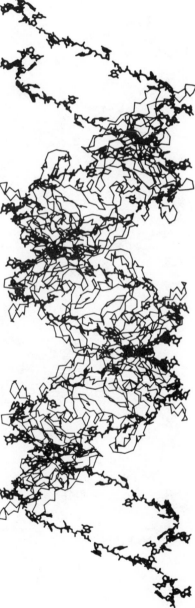

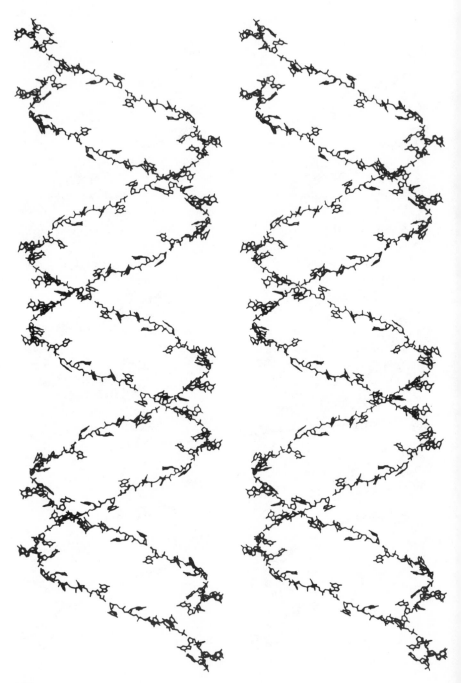

image reconstruction (Gray et al., 1982b). Though locally rather rigid and closely interlocked, it does give rise to a long, reasonably flexible filament. The minor groove of the helix is essentially filled with the extended hydrophobic cores of the protein dimer. The major groove is very broad and allows solvent ready access to the interior of the complex structure, where there is an extensive array of hydrophilic amino acids arising primarily from the N- and C-terminal strands of the individual protein monomers.

As seen in Figure 30, the DNA, though embedded in the binding cleft of the protein and somewhat protected from solvent by protein shielding, does lie nearer the exterior surface of the helix with the gene 5 protein forming the interior core. The "keystone" or wedge shape of the protein dimer is thus ideally suited for the formation of such an arrangement, with the broad DNA binding surface of the protein engaging the nucleic acid and tapering as it proceeds inward to a narrow interior surface that provides the intermolecular contacts required to maintain the helix. It would be difficult to construct a protein molecule more ideally suited in terms of shape and function to engineer its own self-assembly into a helical filament.

The external diameter of the model complex as predicted by the contact analysis is about 93 Å, which is in agreement with several direct measurements of the helix width in electron micrographs and three-dimensional reconstructions as well as that determined by low-angle neutron scattering (see Table I). The repeat distance along the model helix is 90 Å. This is also in agreement with measurements of the gene 5–DNA complex by the techniques cited above. One turn of the model helix will contain 6.4 gene 5 protein dimers, which is consistent with both electron microscopy and neutron diffraction studies. Each gene 5 monomer in the model helix will cover five nucleotides along the DNA single strand. This is again in agreement with investigations that find the protein to bind five nucleotides per monomer, and with our own analysis of the gene 5 DNA binding cleft. The total number of gene 5 protein molecules required to complex the

---

Figure 28. (*Opposite*) Drawing of 2.5 turns of two antiparallel single strands of DNA as they would appear when a part of the gene 5 protein–DNA complex. The radius of the DNA helix is 35 Å, the height of one turn is 90 Å, the helical pitch is 23°, and there are 64 nucleotides per turn. Given that the fd genome contains 6408 nucleotides, then the entire fd DNA would require a complex having 100 turns and therefore a total length of about 9,000 Å.

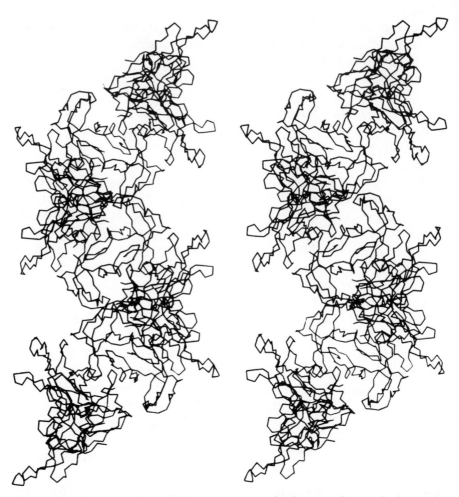

Figure 29. Drawing of the G5BP component of 1.5 turns of the helical complex shown in Figure 27. Note the very open central channel to the assembly where all of the N and C termini of the individual units reside.

6408 nucleotide genome of the fd phage will therefore be about 1280, and given the model helical parameters, this will lead to an overall length for the complex of 0.91 microns. This is the same as the contour length of the gene 5 fd–DNA complex measured from micrographs (Gray et al., 1982b). Thus the model helical structure predicted from the contact analysis appears to conform to essentially all of the physical requirements placed on it by investigations based on other techniques.

Almost all of the reasonable combinations of parameters—those having few prohibited contacts and falling close in structure to the optimum helix—relied upon similar sets of nonbonded atomic contacts at the interfaces and, therefore, the prohibited contacts generally involved subsets of the same atoms. These were confined almost entirely to the tip of the β loop formed by residues Glu 40, Tyr 41, and Pro 42, and occasionally the flanking residues, in apposition to residues 66–70 and residues 48–52 of the adjacent dimer. Because the polypeptide segments 48–52 and 66–70 are rather rigidly fixed in the native protein structure while 40–42 are somewhat flexible and mobile (as evidenced by their thermal parameters, shown in Fig. 5), we believe that the tip of the β loop, residues 40–42, may be of essential importance in regulating the cooperative aggregation that occurs upon DNA binding. It appears to have the potential for conformational variation that would be required to relieve the few prohibited contacts that do still remain in the optimal model structure.

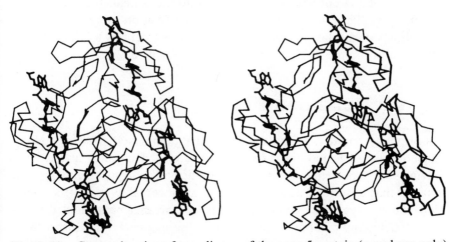

Figure 30. Stereo drawing of two dimers of the gene 5 protein (α-carbons only) with bound pentanucleotides (all atoms) as they would be arranged in a helical configuration based on the optimal parameters derived from the contact analysis. The bound DNA strands have been artificially interrupted every five nucleotides to indicate the juncture point as each strand passes on to the adjacent dimer unit. The protein conformation is the refined native G5BP dimer structure that was used in the contact analysis.

We should point out that Tyr 41 is a residue implicated in direct interaction with DNA (see discussion above regarding DNA binding) and, therefore, would be positionally sensitive to the presence of nucleic acid. Based on considerations independent of this analysis we suggested above that the tip of the β loop, and particularly Tyr 41, do alter their conformation slightly in order to optimally bind DNA, and we have described the modified conformation (see Fig. 24). Indeed, if we assume this minor alteration, which involves stacking of the tyrosine ring on a DNA base, in the present analysis we find that the number of prohibited atomic contacts in the optimal model helix falls to zero.

Thus we propose that in the absence of DNA there exist a few sterically unacceptable atomic contacts that prevent otherwise complementary surfaces from merging and producing aggregate formation. In the process of binding nucleotides the tip of the β loop alters its conformation very slightly in order to optimize interactions between Tyr 41 and a base of the DNA. This, in turn, has the consequence of removing the few blocking interactions, permitting favorable protein–protein interfaces to come into contact concomitant with self-assembly into the helical structure. Hence we envision the 40–41 loop to act as a two-position mechanical switch that allows or disallows cooperative protein interactions to take effect in response to the presence or absence of nucleic acid.

13. THE TOROID MODEL

As previously noted, crystals of the G5BP complexed with oligodeoxynucleotides suggested the existence of a closed circular arrangement having twelve protein monomers. We applied an analysis essentially identical to that described above to deduce an acceptable model for such a dodecamer. Again, we found that only one arrangement yielded a pronounced minimum in our search, and that this aggregate, not surprisingly, closely correlated with a single turn of the helical assembly described above. The arrangement, shown here in Figures 31 and 32, does not form the open turn of a helix but a closed disk structure containing six gene 5 dimers and having point group symmetry 622. Immediately apparent is that although the DNA binding clefts are again aligned, the individual clefts interface in a seriously disjointed manner so that a single DNA strand cannot pass smoothly from one binding site to the next. Even more

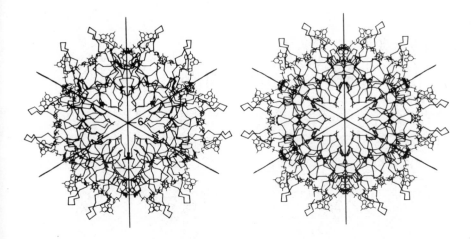

(a)

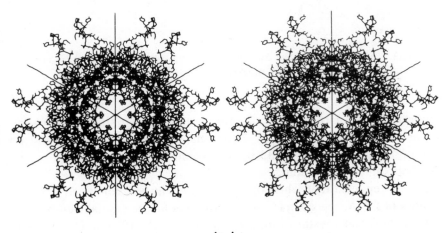

(b)

Figure 31. (a) Stereo drawing of the α-carbon backbone. (b) All atoms of the protein along with all atoms of the oligonucleotides as they appear in the closed toroidal structure derived from the contact analysis. There are six dimers of gene 5 protein in an assembly having 622 point group symmetry. The dyads are provided by the intermolecular twofold axes of the dimers. The DNA binding sites appear continuous across dimer interfaces in these drawings but a reversal of polarity occurs at each junction. Thus the binding of short oligonucleotides would be permitted but not extended polynucleotide chains, consistent with the asymmetric unit of the gene 5 protein oligonucleotide complex crystals.

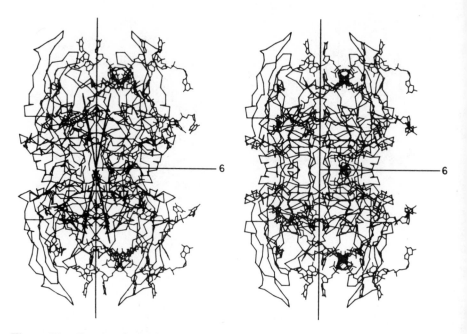

Figure 32.   Stereo view of the toroid complex viewed 90° to that in Figure 31a. The sixfold and dyad axes are indicated.

important, the polarity of the adjacent binding clefts in this toroid assembly are alternately 3'-5' and 5'-3' as one rounds the circle. Hence no continuous DNA strand could possibly bind to the toroid complex, but short oligonucleotides could.

It was reported that when the gene 5 protein was combined with oligomers of DNA, the complexes so formed could be crystallized in a variety of different crystal forms (McPherson et al., 1980b). The capacity to form crystals could not be correlated with oligomer length (up to six) or sequence. The one constant feature of all of these crystals was an asymmetric unit likely composed of six gene 5 protein dimers. The repeated occurrence of this assembly in the complex crystals along with evidence of complex formation induced by oligonucleotides in solution (Rasched and Pohl, 1974) was combined with a number of other findings, including rotation function results, and a model was proposed. This complex structure, whose formation was to be induced by oligonucleotide binding to the protein, consisted of a closed disk of six gene 5 dimers arranged with exact 622 symmetry. The disk was predicted to have a diameter of 100 Å and a height of 90 Å; the top layer of monomers able to bind segments of DNA running 3'-5' and the lower layer segments having opposite polarity (McPherson et al., 1980a).

We believe that the contact analysis employed to deduce the most probable helical complex structure has yielded the structure of the next most favorable arrangement of the protein units as well. This complex, we contend, does in fact occur; it is the aggregate structure formed when oligonucleotides are present rather than linear DNA of greater extent; and it is likely to be the toroid structure that forms the asymmetric unit of the complex crystals.

A particularly satisfying feature of the toroid structure, which in turn reflects the construction of the helix, is the intricate fit and complementarity between adjacent protein units, allowing them to form the densely packed and perfectly contiguous arrangement. This is seen in Figure 33. It is not difficult to imagine how such an array might spontaneously self-assemble if conditions were otherwise favorable or if some small set of requirements were satisfied.

The toroid assembly, like the helical assembly, arises from the cooperative protein–protein interactions that are induced or permitted by the consequences of DNA or oligonucleotide binding. It is therefore consistent with the mechanism suggested above involving a slight conformational change in the β bend containing Tyr 41. A point that should be recognized is that there are two distinctly favorable arrangements of the

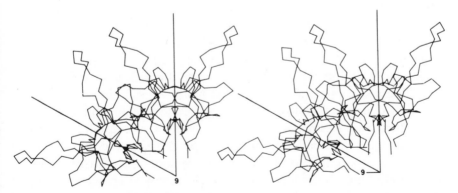

Figure 33. Two G5BP dimers arranged as they would appear in the toroidal aggregate shown in Figures 31 and 32.

gene 5 protein dimers that use very similar but not identical intermolecular contacts. Although the first arrangement, the helix, has clear physiological utility, the purpose of the toroid structure is not immediately apparent and may simply reflect a weakness in specificity of the interacting atomic contact groups. It could also reflect a slight conformational variability in the protein structure that favors, in some instances, one set of contacts over the other depending on the nature of the bound nucleic acid, that is, the oligomer or extended strand. Such variation in aggregate specificity is observed, for example, in the spherical plant viruses where a single type of capsid protein is used to create sixfold as well as fivefold symmetry aggregates (Caspar and Klug, 1962).

14.   RELATIONSHIP OF THE HELIX TO NATIVE CRYSTAL PACKING

An interesting feature of the gene 5 protein–DNA helical complex derived from the contact analysis is its relationship to the arrangement of gene 5 protein molecules found in the C2 monoclinic crystals used for the native structure determination. If the model helical complex seen in Figures 27 and 29 is unwound so that the protein dimers remain a contiguous ribbon, and the helical axis is maintained vertical as in Figure 34, then the angle between the ribbon and the helical axis is 113°. This is rather close to the β angle of the monoclinic crystal, which is about 103°.

A comparison of the contacts between sequential gene 5 dimers in the unwound helix, or ribbon, with the array of gene 5 dimers forming the XZ plane of the monoclinic crystals shows that the relative dispositions of the dimers with respect to one another is nearly the same. The XZ plane of molecules in the monoclinic crystal can almost be generated from the model helix complex by simply aligning the helical axis along the X crystallographic axis and unrolling the helix. As seen in Figure 35, the ribbon of dimers will fall roughly along the Z crystallographic axis with orientations very much the same as is observed in the crystal. Conversely, the helix could be created by rolling up the Z axis of the crystal around the X axis. Again, this result was not a condition explicitly or implicitly incorporated into the model search which employed an analysis only of prohibited atomic contacts. It was a consequence of the optimal helical parameters deduced from the analysis.

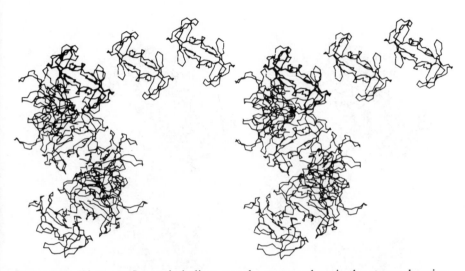

Figure 34. The gene 5 protein helices may be unwound, as in the stereo drawing shown here, to produce a continuous ribbon of dimers. Only the α-carbon chain is shown in this drawing. The angle between the ribbon and the helical axis is 113°.

The implication of this observation is that the gene 5 dimers in the native monoclinic crystal are packed along the Z direction so that in the crystal they display essentially the same interactions with the dimers on either side as they do when a part of the helical complex with single-stranded DNA. The crystal then is, in a sense, composed of unrolled protein helices and uses similar but not quite identical nonbonded contacts to maintain the structure.

This result, in addition to lending some support for the helical complex model presented above, is satisfying from a structural and energetic standpoint as well. Both the crystal and the self-assembling helical complex are minimum-energy structures. One might well expect that an ideal bonding motif designed to fulfill the essentially two-dimensional physiological function of spontaneous helical self-assembly would be incorporated in the formation of a three-dimensional minimum-energy array. Here it does so appear.

We should like to reemphasize that the structure presented here for the gene 5-DNA helical complex is only a model and it has not been visualized in its complete form at atomic resolution. Only the native struc-

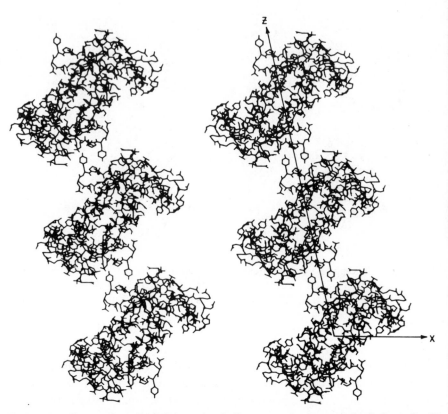

Figure 35.   Stereo drawing of the gene 5 dimers as they are packed along the Z direction in the native monoclinic crystals. This same array may be obtained from the unwound helices shown in Figure 34 by aligning the helical axis along crystallographic X and the ribbon roughly along Z. The intermolecular interactions in both of these two minimum energy structures are quite similar.

ture of the gene 5 protein has been directly determined by X-ray diffraction, hence the model is by no means proven. The model does have one exceptional virtue, however, and that is the extensive variety of methods and techniques that contributed to its creation. Although X-ray diffraction studies of the gene 5 protein and those on DNA provided the underlying structural elements, the interactions and assembly were based on numerous chemical modification studies, NMR, CD, ORD, and UV spectroscopy, low-angle neutron scattering, electron microscopy, and a host of other biochemical and biophysical experiments. Thus, this particular model serves as an especially satisfying synthesis and correlation of work from many laboratories.

REFERENCES

Alberts, B. M., Amodio, F. J., Jenkins, M., Gutmann, E. D., and Ferris, F. L. (1968), *Cold Spring Harbor Symp. Quant. Biol.* **33**, 289–305.

Alberts, B., Frey, L., and Delius, H. (1972). *J. Mol. Biol.* **68**, 139–152.

Alma, N. C. M., Harmsen, B. J. M., Hull, W. E., van der Marel, G., van Boom, J. H., and Hilbers, C. W. (1981). *Biochemistry* **20**, 4419–4428.

Alma-Zeestraten, N. C. M. (1982). PhD. Thesis., University of Nijmegen, The Netherlands, Dept. Biophysical Chemistry.

Anderson, R. A., Nakashima, Y., and Coleman, J. E. (1975a). *Biochemistry* **14**, 907–917.

Anderson, R. A., Nakashima, Y., and Konigsberg, W. (1975b). *Int. Symp. Photobiol.*, Williamsburg, Virginia.

Brayer, G. D., and McPherson, A. (1983a). *J. Mol. Biol.* **169**, 565–596.

Brayer, G. D., and McPherson, A. (1984). *Biochemistry* **23**, 340–349.

Caspar, D. L. D., and Klug, A. (1962). *Cold Spring Harbor Symp. Quant. Biol.* **27**, 1.

Cavalieri, S. J., Neet, K. E., and Goldthwait, D. A. (1976). *J. Mol. Biol.* **102**, 697–711.

Coleman, J. E., Anderson, R. A., Ratcliffe, F., and Armitage, I. M. (1976). *Biochemistry* **15**, 5419–5430.

Coleman, J. E., and Armitage, I. M. (1978). *Biochemistry* **17**, 5038–5045.

Coleman, J. E., and Oakley, J. L. (1980). *Crit. Rev. Biochem.* **7**, 247–389.

Cuyper, T., Van der Ouderaa, F. J., and De Jong, W. W. (1974). *Biochem. Biophys. Res. Commun.* **59**, 557–564.

Day, L. A. (1973). *Biochemistry* **12**, 5329–5339.

Denhardt, D. T. (1975). *Crit. Rev. Microbiol.* **4**, 161–223.

Felsenfeld, G., Sandeen, G., and von Hippel, P. H. (1963). *Proc. Natl. Acad. Sci. USA* **50**, 644–651.

Franklin, R. E., and Gosling, R. G. (1953). *Nature (London)* **17**, 740–741.

Garssen, G. J., Hilbers, C. W., Schoenmakers, J. G. G., and van Boom, J. H. (1977). *Eur. J. Biochem.* **81**, 453–463.

Garssen, G. J., Kaptein, R., Schoenmakers, J. G. G., and Hilbers, C. W. (1978). *Proc. Natl. Acad. Sci. USA* **75**, 5281–5285.

Garssen, G. J., Tesser, G. I., Schoenmakers, J. G. G., and Hilbers, C. W. (1980). *Biochim. Biophys. Acta* **607**, 361–371.

Gray, D. M., Gray, C. W., and Carlson, R. D. (1982a). *Biochemistry* **21**, 2702–2713.

Gray, C. W., Kneale, G. G., Leonard, K. R., Siegrist, H., and Marvin, D. A. (1982b). *Virology* **116**, 40–51.

Hendrickson, W. A., and Konnert, J. H. (1980). *Acta Crystallogr.* **A36**, 344–350.

Herrick, G., and Alberts, B. (1976). *J. Biol. Chem.* **251**, 2133–2141.

Hilbers, C. W., Garssen, G. J., Kaptein, R., Schoenmakers, J. G. G., and van Boom, J. H. (1978). In *Nuclear Magnetic Resonance Spectroscopy in Molecular Biology,* B. Pullman, Ed., D. Reidel, Holland, pp. 351–364.

Konnert, J. H. (1976). *Acta Crystallogr.* **A32**, 614–617.

Lica, L., and Ray, D. S. (1977). *J. Mol. Biol.* **115**, 45–59.

Marvin, D. A., and Hohn, B. (1969). *Bacteriol. Rev.* **33**, 172–204.

Mazur, B. J., and Model, P. (1973). *J. Mol. Biol.* **78**, 285–300.

Mazur, B. J., and Zinder, N. D. (1975). *Virology* **68**, 490–502.

McPherson, A. (1976). *J. Biol. Chem.* **251**, 6300–6303.

McPherson, A., Jurnak, F. A., Wang, A. H. J., Kolpak, F., Rich, A., Molineux, I., and Fitzgerald, P. M. D. (1980a). *Biophysical J.* **10**, 155–173.

McPherson, A., Molineux, I., and Rich, A. (1976). *J. Mol. Biol.* **106**, 1077–1081.

McPherson, A., Wang, A. H. J., Jurnak, F. A., Molineux, I., Kolpak, F., and Rich, A. (1980b). *J. Biol. Chem.* **255**, 3174–3177.

Nakashima, Y., Dunker, A. K., Marvin, D. A., and Konigsberg, W. (1974). *FEBS Lett.* **43**, 125.

O'Connor, T. P., and Coleman, J. E. (1982). *Biochemistry* **21**, 848–854.

O'Connor, T. P., and Coleman, J. E. (1983). *Biochemistry* **22**, 3375–3380.

Oey, J. L., and Knippers, R. (1972). *J. Mol. Biol.* **68**, 125–138.

Paradiso, P. R., and Konigsberg, W. (1982) *J. Biol. Chem.* **257**, 1462–1467.

Pratt, D., Laws, P., and Griffith, J. (1974). *J. Mol. Biol.* **82**, 425–439.

Pretorius, H. T., Klein, M., and Day, L. A. (1975). *J. Biol. Chem.* **250**, 9262–9269.

Rasched, I., and Pohl, F. M. (1974). *FEBS Lett.* **46**, 115–118.

Ray, D. S. (1977). *Compr. Virol.* **7**, 105–178.

Salstrom, J. S., and Pratt, D. (1971). *J. Mol. Biol.* **61**, 489–501.

Schaller, H., Beck, E., and Takanami, M. (1978). In *The Single-Stranded DNA Phages,* Cold Spring Harbor Laboratories, Cold Spring Harbor, New York, pp. 139–163.

Thiessen, W. E., and Levy, H. A. (1973). *J. Appl. Crystallogr.* **6**, 309–346.

Torbet, J., Gray, D. M., Gray, C. W., Marvin, D. A., and Siegrist, H. (1981). *J. Mol. Biol.* **146**, 305–320.

Venkatachalam, C. M. (1968). *Biopolymers* **6**, 1425–1436.

9

Structure of Bovine Pancreatic Ribonuclease by X-Ray and Neutron Diffraction

ALEXANDER WLODAWER
National Measurement Laboratory
National Bureau of Standards
Washington, DC

and

Laboratory of Molecular Biology
National Institute of Arthritis, Diabetes, and
Digestive and Kidney Diseases
Bethesda, Maryland

CONTENTS

1. INTRODUCTION

Bovine pancreatic ribonuclease (EC 3.1.27.5) has been an object of intensive physical-chemical studies for well over 50 years. Three properties of this enzyme contributed to its popularity. First, it could be easily and inexpensively isolated in large quantities. More than a kilogram of purified protein became available to researchers in the early 1940s, providing, for the first time, an opportunity for a number of laboratories to study a pure, uniform, and well-characterized enzyme. Second, ribonuclease was successfully crystallized (Kunitz, 1940) and the crystals diffracted to 2 Å resolution even under air-dried conditions (Fankuchen, 1941). Seven different crystal forms were grown initially (King et al., 1956), and more became available later. Third, Richards and Vithayathil (1959) found that, when the native enzyme (RNase A) was subjected to limited proteolysis using subtilisin, the resulting RNase S consisted of two protein chains which could be reversibly dissociated. The complex of two polypeptide fragments was fully active but the separate chains were inactive. This property was crucial for a variety of later experiments.

For these and other reasons, RNase has been utilized in a number of pioneering experiments. It was the first enzyme and only the second protein whose sequence was determined (Smyth et al., 1963; Potts et al., 1962). It provided the first NMR spectrum of a protein (Saunders et al., 1957). It is a standard protein in the investigation of hydrogen exchange (Schreier and Baldwin, 1976; Rosa and Richards, 1979, 1981). And it was the fourth protein and the third enzyme (after myoglobin, lysozyme, and carboxypeptidase) whose structure was solved by X-ray diffraction. Initial reports came from three groups. The structure of RNase A was solved at 5.5 Å resolution by Avey et al. (1967) and at 2.0 Å by Kartha et al. (1967). RNase S structure was reported at 6 Å and 3.5 Å (Wyckoff et al., 1967b, 1967c). The latter work was later extended to 2.0 Å (Wyckoff et al., 1970), while the 2.5 Å structure of RNase A was presented by Carlisle et al. (1974). Altogether, at least 11 laboratories published crystallographic studies on the structure of RNase and of its various modifications and complexes.

The chemical and structural properties of RNase have been extensively reviewed by Richards and Wyckoff (1971) and more recently by Blackburn and Moore (1982). This review, of a more limited scope, discusses our current knowledge of both the structure and the mechanism of action of this enzyme, based primarily on the data obtained by X-ray and neutron diffraction of single crystals. Discussion of structural properties of RNase elucidated by other techniques will be rather fragmentary. Moreover, we will not consider the properties of bacterial ribonucleases (which were subjects of recent structural investigations) because these were shown to be structurally unrelated to the class of pancreatic ribonucleases (Mauguen et al., 1982; Heinemann and Saenger, 1982; Nakamura et al., 1982).

2. PHYSICAL PROPERTIES AND THE PHYSIOLOGICAL ROLE OF RNASE

The reaction catalyzed by pancreatic RNase is shown in Figure 1. The substrate for the enzyme is single-stranded RNA, and RNase is very specific in its requirement for a pyrimidine base on the 3 ' side of the phosphodiester linkage to be hydrolyzed. Either a pyrimidine or a purine base can be present on the 5' side. In the first step of the catalytic reaction, the phosphate initially involved in a 3'-5' bridge is converted to a pentacoordinated intermediate and then to a 2'-3' cyclic monophosphate. In the second step, the cyclic phosphate is hydrolyzed to a 3' monophosphate ester, again via a pentacoordinated intermediate. Although the pathway for catalysis has been well characterized for more than 20 years, the details of the relationship between structure and function are not yet unambiguous, as is discussed later in this review.

Little is known concerning the physiological significance of the enzyme. The physiological role of pancreatic RNase was initially considered to be quite limited. The abundance of RNase in the digestive tract of ruminants was explained as being caused by the need to reutilize the components of bacterial RNA rather than to digest the dietary RNA (Barnard, 1969). This opinion may have been oversimplified, especially in view of the presence of related enzymes in many tissues (Blackburn and Moore, 1982). Nevertheless, RNase is remarkably well characterized for an enzyme that does not appear to be physiologically crucial, but this is a legacy of its historical utilization as a "standard" enzyme.

The physical properties of bovine pancreatic RNase have been summarized by previous reviewers (Richards and Wyckoff, 1971; Blackburn and Moore, 1982). Briefly, bovine RNase A consists of a single polypeptide chain composed of 124 amino acids, with a molecular weight (cal-

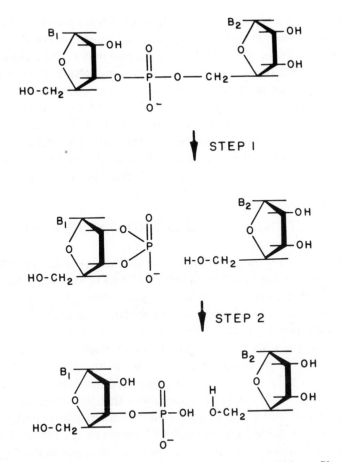

Figure 1. The chemical reaction catalyzed by pancreatic RNase. Site B_1 (left) is always occupied by a pyrimidine, while either a pyrimidine or purine may occupy site B_2 (right).

culated on the basis of the sequence) of 13,680. Partial specific volume of the protein is 0.703 ml/g, the radius of gyration is 18.3 Å, and the sedimentation coefficient is 1.78 S. The protein is basic, with an isoionic point of 9.6 (Richards and Wyckoff, 1971).

RNase A has been subjected to a variety of proteolytic and chemical modifications in order to elucidate the structure-function relationship. The most important result was the discovery of RNase S, a product of the cleavage of the peptide bond between residues 20 and 21 by subtilisin

(Richards and Vithayathil, 1959). Although RNase S is less stable than the parent enzyme (as indicated by its lower transition point for thermal unfolding), its enzymatic activity as measured under identical assay conditions is undiminished. The ease of separating RNase S into two completely inactive components (S peptide consisting of amino acids 1–20 and S protein encompassing the remainder of the molecule) and the fact that the residues belonging to the active site are distributed between the two polypeptides make this enzyme system extremely valuable in monitoring the catalytic activity of semisynthetic modifications in which individual amino acids have been replaced (Blackburn and Moore, 1982).

3.　STRUCTURAL WORK ON RIBONUCLEASE

The structure of the native ribonuclease (RNase A) and of its proteolytic modification (RNase S) has been solved at high resolution (2 Å or better) in a number of laboratories. In addition, glycosylated ribonuclease (RNase B) has also been crystallized, and its structure is under investigation. Although the initial reports of the structure of RNase A and S appeared almost simultaneously in 1967, the atomic coordinates of only the latter modification were published (Wyckoff et al., 1970; Richards and Wyckoff, 1973), and for almost a decade, they provided the only structural model of the enzyme. Although it was anticipated that the structures of RNase A and S would be very similar, a rigorous comparison confirming that assumption was made only recently (Wlodawer et al., 1982a). Nevertheless, we will discuss the results concerning the structure of the three forms of RNase separately, starting with the structure known the longest, namely RNase S.

3.1.　The Structure of RNase S

Several crystal forms of RNase S have been grown, and two have been studied in detail. The original structure was obtained using type Y crystals. These crystals were grown by dissolving the protein in 6 M CsCl, 0.1 M acetate buffer, pH 6.1, to the concentration of 75 mg/ml. Ammonium sulfate (80% saturated) was used as a precipitant (Wyckoff et al., 1967b). Form Y crystals belong to the space group P3$_1$2$_1$ and have unit cell parameters $a = b = 44.5$ Å, $c = 97.3$ Å. Each of the six asymmetric units contains a single molecule of RNase. Crystals used for data collection were usually transferred in stages to 75% saturated ammonium sulfate, pH 5.5. The cesium chloride was washed out in order to reduce the background caused by the scattering of the heavy atoms. The data used in the original studies were collected on a four-circle diffractometer, with crystals mounted in flow cells (Wyckoff et al., 1967a). In such cells the crystals are immobilized in a plastic or glass capillary with cotton fibers, and it is possible for the solution to flow in a continuous manner. The use of flow cells facilitated the search for heavy atom derivatives and made the studies of the binding involving substrate and product analogs easier.

The structure of the native RNase S has been reported at 6 Å, 3.5 Å, and 2.0 Å resolution (Wyckoff et al., 1967b, 1967c, 1970). Isomorphous derivatives have been prepared using uranyl acetate, tetracyanoplatinate, and dichloroethylenediaminoplatinate(II). All compounds except the latter produced multiple sites. The figure of merit was 0.97 at 10 Å, 0.8 at 3 Å, and 0.6 at 2 Å. The rms error in electron density was estimated as 0.23 electrons/$Å^3$ for the 2 Å map. The map was easily interpretable, and a list of atomic coordinates was provided. These coordinates were later refined by Fletterick and Wyckoff (1975), using a real-space algorithm. This data set (designated 6D) was widely distributed, both in printed form (Richards and Wyckoff, 1973) and by the Protein Data Bank (Bernstein et al., 1977), on computer-readable media. The model was further refined using an automated difference Fourier method (Powers, 1976). The resulting structure (designated 7B) had acceptable rms deviations of bond lengths from ideality (0.042 Å) but had large deviations of atomic positions from planarity (0.074 Å). The crystallographic R factor was 0.31. Further refinement of that model using the restrained least-squares procedure of Hendrickson and Konnert (1981) lowered the R factor to 0.233 and improved the geometry (Taylor et al., 1981), but the resulting coordinates were not published. A more careful refinement using least-squares methods is under way (Wyckoff, personal communication).

Crystals of form Z of RNase S belong to the monoclinic space group C2, $a = 101.5$ Å, $b = 32.0$ Å, $c = 69.6$ Å, $\beta = 90.8°$. These crystals were grown from 65% saturated ammonium sulfate at pH = 7.2, and before collection of the data, they were transferred to the same standard buffer that was used for form Y crystals. In form Z, each asymmetric unit contains two molecules of ribonuclease. The structure was solved using multiple isomorphous replacement, first at 6 Å resolution (Mitsui and

Wyckoff, 1975) and later at 4 Å (Torii et al., 1978). They reported that the structure of the two independent molecules (ZA and ZB) was very similar, although their environment was quite different due to crystal packing.

As noted previously, RNase S can be reversibly dissociated into S peptide (residues 1–20) and S protein (21–124). Pandin et al. (1976) reported that both the reconstituted RNase S and a semisynthetic RNase S, in which the native S protein was associated with a synthetic S peptide (obtained by solid phase methods), could be crystallized. The crystals were isomorphous with the Y form of RNase S and could not be distinguished by either their appearance (Fig. 2) or diffraction pattern. Other semisynthetic forms of RNase S were also crystallized. Taylor et al. (1981) reported the structure of an inactive analog containing a synthetic S peptide with residues 1–16 only, in which 4-fluoro-*L*-histidine replaced His 12. The structure was virtually identical with that of the native RNase S (since residues 17–20 are very poorly delineated in the native enzyme), and a peak corresponding to the position of fluorine was used to assign absolute orientation to His 12. Another semisynthetic RNase S studied by X-ray diffraction contained ornithine instead of Arg 10 (Valle et al., 1977).

RNase S can easily form complexes with mono- and dinucleotides. Such complexes are usually prepared by soaking native crystals in suitable media previous to X-ray data collection or by using flow cells. A number of complexes have been studied by crystallographic techniques. These include RNase complexes with mononucleotides 3'-AMP, 5'-AMP, (3'-5')-cyclic AMP, ATP, and 3'-UMP (Richards et al., 1971); with substrate analogs UpcA (Richards and Wyckoff, 1973), 2'-F-UpA (Pavlovsky et al., 1978), C(2'-5')A (Wodak et al., 1977); and with other nucleotides such as A(3'-5')C (Mitsui et al., 1978), 4-thiouridylic acid (Torii et al., 1978), and pTp (Iwahashi et al., 1981). These structures are discussed in Section 5.5.

3.2. The Structure of RNase A

All studies of the structure of RNase A utilized monoclinic crystals belonging to the space group $P2_1$ (King et al., 1956). The crystals used in different studies were grown from ethanol, methanol, MPD, or *t*-butanol. Crystals could also be grown from fully deuterated RNase A (Bello and

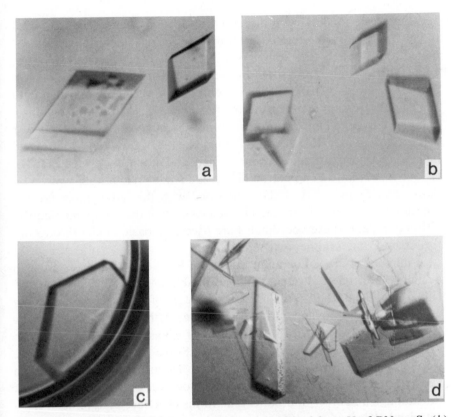

Figure 2. Crystals of RNase. (*a*) Trigonal crystals of form Y of RNase S. (*b*) Crystals of semisynthetic RNase S', in which the S peptide was prepared by solid state technique, but had normal sequence (Pandin et al., 1976). (*c*) Large crystal (length 8 mm) of the monoclinic form of RNase A, soaked in a deuterated solution in preparation for neutron experiments. (*d*) Orthorhombic crystals of RNase B, grown by Rubin et al. (1982).

Harker, 1961). Although the reported unit cell parameters differ by up to a few percent, it is obvious that all of these crystals are very similar. The unit cell parameters for the crystals grown from *t*-butanol are $a = 30.18$ Å, $b = 38.4$ Å, $c = 53.32$ Å, $\beta = 105.85°$, and each asymmetric unit contains one protein molecule.

The main chain tracing for RNase A was first published by Avey et al. (1967), on the basis of 5.5 Å data, and by Kartha et al. (1967), who used data extending to 2.0 Å. The reasons for the serious discrepancies between these models were discussed by Kartha (1967). The folding of the main chain as determined by Kartha was very similar to that determined for RNase S (the comparison was made by Dickerson and Geis, 1969). More details of the structure were provided by Carlisle et al. (1974) on the basis of 2.5 Å data, but the coordinates were not published until Wlodawer (1980) completed the refinement at 2.5 Å resolution. Subsequently, the room temperature structure was refined at 2.0 Å (Wlodawer et al., 1982a) and at 1.45 Å (Borkakoti et al., 1982). Low-temperature and high-temperature structures were refined by Gilbert and Petsko (1984). RNase A was used in a neutron diffraction study (Wlodawer, 1980; Wlodawer and Sjölin, 1981, 1983). Published with the latter paper are the coordinates resulting from a joint refinement with the X-ray and neutron diffraction data; these coordinates are used as a basic model throughout this chapter unless otherwise indicated.

A number of crystallographic studies of RNase A complexes are also available. Kartha et al. (1968) studied the binding of phosphate ion and Gilbert and Petsko (1984) reported the structure of RNase complexed with a transition state analog, uridine vanadate, with d(CpA), 2'-UMP, 3'-UMP, and 2'-3'-cyclic CMP. Complexes of RNase A with 2'-CMP, with cytidine-N(3)-oxide 2'-phosphate and with 8-oxo-guanosine 2'-phosphate were studied by Borkakoti (1983a, 1983b). Neutron studies of a uridine vanadate complex have been reported (Wlodawer et al., 1983).

Although a number of other crystal forms of RNase A have been characterized (King et al., 1956), no structural studies utilizing them have been published. The form which appears to be the most interesting was grown at high pH by Martin et al. (1976). This form of RNase A is isomorphous with the W form of RNase S (space group $P3_121$, $a = b = c = 64.4$ Å). A comparison of these two modifications having identical crystal lattice contacts could be of interest, but the results have not been published yet. Another interesting form of RNase available in the W crystal form, that of the semisynthetic complex of residues 1 to 118 and 111 to 124, was reported by Sasaki et al. (1979). Doscher et al. (1983) crystallized a similar, catalytically defective complex in which Phe 120 was replaced by leucine.

3.3. The Structure of RNase B

RNase B differs from RNase A by being glycosylated at a single site (Asp 34; Plummer et al., 1968). Monoclinic crystals have been grown by Brayer and McPherson (1982). They belong to space group $P2_1$, $a = 51.4$ Å, $b = 75.6$ Å, $c = 31.0$ Å, $\beta = 108°$, with two molecules in each asymmetric unit. A different crystal form was grown under virtually identical conditions by Rubin et al. (1982). These crystals are orthorhombic (space group $P2_12_12_1$, $a = 59.2$ Å, $b = 56.0$ Å, $c = 81.0$ Å), with two molecules in each asymmetric unit. The investigation of the structure is being continued by both groups.

Brayer and McPherson (1982) have also reported crystallization of complexes of RNase B with tetra- and hexadeoxyphosphoadenosine. The investigation of the structure of these crystals, which are isomorphous between themselves but not with the native enzyme, is under way.

4. EXPERIMENTAL TECHNIQUES

Studies of the structure of ribonuclease have led to the development of a number of tools, which were also found useful in the investigation of other proteins. These included innovations in crystal mounting, in methods of integrating reflection intensities, and in the approach to preparing heavy atom derivatives and nucleotide complexes. RNase A was also one of the first proteins to be studied by neutron diffraction, and we briefly discuss some relevant properties of this technique.

4.1. Crystal Mounting

A large fraction of the data discussed in this chapter was obtained using RNase crystals mounted in flow cells. This method of crystal mounting was introduced by Wyckoff et al. (1967a) and was subsequently modified (Wyckoff et al., 1970). Briefly, the crystal is placed in a polyethylene tube or in a quartz capillary open on both ends and is immobilized by surrounding cotton fibers or Sephadex. Tubes with a smaller diameter are inserted below and above the crystal, and the whole assembly is attached to a brass yoke, which in turn is inserted into a standard goniometer head. One end of the inner tubing is placed into a reservoir containing the soaking solution, while the other empties into a waste container. The rate of flow can be controlled by changing the height of the reservoir. Almost all of the data collected on the Y form of RNase S crystals were collected using the flow cells described above and a modified yokeless cell has been used to collect the diffraction data for RNase A (Gilbert and Petsko, 1984).

Recently it was shown that mounting crystals in the mother liquor retards radiation damage caused by an X-ray beam (Narayana et al., 1982). This protection may have been important in the studies of RNase.

A different challenge in crystal mounting was presented by the requirements of the neutron data collection. The length of time required for data collection (several months) necessitated the prevention of any crystal movement and even the slowest evaporation of the mother liquor. In addition, due to the conflicting requirements imposed by the crystal morphology and the optimization of measurements, crystals had to be placed with their shortest dimension parallel to the axis of the tube. These problems have been solved by Wlodawer (1980) by placing a crystal on the flat bottom of a quartz tube, surrounding it with mother liquor, and immobilizing it with quartz wool. The tube was packed with a thick layer of silicone grease, sealed with dental wax, and attached to a brass pin with epoxy glue, which provided the final vapor barrier. A crystal mounted in this manner showed no changes in the diffraction intensities for more than two years of intermittent data collection.

4.2. Low-Temperature and High-Temperature Crystallography

One of the objectives of the extensive studies of the interaction of RNase A with substrate, product, and transition state analogs was to investigate the details of binding under conditions most similar to the physiological ones. Substrate–enzyme complexes can often be formed and studied at cryogenic temperatures, when the activity of the enzyme is negligibly low. It is therefore possible to observe the interactions with real substrate rather than with the analogs. Gilbert and Petsko (1984) employed a device described by Marsh and Petsko (1973) to study RNase A crystals in temperatures ranging from $-32°C$ to $+47°C$. The elevated temperatures were found useful in the investigation of the flexibility of the enzyme (Gilbert et al., 1982).

4.3. Integration Of Diffraction Intensities

An efficient way of obtaining integrated intensities was applied in the data collection for RNase S. This limited step-scan technique was introduced by Wyckoff et al. (1967a) in order to speed up the process of data collection and to minimize the background. The data are collected by measuring a limited number of points in the vicinity of the peak of a reflection at small intervals of the most sensitive parameter (usually ω). A fraction of these points, approximately 50%, is ignored, while the sum of the remaining ones is taken as the count. This procedure guarantees that a larger fraction of the total measurement time is spent on the most intense part of the peak, unlike the case where the scan covers the entire peak. Moreover, by summing the highest contiguous integrated intensity, it is possible to compensate for the errors in the determination of the orientation matrix. The scan range can also be extended if the peak is found at the limit. The backgrounds can be measured as either a function of 2θ only or as a function of 2θ and of other diffractometer angles (Krieger et al., 1974).

Another problem in the integration of diffraction data was encountered in the course of investigations of RNase by neutron diffraction. The background, primarily due to the incoherent scattering of neutrons by hydrogen atoms, is relatively much higher than the background encountered in X-ray diffraction experiments. As a consequence, it is difficult to differentiate between the signal and noise for the weaker reflections, which comprise a large fraction of a total data set. A technique enabling precise estimation of the extent of reflections was introduced by Sjölin and Wlodawer (1981), specifically for processing RNase data. Although this "dynamic mask" procedure was developed primarily for a diffractometer equipped with a linear detector, it appeared to be applicable for more general use. It has since been applied to the output of four-circle diffractometers, to precession and oscillation films, and to processing the data collected with a neutron area detector (Wlodawer and Sjölin, 1982a).

Reflections to be integrated by the dynamic mask procedure are considered to reside in boxes large enough to contain the whole peak and possibly parts of some other reflections as well. Boxes can be linear, two-dimensional or three-dimensional, depending on the actual technique of data collection. In the case of a linear detector or films, reflections reside in two-dimensional boxes. Because noise may prevent the direct determination of the extent of reflections, it is necessary to apply a smoothing algorithm to suppress the noise and emphasize the actual data. This can be done, because the reflection is contiguous while the noise is random. An actual smoothing algorithm depends on the experimental details, such as the number of pixels contained in a reflection box and the average count per pixel. In the case of two-dimensional data (Sjölin and Wlodawer,

1981), each point was averaged with its eight neighbors, using different weights for different classes of points. The initial estimate of the background and its variance was obtained either from the intensities of points unlikely to contain any parts of the peak or on the basis of a universal background curve, measured during data collection. Once the data had been smoothed, a "statistical filter" was applied, so the data belonging to the peak and to the background could be distinguished. The distribution of intensities corresponding to the tentative background area could be checked for points clearly exceeding the average value, because such points could belong to other reflections. This procedure would then be cycled until convergence was reached. Actual integration was performed by summing up the points within the peak boundary, disregarding, by then, the smoothed data. Reflections too weak to indicate their own masks were integrated by using averaged masks calculated for a number of medium reflections in the same area of reciprocal space. This procedure led to a significant improvement of the neutron diffraction data compared with what could be obtained with simple box integration methods.

4.4. Neutron Diffraction

Application of the technique of neutron diffraction to studies of single crystals of protein has recently been reviewed by Wlodawer (1982). This technique is complementary to the X-ray diffraction and can yield information not readily obtainable by other means. This includes the possibility of differentiating N from C or O atoms, which in turn makes it possible to fix the orientation of histidine, glutamine, and asparagine side chains (Fig. 3). This feature is of particular importance in the studies of RNase, because two histidines are involved in its enzymatic activity. It is also possible to distinguish between exchangeable and nonexchangeable amide hydrogens. With neutron diffraction studies of RNase A, the kinetic hydrogen exchange studies of Schreier and Baldwin (1976) and Rosa and Richards (1979, 1981) were complemented. Hydrogen bonds can be observed directly and need not be inferred from the positions of other atoms. Solvent structure is more visible than with X-ray methods.

Conditions for growth of large crystals of RNase A were established by Norvell (1976, unpublished results), and the procedure was summarized by Wlodawer (1980). Crystals were grown by batch technique using 43% t-butanol as a precipitant and as much as 100 mg of protein for a

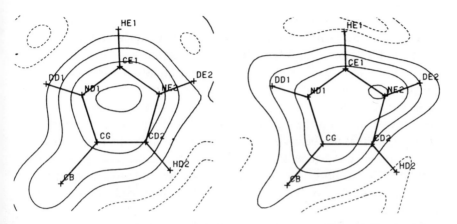

Figure 3. Comparison of X-ray (left) and neutron (right) difference Fourier maps for a side chain of histidine. Both are "fragment $F_0 - F_c$" maps, calculated after subtracting the contribution of the atoms belonging to His 12 side chain from the calculated structure factors. The coordinates are taken from a model based on joint X-ray and neutron refinement of RNase A (Wlodawer and Sjölin, 1983). The orientation of His 12 can be determined unambiguously only on the basis of the neutron map, because the electron density is symmetrical.

single experiment. The largest crystals reached a volume of 100 mm^3; the crystals used for data collection were between 25 and 35 mm^3. Crystals were soaked in fully deuterated synthetic mother liquor for about six months prior to the collection of the data, in order to remove practically all exchangeable hydrogens. Although the principal aim of this procedure was to lower the background caused by the incoherent scattering of neutrons by hydrogen atoms, information about amide hydrogen exchange became available as a by-product. Neutron diffraction data were collected using the flat-cone diffractometer at the National Bureau of Standards reactor (Prince et al., 1978). The neutrons were monochromatized by reflection from a 50 × 100 mm graphite crystal, and the chosen wavelength was 1.68 Å. Diffraction data extending to 2Å resolution were collected for the native RNase A and for a complex with uridine vanadate.

4.5. Structure Refinement

Several refinement techniques have been utilized in the investigation of the structure of native RNase and of its complexes with nucleotides. The coordinates of RNase S presented by Richards and Wyckoff (1973) and deposited in the Protein Data Bank as the set 6D resulted from a refinement by Fletterick and Wyckoff (1975). The authors searched for the steepest gradient in the fit between the coordinates and the multiple isomorphous replacement (m.i.r.) electron density map. The real space method can improve the fit of the coordinates to the map but does not impose restraints on the geometry of the resulting model. A more complete refinement has been achieved by Powers (1976); he used four cycles of automated difference Fourier, with intermediate cycles of nonrigid structural constraints. Although this technique leads to a clear improvement in the quality of the model, it is not as powerful as modern least-squares methods.

An energy refinement program has been used in the investigation of the structure of cytidylyl-(2'-5')-adenosine(C_2,P_5,A) when bound to RNase S (Wodak et al., 1977). The protein was assumed to be in its native conformation, except for residues 41, 45, 67, 69, 71, and 119, which were allowed to vary together with the dinucleotide. The refinement minimized the conformatinal energy of the protein-inhibitor complex, with only nonbonded and Coulombic energy terms being included in the calculations. This procedure improved the interpretation of a rather noisy difference Fourier map. A different method of refining the structure of a dinucleotide was used by Pavlovsky et al. (1978) in their investigation of 2'-F-UpA bound to RNase S. The atomic coordinates of the dinucleotide were refined by searching for the best fit to electron density. The refinement included optimization of orientational and positional parameters for 2'-F-UpA as a whole as well as the conformational angles of rotation around interatomic bonds. Changes in conformation were constrained by standard values of interatomic distances and bond angles, with the state of ribose rings determined by a sequential trial of different possible standard conformations.

All other refinement efforts on RNase involved restrained least-squares refinement algorithms. A program system developed by Hendrickson and Konnert (1981) was used by Taylor et al. (1981) to refine the native structure of RNase S, by Wlodawer et al. (1982a) to provide the first highly refined structure of RNase A, by Gilbert and Petsko (1984) in the investigation of RNase A at low temperatures and in the presence of nucleotides, and by Wlodawer and Sjölin (1983) for the joint refinement of RNase with X-ray and neutron data.

A joint refinement procedure simultaneously utilizing both X-ray and neutron data, with both types of diffraction data entered into the same refinement matrix, was developed by Wlodawer and Hendrickson (1982). The program used for this purpose was a modification of the standard least-squares refinement algorithm of Hendrickson and Konnert (1981). The major tasks in implementing the joint refinement procedure involved the incorporation of hydrogen and deuterium atoms, the introduction of special classes of nonbonded contacts for H atoms involved in hydrogen bonds, the calculation of structure factors and their derivatives based on the appropriate scattering factors, and the inclusion of a separate scale factor refinement. Because the X-ray and neutron data were entered into the refinement with separate weights, it was easy to change the relative importance of separate classes of observations. The main advantage of the joint refinement procedure is that it fits a single model to both types of observations, unlike separate neutron refinement methods, which usually lead to serious departures of the neutron models from their X-ray counterparts.

The occupancies of amide hydrogens in RNase were calculated using the joint refinement program with only the neutron diffraction data, after constraining the coordinates and temperature factors for all atoms to the starting values. Although the occupancies of all atoms were allowed to vary in an unrestrained manner, the standard deviations in the occupancies of amide hydrogens were twice as large as those for nonexchangeable atoms (0.44 compared with 0.21, Wlodawer and Sjölin, 1982b). Standard deviations for atoms other than amide hydrogens were used to estimate the error in the determination of occupancies, which was about 15%. In the final run, a maximum occupancy of 1.0 was enforced for all atoms other than amide hydrogens, for which the occupancies were not limited. Initially all amide hydrogens were assigned as deuterium with an occupancy of 1.0; in the ideal case, the occupancies of nonexchangeable hydrogens should become -0.55 (based on the magnitude and the sign of scattering length of H and D). As a result of such refinement, amide hydrogens were assigned as protected (less than 20% D), partially protected (20–60% D), or unprotected (over 60% D).

A different least-squares refinement algorithm was utilized by Borkakoti et al. (1982) in their refinement of RNase A at 1.45 Å resolution. Although the details of the computational procedure differed from the ones described above, the general principles involving the utilization of stereochemical information in the refinement remained the same. A comparison of the structure of RNase A resulting from both refinement programs is under way (Wlodawer, Borkakoti, and Moss, unpublished).

5. RESULTS

5.1. The Sequence of RNase

The sequence of RNase has been independently determined by groups at Rockefeller University (Smyth et al., 1963, and references therein) and at the National Institutes of Health (Potts et al., 1962, and references therein). It is shown in Figure 4. A large number of sequences of RNases from different species have subsequently been published, and it is obvious that a considerable part of the sequence is conserved throughout the evolutionary development. It seems probable, therefore, that the three-dimensional structure of RNase is also preserved during evolution. Because the ribonucleases from other species have not been studied by diffraction techniques, no direct proof is available yet. (A number of other enzymes with similar activity have now been studied by X-ray diffraction, but they are quite distant from the bovine pancreatic RNase). The sequence itself has no unusual features, and all amino acids other than tryptophan are present. Of the four histidines found in the bovine enzyme, three (12, 48, and 119) are strictly conserved; His 105 is not. Other conserved amino acids include Lys 41, which is implicated in the enzymatic activity of the enzyme. All eight half-cystines present in RNase (forming disulfide bonds between residues 26–84, 40–95, 65–72, and 58–110) are found in the enzyme from all sources.

5.2. Secondary Structure

The secondary structure of RNase A has been discussed in considerable detail by Wlodawer et al. (1982a), Wlodawer and Sjölin (1983), and Borkakoti et al. (1982, 1983). Both groups are in good agreement on the features of the hydrogen bonded network involving main-chain atoms. A large part of the RNase molecule consists of regular structural elements such as α helices and β-pleated sheets, with an extensive network of hydrogen bonds stabilizing the molecule. The diagram of such interactions

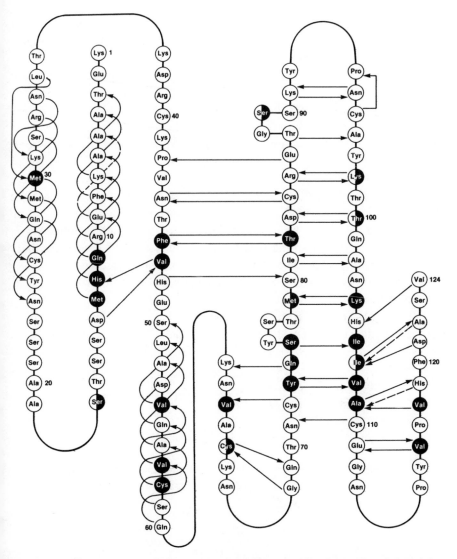

Figure 4. The sequence of RNase A and the diagram of hydrogen bonds involving the atoms in the main chain. Bonds shorter than 3.15 Å are solid, those between 3.15 Å and 3.35 Å and with expected angles are dashed. Filled circles correspond to the amide hydrogens observed in neutron experiments as fully protected, while half-filled circles denote partially protected amides (Wlodawer and Sjölin, 1983).

is shown in Figure 4, and the only differences between this model and that of Borkakoti et al. (1982) are the lack of a hydrogen bond between 31N-27O in the helix H1, between 61N-74O and 72N-63O in the β sheet, and an extra bond between 94N and 91O. These differences do not influence the identical interpretation of the secondary structure and are possibly artifacts of the refinement. The assignment of the bonds in RNase S is virtually identical with that in RNase A (Richards and Wyckoff, 1973).

Three helical regions can be found in a molecule of RNase A, as shown in Figure 5. If a helix is defined as extending from the first hydrogen bond donor to the last acceptor, the helices consist of residues 3 to 13 (helix H1), 24 to 34 (helix H2), and 50 to 60 (helix H3). The torsion angles for the residues involved in the helix H1 are close to the ideal values for the α helix throughout its length. Helix H2 is a distorted α helix, with residue 34 forming a hydrogen bond characteristic for a 3_{10} helix. Helix H3 is even more distorted than the first two. Residues 50 to 55 form a partially distorted α helix, while 56 to 60 progressively distort toward a 3_{10} helix.

A large part of the RNase molecule consists of β structure. The backbone of the molecule consists of a pair of antiparallel β strands of residues 71–92 and 94–110. Each of the two strands is partly twisted, and each can be best described as a letter V, with residues 77 and 104 at the apices.

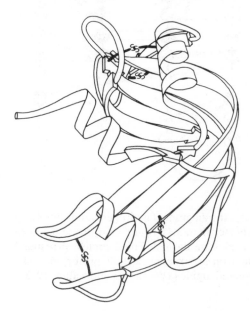

Figure 5. Schematic diagram of the backbone of RNase A, showing the elements of the secondary structure and the disulfide bonds (drawn by Jane Richardson).

Residues 76–78 and 88–89 bulge out and do not participate in the hydrogen bonding characteristic for the β structure. In each part of the V, another shorter β strand completes the sheet. These additional strands consist of residues 41–48 and 118–123, the latter considerably distorted. Residues 61–65 form a short fourth strand on one side of the V. All of the β strands are antiparallel.

5.3. Tertiary Structure and Intramolecular Interactions

The agreement on the features of the tertiary structure of RNase is also very good between different models. The geometry of bends in RNase A has been discussed in detail by Borkakoti et al. (1983). The bends present in the structure have been analyzed as postulated by Lewis, Momany, and Scheraga (1973). Of the eight bends identified by the authors, five belong to type IV (15–18, 16–19, 36–39, 65–68, 75–78), one to type III (87–90), and two to type VI (91–94 and 112–115). The geometry of some of these bends is at variance with their counterparts in RNase S, but this may be due to the incomplete refinement of the structure of RNase S rather than to real differences between the two forms of the enzyme.

Four proline residues are present in ribonuclease. It has been agreed by all investigators that two of them (Pro 42 and 117) are in *trans* configuration, while the other two (Pro 93 and 114) are *cis*. Such *cis* proline configurations have been found in a number of proteins and are not considered unusual. Both *cis* prolines are involved in sharp turns between β regions.

Four disulfide bonds are present in each molecule of RNase. Three of these bridges (26–84, 40–95, 58–110) are left-handed ($\chi_3 \cong -90°$) and one (65–72) is right-handed ($\chi_3 \cong 90°$), being part of a very tight local loop containing only seven residues. The location of the disulfide bonds in the structure can be seen in Figure 5.

A complete list of hydrogen bonds involving both the main-chain and side-chain atoms was published by Wlodawer and Sjölin (1983). The information provided by Borkakoti et al. (1983) is in substantial agreement concerning the number of bonds made by each residue, but not all of the interactions have been listed. The descriptions of the bound solvent, however, are somewhat different. Even though the number of assigned water sites is virtually identical (128 by Wlodawer and Sjölin, 1983, and 123 by Borkakoti et al., 1984), only 58 of them were found in similar locations (distance less than 1 Å). It is not yet certain whether the discrepancy is due to true differences between the structures (different solvent, deuteration, etc.), or whether it was caused by artifacts in the refinement. The problem is now under investigation (Wlodawer, Borkakoti, and Moss, unpublished).

The number of hydrogen bonds between the neighboring molecules in the crystal is small. Only four direct contacts between the peptide atoms can be found. The terminal amino group of Lys 1 makes a bond to the carbonyl oxygen of Lys 37, NH of Ser 23 bonds to O of Ser 15, NH1 of Arg 39 to OE1 of Glu 111, and NH of Thr 70 to O of Asp 38. Eight water molecules are involved in intermolecular bridges, and two more bridges involve more than one water molecule. The small number of intermolecular hydrogen bonded interactions may be responsible for the considerable enzymatic activity of the crystalline protein. It may also be an important factor in the interpretation of the results of hydrogen exchange, because it is unlikely that so few interactions could seriously modify the flexibility and accessibility to solvent of different parts of the secondary structure of RNase.

Two crystallographically accessible phenomena can be used to monitor the flexibility of the polypeptide chain in RNase. Both Wlodawer et al. (1982b) and Borkakoti et al. (1984) have plotted the averaged temperature factors for side chain and main chain atoms of each residue. A strong correlation can be found between the temperature factors and the secondary structure, with the residues involved in the helical and β sheet hydrogen bonding being less flexible. The temperature factors by themselves are not, however, a sufficient measure of chain flexibility, since high-temperature factors may also be caused by static disorder and it is sometimes difficult to distinguish whether they are due to local vibrations or rather to concerted motions of a whole region. Hydrogen exchange, on the other hand, can provide information about long-range flexibility in different regions of the protein as well as about solvent accessibility. RNase was indeed one of the first proteins subjected to such studies using neutron diffraction (Wlodawer and Sjölin, 1982b). The advantage of neutron diffraction investigation over the chemical or NMR methods lies in the high level of confidence that can be placed in assigning individual amide hydrogens to be either protected or exchanged. However, neutron diffraction does not provide any direct kinetic information about the course of the exchange process.

In view of the relatively large errors in the estimates of individual occupancies, the protected amides were divided into two classes. Nineteen amides were found to be fully protected (more than 80% H) and eight were partially protected (40–80% H). The protected amides are marked in Figure 4, which also shows the correlation between the position of the protected amides within the peptide chain and between the elements of secondary structure. A vast majority (23 out of 27) of the protected amide hydrogens form hydrogen bonds to main-chain oxygen atoms in either helical or β sheet secondary structures. However, participation in such bonds is not a guarantee of protection, because almost two-thirds of the hydrogens involved in the main-chain hydrogen bonding are fully exchanged. Of the remaining protected hydrogens, one is involved in an unambiguous hydrogen bond to the side-chain group and one forms a bond to a carbonyl oxygen in irregular secondary structure.

The distribution of protected amides in the RNase structure is not uniform (Fig. 6). Of two distinct areas surrounding the cleft near the active site, one appears to be much more flexible than the other. In particular, the part of the β sheet containing residues 63, 65, 73–75, 79, 104, 106–109, 116, and 118 appears to be more protected from exchange than any

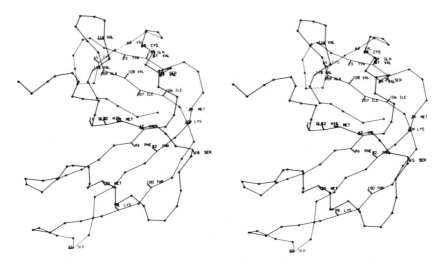

Figure 6. Stereo diagram showing positions of all amide nitrogens in the RNase molecule as well as those amide hydrogens that were found to be at least partly protected against exchange (Wlodawer and Sjölin, 1983).

other region in the protein. Only the central part of the β sheet found in the other half of the molecule contains protected amides (46, 82, 98, and 100). Each of the helices contains at least one protected amide, but three amides in a row are protected only in the first helix. These amides (11–13) are located at the carbonyl end of the helix, near the active site cleft. At this stage of investigation, the reasons for the differences in the protection of some of the amides found in rather similar environments are not understood.

5.4. Active Site of RNase

The active site of RNase is formed by the side chains of His 12, Lys 41, Val 43, Asn 44, Thr 45, His 119, Phe 120, Asp 121, and Ser 123 (Richards and Wyckoff, 1973). The relevant part of the molecule is shown in Figures 7 and 8. Of these residues, His 12 and 119 are directly responsible for the catalytic activity of the enzyme, and Lys 41 is probably involved also.

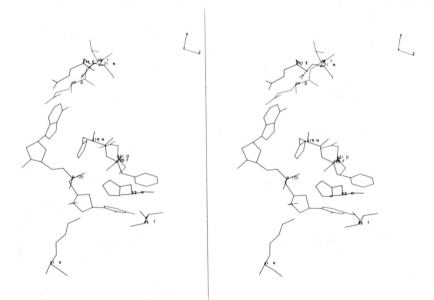

Figure 7. Active site of RNase A. Diagram shows selected parts of a polypeptide chain of RNase and of the substrate analog d(CpA) bound in the active site. The coordinates were refined by Gilbert and Petsko (1984) using X-ray diffraction data extending to the resolution of 1.5 Å.

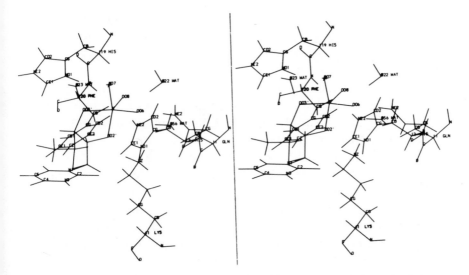

Figure 8. Uridine vanadate interacting with the selected residues that form the active site of RNase (Gln 11, His 12, Lys 41, His 119, and Phe 120). This complex is similar to the substrate in transition state (Wlodawer et al., 1983).

The role of the other side chains mentioned above is limited to assisting in the binding and orienting of the substrate. Wyckoff and Richards (1973) labeled the parts of the active site R_1, B_1, p_1, R_2, and B_2. Site p_1 was unambiguously assigned the position occupied by the inorganic phosphate (or sulfate) in the native structure. The exact environment of the other sites has not been described in much detail, but the model of binding of the substrate analog, d(CpA) (Fig. 7) clearly shows the interactions between the protein and the substrate. The site B_1 is always occupied by a pyrimidine (with its ribose in the site R_1), while the site B_2 can be occupied by either a pyrimidine or a purine. A base occupying the site B_1 makes hydrogen bonds to OG1 and N of Thr 45, as well as van der Waals contacts to the side chain of Phe 120. The ribose at R_1 rests against Lys 41 and Phe 120 but is not hydrogen bonded to the protein. The environment of the site B_2 includes the side chains of Glu 111, Glu 69, Asn 71, and Ala 109, with hydrogen bonds made to NE2 of Gln 69 and OE1 of Glu 111. The ribose at R_2 is adjacent to His 119 but makes no hydrogen bonds to the protein.

The details of the interactions between the phosphate at p_1 and the histidines 12 and 119 were not clearly determined until quite recently.

The reason is that the electron density for a histidine side chain is symmetric and the assignment of nitrogen and carbon atoms in the imidazole ring is indirect (Fig. 3). In addition, four positions of His 119 were postulated in the structure of RNase S (Richards and Wyckoff, 1973), even though the coordinates for only one of them (site IV) were listed. Site I of His 119 corresponded to the position of the sulfate and was probably in error, while sites II and III might have also been artifacts. In the refined structure of RNase A (Wlodawer et al., 1982b), the position of His 119 was very clear and probably corresponded to site IV in RNase S. The absolute orientations of His 12 and His 119 were determined in a neutron investigation (Wlodawer and Sjölin, 1981, 1983). It was shown that the nitrogen NE2 of His 12 forms a hydrogen bond to the oxygen O1 of the phosphate, while the nitrogen ND1 of His 119 binds to the O3'. Both OE1 and Glu 11 and NZ of Lys 41 seem to be within hydrogen bonded distances of O2. The orientation of His 12 was further confirmed in a study of a semisynthetic RNase S, in which His 12 was substituted by 4-F-His (Taylor et al., 1981). The position of the fluorine atom was visible in a difference Fourier map and was in agreement with the results of a neutron study. Borkakoti et al. (1982) observed partial disorder of His 119, with site IV occupied 80% while another site was 20% occupied. In the absence of equivalent data on RNase S, it is impossible to determine if the alternate site corresponds to sites II or III or whether it is entirely different. The question of disorder of His 119 will have to be resolved in the future, in view of the conflicting data presented in different publications.

Another important residue presenting the investigators with many problems was Lys 41. The density for its side chain was not seen in the maps for the native RNase S, and for a while it was assumed that Lys 41 is subjected to very large shifts upon nucleotide binding. Wlodawer and Sjölin (1983) located a large peak in the vicinity of the phosphate in the neutron map of RNase A and refined the position of Lys 41 on the assumption that the peak corresponded to the terminal ND_3 group. Following the refinement, both X-ray and neutron maps were quite clear and confirmed that Lys 41 was indeed bound to the O4 of the phosphate even in the native RNase. The temperature factor was, however, quite high ($B = 20 \text{ Å}^2$), which indicated that Lys 41 was very flexible. Similar temperature factors are also observed in the presence of substrates and product analogs, with the notable exception of the transition state analog uridine vanadate. When uridine vanadate is bound to RNase, the average temperature factor for Lys 41 is lowered to 8 $Å^2$ (Wlodawer et al., 1983). It forms a stronger hydrogen bond to O2′ of the cyclic vanadate than to the equivalent oxygen of the substrate (or to the inorganic phosphate). The stabilization of Lys 41 in the transition state complex reinforces the assumption that this residue plays a direct role in the catalytic mechanism.

5.5. Interaction with Inhibitors and Other Molecules

A large number of complexes of RNase A and S with substrate, product, and transition state analogs have been studied crystallographically. Some of the complexes were studied at high resolution, and the atomic coordinates have been published. In other investigations, the resolution employed was much lower and the positions of the ligands have been found in only very general terms. Figure 9 shows a number of mono- and dinucleotides bound to RNase and summarizes the results of a number of investigations. Some comments about the preparation of this figure are in order. The α-carbon backbone of RNase and the side chains of His 12, His 119, and of Lys 41 were taken from the model based on joint X-ray–neutron refinement by Wlodawer and Sjölin (1983). For the nucleotide complexes with RNase S, the coordinates were rotated into the RNase A cell as described by Wlodawer et al. (1982a). Such a procedure may lead to some loss of precision in the presentation, but it facilitates the preservation of the uniformity of views. Because the complexes with RNase S were not refined using least-squares procedures, the details of the interactions with the protein were not established with high accuracy anyway. Figure 7, on the other hand, shows the details of the interaction between the analog d(CpA) and RNase A, and in that case, all coordinates were taken from the highly refined X-ray structure of the complex, preserving the orientation which was used to generate Figure 9.

A number of dinucleotides have been investigated that closely resemble the substrate but for some reason cannot be cleaved by RNase. One of the first to be studied was UpcA, an analog to the usual dinucleoside phosphate UpA which differs in that the 5′-oxygen atom of the ribose attached to the adenine base has been replaced with a methylene group. The phosphorus–carbon bond cannot be cleaved by RNase, but the stereochemistry in the vicinity of the phosphorus atom has not been drastically altered. The structure of this complex was studied at 2 Å resolution by Carlson (1976) and was discussed by Richards and Wyckoff (1973) and

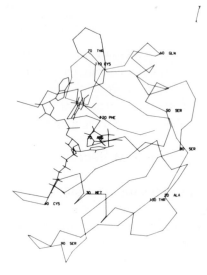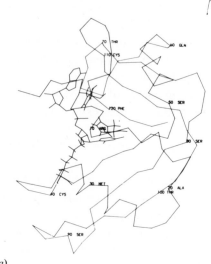

(a)

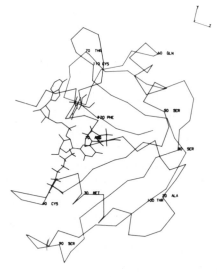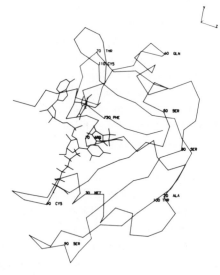

(c)

Figure 9. Structures of substrate, transition state, and product analogs. The atomic coordinates of these compounds are superimposed on the α-carbon tracing of the protein main chain taken from the structure refined by Wlodawer and Sjölin (1983). (a) Unrefined coordinates of a dinucleotide UpcA (Richards and Wyckoff, 1973). (b) Partially refined coordinates of 2′-F-UpA; note *syn* conformation of the adenine base (Pavlovsky et al., 1978). (c) Partially refined coordinates of

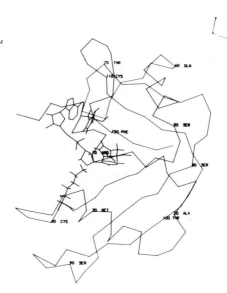

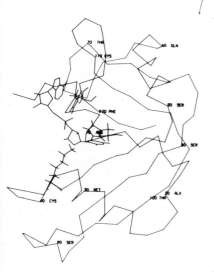

(b)

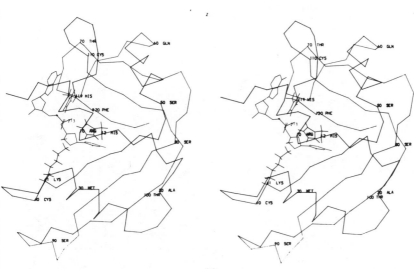

(d)

C(2′-5′)A (Wodak et al., 1977). (d) Highly refined coordinates of d(CpA). On the following pages: (e) The coordinates of 2′-UMP. (f) The coordinates of 3′-UMP. (g) The coordinates of (2′-3′)-cyclic CMP. Models d–g are based on the refinements reported by Gilbert and Petsko (1984). (h) The coordinates of uridine vanadate, a transition state analog (Wlodawer et al., 1983).

421

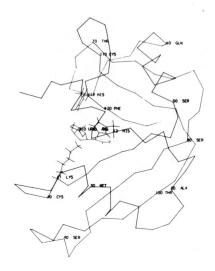
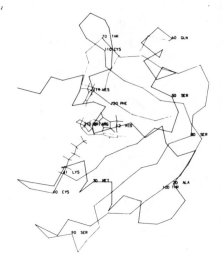

(e)

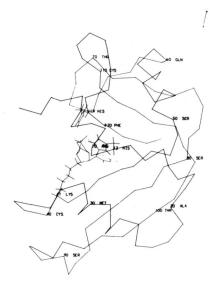
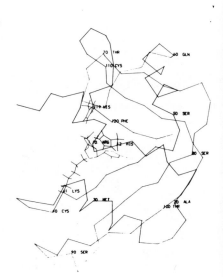

(g)

Figure 9. (Continued)

422

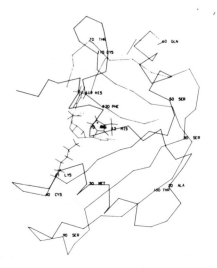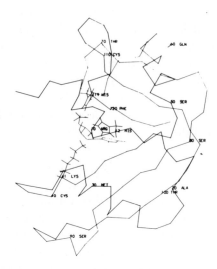

(f)

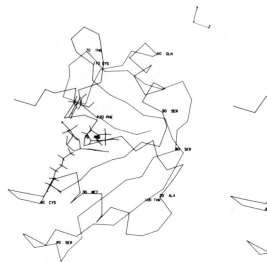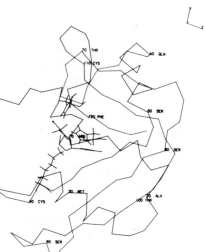

(h)

Figure 9. (Continued)

423

Wyckoff et al. (1977). The coordinates of the unrefined structure were published by Richards and Wyckoff (1973) and were used in drawing Figure 9a. It can be seen that the dinucleotide is highly extended, with the tips of the bases 14 Å apart. The sugar-base conformation is *anti* for both nucleotides. The details of the interactions with the protein are similar to those previously discussed for d(CpA). Richards et al. (1970) and Wyckoff et al. (1977) have suggested that Thr 45 is an ambivalent hydrogen bond donor-acceptor, facilitating the binding of both U and C at the site B_1.

The structure of a complex with another very similar substrate analog, 2'-F-UpA, has been studied at 2.5 Å resolution by Pavlovsky et al. (1978). The properties of this dinuleotide (2'-deoxy-2'-fluoro uridylyl-(3'-5')-adenosine) were expected to be even closer to those of the actual substrate than in the case of UpcA, because the substitution of 2'-OH by 2'-F produces no significant changes in the conformational state of the ribose. The atomic coordinates of 2'-F-UpA were refined by searching for the best fit of the atomic model to the electron density calculated for a difference Fourier map, with the conformation of the ribose rings determined by sequential trials of different possible standard conformations. The studies were performed at pH 5.5 and 7.2 respectively, with no apparent differences. The results (Fig. 9b) are in good agreement with those for UpcA, except that the sugar-base conformation for the adenosine is *syn* rather than *anti*. This result is at variance with other studies and may indicate an error in the map fitting. If the difference in the conformation were real, it would complicate the interpretation of the relationship of the analogs and the real substrates because the conformation of the analogs are very similar to the actual substrates.

Wodak et al. (1977) published the coordinates of $C_{2,P5,}A$ bound to RNase S (Fig. 9c). These coordinates were refined by an energy minimization procedure using 2 Å data and are in general agreement with the results obtained using UpcA, except in the vicinity of the phosphorus, where the 2' linkage introduces some differences in stereochemistry. The binding of $C_{2,P5,}A$ affected the positions of Lys 41, Thr 45, Gln 69, Asn 71, and His 119 in the enzyme.

Gilbert and Petsko (1984) refined the structure of RNase A with d(CpA) bound in the active site, using 1.5 Å data collected at $-32°C$. This study represents the best refined structure of a dinucleotide with RNase and is shown in some detail in Figures 7 and 9d. The interactions between the dinucleotide and the enzyme were discussed previously.

Other dinucleotides with appreciably different binding properties were studied in complexes with the Z form of RNase S. Mitsui et al. (1978) studied low resolution difference Fourier maps of ApC and found an indication of a binding of cytosine at the B_1 and R_1 sites of both molecules forming the asymmetric unit. A site for the adenosine was detected only in one of the RNase molecules. It was assumed that the other adenosine was disordered and that the binding of the first one was facilitated by the crystal packing. This experiment confirmed the high specificity of the B_1 site for the pyrimidine nucleotides.

The structures of a large number of mononucleotides complexed to RNase have been studied. As expected, mononucleotides such as 3'-CMP and 3'-UMP can be found occupying the B_1 and R_1 sites, while 3'-AMP and 5'-AMP were located in sites B_2 and R_2 (Richards and Wyckoff, 1971). Iwahashi et al. (1981) investigated a complex of thymidine-(3'-5')-diphosphate (pTp) with RNase S (Z form) and concluded that they found evidence for a specific binding site, p_0, for the 5'-phosphate group. The structures of 2'-UMP and 3'-UMP bound to RNase A (Gilbert and Petsko, 1984) are shown in Figures 9e and 9f, respectively.

A particularly interesting complex is prepared by reacting RNase with uridine vanadate, a strong inhibitor. Lindquist et al. (1973) postulated that the vanadium atom should be pentacoordinated in such a complex and resemble the phosphorus in the transition state of the cyclization. Alber et al. (1983) studied the structure of this complex using X-ray diffraction and Wlodawer et al. (1983) utilized both X-ray and neutron 2 Å diffraction data in their investigations. The preliminary results of the latter studies are shown in Figures 8 and 9h. It is clear that the O3' lies in the basal plane, while the O2' is in the apical position, as postulated by Lindquist et al. (1973). Details of interaction with the protein differ, however, from the predictions. Alber et al. (1983) placed the ε-amino group of Lys 41 in the vicinity of an equatorial oxygen of the intermediate; this was disputed by Wlodawer et al. (1983). In the combined X-ray–neutron refinement of the structure of an RNase complex with uridine vanadate, these authors observed that NZ of Lys 41 was only 2.79 Å from the 2'-oxygen of the ribose and that the distance to the nearest basal oxygen was 3.38 Å. At the same time, NE2 of His 12 was observed at a distance of 2.62 Å from one of the equatorial oxygens, with an indication of the presence of a proton between these two atoms. This proton appeared to be bound to the imidazole. The other basal oxygen was making a hydrogen bond with an amino group of Gln 11. The distance from NE2 of His 12 to the 2′-oxygen was 3.22 Å, but the angle was incorrect for a postulated hydrogen bond. Nitrogen ND1 of His 119, on the other hand, was found in its expected position 3.03 Å from the apical oxygen opposite O2′. This oxygen was predicted by Lindquist et al. (1973) to represent a coordinated water molecule, but it was found to be very similar in appearance to the other oxygens, with no sign of the deuterium atoms in the neutron map. The oxygen was bonded, however, to a water molecule. The implications of this observed geometry remain unclear, and it may be necessary to reinterpret the existing models to fit the new observations.

Allewell et al. (1973) investigated the structure of an ε-41 dinitrophenyl RNase S, a covalently modified inactive form of the enzyme. The binding of DNP does not obstruct the active site, thus confirming the importance of a free Lys 41 for the catalysis. The binding of 3′-pyrimidine nucleotides was found to be unaffected by DNP.

Allewell and Wyckoff (1971) studied the binding of Cu(II) to RNase S. The intramolecular copper binding sites occurred near His 119, His 105, and Glu 86 and near the sulfate bound at the active site. In addition, several intermolecular sites were also identified.

The nature of the ligand bound in the active site of the native RNase is controversial. This ligand was uniformly assumed to be sulfate in the studies of RNase S, particularly because the crystals were grown from solutions of ammonium sulfate (Richards and Wyckoff, 1971). Kartha et al. (1968) studied the structure of the crystals of RNase A grown in the presence of either phosphate or arsenate, with a clear density seen in a difference Fourier map calculated using these two data sets. Nevertheless, no chemical analyses were performed to verify that the ligand was indeed a phosphate. Similarly, Wlodawer et al. (1982a) have assumed that a large peak found in the active site of RNase A corresponds to a phosphate ion, without any independent confirmation of this interpretation. Gilbert et al. (1982), on the other hand, unambiguously identified the ligand as sulfate using laser Raman spectroscopy. In that case, the crystals were grown from deionized water, without any buffer, so the sulfate must have remained bound throughout the protein purification procedure. Similar results were reported by Borkakoti et al. (1982), who ruled out the presence of a phosphate using the energy dispersive analysis by X-rays (EDAX) and who also used laser Raman spectroscopy to identify the sulfate. Whether the sulfate is still bound even if a phosphate buffer is used in crystallization is not known. In either case, the presence of a large anion in the active site of native RNase always caused difficulties in the interpretation of difference Fourier maps for the nucleotides, because no density could be observed at site p_1.

5.6. Critical Evaluation of the Accuracy of Results

The results of structural studies described in this review can be broadly divided into three classes. The structure of the Z form of RNase S has been solved at a resolution of 4 Å and was not refined. As a consequence, nucleotide binding studies can only indicate general features of the binding sites but cannot be used for making detailed descriptions of the protein–substrate interactions. The structure of the Y form of RNase S is known in more detail because 2.0 Å data were used and partial refinement was completed by Powers (1976). The R factor for that refinement was 0.31, which indicates that the atomic positions were determined with an estimated error of about 0.35 Å. Much larger errors were probably present in some parts of the structure, particularly on the surface. It can also be assumed that the position of Lys 41 may have been in error, and the exact orientation of His 119 may be questionable. Least-squares refinement of Powers's model by Taylor et al. (1981) led to an immediate reduction of the R factor to 0.23, indicating that, by and large, the RNase S model was close to being correct.

Wlodawer et al. (1982a) estimated by the method of Luzzati (1952) that the coordinates of RNase A were determined with an r.m.s. error of 0.17 Å for the refined X-ray structure at 2.0 Å resolution. Nevertheless, several side chains (including Lys 41) had to be completely repositioned during the subsequent joint X-ray–neutron refinement (Wlodawer and Sjölin, 1983). It should also be pointed out that, at an intermediate step of the X-ray refinement ($R = 0.20$), a stretch of four residues was shifted out of register, indicating that even the structures characterized by comparatively low R values can contain very serious errors. The structure of the solvent, in particular, is probably still subject to considerable modifications, as shown by the lack of agreement on the positioning of the majority of water molecules between the structures refined by Wlodawer and Sjölin (1983) and by Borkakoti et al. (1984). Even for the best refined structure of RNase, the R factor is still considerably higher than 0.12–0.13 obtained for several proteases (Fujinaga et al., 1982), indicating the presence of uncorrected errors. The results of high resolution studies of the complexes of RNase with inhibitors (Gilbert and Petsko, 1984) are probably accurate in the vicinity of the active site (an area to which the authors paid most of their attention), but considerable errors may still be present elsewhere ($R > 0.22$). The degree of hydrogen–deuterium exchange found in the neutron investigations was probably accurate to about 15% (Wlodawer and Sjölin, 1982b).

The results of studies on the binding of inhibitors to RNase clearly show the limitations inherent in the interpretation of unrefined or partly refined X-ray structures, as well as the exaggerated faith placed in such structures by noncrystallographers. A typical example is provided by the interpretation of the coordinates of the RNase–UpcA complex, published by Richards and Wyckoff (1973) with clear warnings about the lack of their refinement. These coordinates were subsequently used by Holmes et al. (1978) as a starting point of a computer simulation of RNase action, with little regard being paid to rather large errors expected in the published structure. It is not surprising that the recalculated structure of minimum energy deviated seriously from the X-ray model. Similarly, after Pavlovsky et al. (1978) published the results of their investigation of the 2'-F-UpA complex, these results were interpreted by Arus et al. (1981) as a proof of the existence of a hydrogen bond between His 12 and the 2'-oxygen of the ribose. This was a serious overinterpretation, due to the uncertainties in the partially refined X-ray model and in particular, to the previously discussed ambiguity in assignment of the orientation of imidazole side chains on the basis of X-ray data. A mechanism devised without any reference to the X-ray structure can be even less convincing (Rübsamen et al., 1974). Imprecise knowledge of the details of the interaction of the substrate with the catalytically active groups in the enzyme may also present an impediment in the design of artificial, nonprotein enzymes (Breslow, 1982). In general, the interpretation of difference Fourier maps points to the locations of the substrate, but the nature of the specific bonding between the substrate and the enzyme, such as the details of bond length and angles, should be treated with some skepticism. Such information is, however, usually trustworthy if based on well refined high resolution structures, such as the structures of native RNase A and of the inhibitor complexes described in Section 5.5.

6. MECHANISM OF ACTION

The mechanism of action of RNase has been under investigation for more than twenty years, but some questions about the details are still unresolved. Most of the results of these studies were summarized by Richards and Wyckoff in 1971. Little new information appeared until very recently, as evidenced by the review of Blackburn and Moore (1982). The model currently finding the widest acceptance is the in-line mechanism of Usher et al. (1970, 1972). Other models discussed by Richards and Wyckoff (1971) are mostly of historical interest and are not reviewed here.

The first step of the reaction catalyzed by RNase is a transphosphorylation of the 3'-5' linkage to give an oligonucleotide terminating in a pyrimidine (2'-3')-cyclic phosphate. The 2'-OH group on the 3'-ribose becomes at least partially deprotonated, making it a nucleophile. During or after deprotonation, the phosphate group moves towards the 2'-oxygen and forms a trigonal bipyramid transition state or intermediate. Usher et al. (1972) have conclusively shown that this step proceeds by an in-line mechanism. This was accomplished by using RNase to catalyze a reverse reaction, the synthesis of the dinucleoside phosphorothioate Up(S)C from the crystalline isomer of uridine-(2'-3')-cyclic phosphorothioate and cytidine. The stereochemistry of the groups around the phosphate in the product was fixed by the choice of isomer and by the geometry of reaction. In the in-line mechanism, cytidine would attach to the phosphorus on the side opposite to the 2'-oxygen, while the adjacent mechanism would involve the reaction on the side of the 2'-oxygen, followed by pseudorotation. The product, dinucleoside phosphate, was subsequently treated with base to reform cyclic phosphorothioate by a reaction known to be in-line. If the mechanism of the enzymatic reaction were adjacent, the product of this reaction would have to be the stereoisomer opposite to the original substrate. If the mechanism of RNase catalysis were in-line, both the original substrate and the end product would have to be identical. Because the identity of the product of enzymatic reaction and of the substrate of base catalysis was experimentally confirmed, the in-line mechanism was proved for the reaction under study. By the principle of microscopic reversibility, the geometry of the reverse reaction must be the same as the geometry of the forward reaction, so that the conclusions drawn on the basis of this experiment were valid for either direction.

The breaking of the bond between the phosphate and the 5'-oxygen concludes the first step of the catalysis and results in a cyclic-(2'-3')-phosphate, which in turn becomes a substrate in the second step. The cyclic phosphate is hydrolyzed to produce a 3'-nucleotide. Because a nucleotide is obviously not equivalent to a water molecule, it is conceivable that the two steps could follow different pathways. The geometry of the second step was studied by Usher et al. (1970). In their investigation, an isomer of the previously mentioned uridine-(2'-3')-cyclic phosphorothioate was hydrolyzed at pH 7 by RNase in water enriched in ^{18}O, and the resulting monoester was recycled by treatment with diethyl phosphorochloridate, yielding a mixture of both isomers of the starting compound. Because the nonenzymatic ring closure is known to be in-line, it was predicted that the in-line mechanism of the ring opening reaction would lead to the recovery of the original stereoisomer lacking ^{18}O and the opposite isomer would contain one equivalent of ^{18}O from the water. This was indeed found experimentally, and thus it is now accepted that the mechanism of both catalytic steps is in-line. When the mechanism of action (in-line vs. adjacent) is discussed, it is assumed that there exists a transition state (or intermediate) in which the phosphorus is pentacoordinated, with the 2'-oxygen occupying the apical position. This is necessitated by the preference rules, which indicate that the attacking and leaving groups must be apical, as well as by the geometric considerations, which place 3'-oxygen in a basal position in order to relieve the strain on the five-membered ring. Much more strain would be present if both oxygens were basal and were involved in making a trigonal structure. Lindquist et al. (1973) noted that stable compounds of vanadium(IV) and vanadium(V) are often pentacoordinated and suggested that uridine vanadate, a potent inhibitor of RNase, might closely resemble the substrate portion of the transition state. This prediction was confirmed by both the X-ray (Gilbert and Petsko, 1984) and neutron (Wlodawer et al., 1983) studies of the RNase A-uridine vanadate complex.

Although Usher et al. (1970, 1972) have elucidated the in-line mechanism using synthetic chemistry, they were careful not to implicate any particular groups in the enzyme as the actual base and acid required for catalysis. Instead, such assignments relied mostly on NMR and diffraction data for their elucidation. A very elegant model was presented by Roberts et al. (1969), and these geometrical arrangements were later extensively quoted (Lindquist et al., 1973; Deakyne and Allen, 1979; Umeyama et al., 1979). In this model, His 12 is located in the vicinity of the 2′-oxygen of the pyrimidine ribose and His 119 is adjacent to the 5′-oxygen of the other base. It was postulated that Lys 41 formed a salt bridge to a free oxygen on the phosphorus. In the initial stage of the reaction, His 12 in the base form would abstract a proton from the 2′-oxygen, which would become a nucleophile and attack the phosphorus, forming a pentacoordinated intermediate. Each of the nonesterified oxygens of the intermediate has a full formal charge, thus strengthening the interaction with His 119 and Lys 41. Because the leaving (5′) group is apical, it can leave directly, without any need for pseudorotation. It receives the proton from His 119, which acts as a general acid. The exact mechanism of this proton transfer was not formulated for lack of data, and it is uncertain even today.

The second step of the reaction, the hydrolysis of the cyclic phosphate, was predicted to be an exact reversal of the first step. His 119 in its base form would abstract a proton from a water molecule, which would then attack the phosphorus opposite to the 2′-oxygen. His 12, acting as an acid, would then protonate the 2′-oxygen, thus converting the cyclic nucleotide to a 3′-form.

Although other possible geometries of the active site and alternate mechanisms have been considered, none was found to be as plausible as the one described above. The geometry described here was utilized by Breslow (1982) in his synthesis of an artificial enzyme that would catalyze the cleavage of RNA. Such a molecule was indeed synthesized, and it was determined that it was capable of producing a scission in RNA every 20 minutes or so in a 20 mM solution of the compound.

The role of histidines in the catalytic mechanism of RNase is comparatively straightforward. The same is not true for Lys 41. Initially it was in question whether this side chain was necessary at all, but this has now been resolved. Alber et al. (1983) discussed the remarkable change in the flexibility of Lys 41 in the presence and absence of the transition state analog, uridine vanadate. The temperature factor of its side chain is about 20 Å^2 in the free enzyme and in the complexes with the substrate and product analogs, but it is reduced to 8 Å^2 in the presence of the transition state analog (Wlodawer et al., 1983). Furthermore, the temperature-dependence of the temperature factors indicated that they are due to real thermal motions and not to static disorder. It was believed that the rigid Lys 41 stabilizes the transition state, whereas the flexible side chain allows its formation (as well as decomposition). A rigid catalyst would not be able to bind both a tetrahedral substrate (and product) and a trigonal bipyramidal intermediate with the same efficiency, or it would hinder rearrangement occurring during the reaction.

The main difficulty in accepting the mechanism postulated above derives from the observation of Wlodawer et al. (1983) that the positions of His 12 and Lys 41 required for the efficient transfer of protons seem to be inverted. It is still possible that the role of the residues is the same, but the simple scheme of how the protons are transferred will probably have to be modified, and more theoretical studies of the subject are needed.

7. COMPARISONS WITH THE RESULTS OBTAINED WITH OTHER METHODS

The results described in this review have been obtained primarily in X-ray and neutron diffraction experiments but many other techniques have also been utilized in the studies of the structure of RNase. The results obtained using NMR spectroscopy are of particular interest, because this technique can yield information about the changes of the structure caused by pH titrations or the interactions with substrate analogs, thus enhancing the static picture emerging from diffraction experiments. NMR results have been extensively reviewed by Karpeisky and Yakovlev (1981), by Blackburn and Moore (1982), and by Cohen et al. (1983). Many NMR experiments have been performed in order to investigate the resonances of the four histidines present in RNase and, in particular, the resonances assigned to CE1 protons (Meadows et al., 1967). Each of the resonances titrated with a specific pH value and exhibited distinct properties. However, correlation of these resonances with specific residues presented some difficulties. In particular, the original assignments of the resonances corresponding to His 12 and His 119 (Meadows et al., 1968) had to be reversed later (Markley, 1975; Shindo et al., 1976; Karpeisky and Yakovlev, 1981). In the final assignment, the rates of deuterium (or tritium) exchange observed for these resonances corresponded to the degree of accessibility indicated by the structure based on diffraction data, namely His 105 > His 119 > His 12 > His 48, and to the ease of carboxymethylation of the active site histidines, namely His 119 > His 12 (Cohen et al., 1983). Cohen and Shindo (1975) have found that the minor inflections observed in the titration curves of His 12 and His 119 at low pH became more prominent with an increase in the concentration of the inorganic phosphate. These results were interpreted as an indication of a direct interaction between the phosphate and both histidines even before conclusive proof became available in the diffraction work. Jentoft et al. (1981) found that the resonance due to methylated Lys 41 is perturbed by a nearby histidine, which they assigned to be His 12. They postulated that these two residues must interact, and this conclusion was later confirmed by Wlodawer et al. (1983). Their suggestion that Lys 41 may also interact with the 2'-oxygen of the ribose and that it may aid in its deprotonation is also in agreement with the latest results of the diffraction studies of the complex of RNase A with uridine vanadate.

Antonov et al. (1978) studied the binding of 2'-F-UpA to RNase, using both proton and ^{31}P NMR. On the basis of the nuclear Overhauser effect, they have concluded that both bases of the dinucleotide had to be in *anti* conformation. This result was contradictory to the one obtained by X-ray diffraction and published by the same group (Pavlovsky et al., 1978) but was in good agreement with other diffraction results, as discussed earlier in this review.

Cohen and Wlodawer (1982) discussed how some results of the NMR studies of RNase were overinterpreted in order to show direct interaction between His 12 and His 119; no such interaction was subsequently observed in any of the refined structures. Similarly, NMR failed until recently to prove the interaction between His 119 and Asp 121 in an unambiguous way (Blackburn and Moore, 1982), whereas this hydrogen bond was firmly established on the basis of the diffraction data (Wlodawer and Sjölin, 1981). The conclusion of Cohen and Wlodawer (1982) was that NMR spectroscopists should be very cautious while trying to interpret crystallographic data, especially if the structure studied by diffraction techniques was not refined.

Other methods were also found to be useful for comparing the structure in solution and in the crystalline state. Timchenko et al. (1978) studied the large-angle diffuse X-ray scattering of RNase solutions and concluded that the structures in the crystal and in solution do not show gross differences. Bello and Nowoswiat (1965) studied the activity of crystalline RNase by performing enzyme assays and concluded that the crystals suspended in 75% 2-methyl-2,4-pentanediol retain full enzymatic activity.

RNase S has been utilized in experiments aimed at elucidating the flexibility of the polypeptide chain. Schreier and Baldwin (1976) used tritium exchange to establish that four amide protons belonging to the N-terminal α helix are more highly protected from exchange than the rest of the S peptide. This result is in good agreement with the neutron studies of hydrogen exchange (Wlodawer and Sjölin, 1982b). Rosa and Richards (1979, 1981) enhanced the resolution of the technique of tritium exchange by separating the digested peptides on a high pressure liquid chromatography column under conditions in which no further exchange was likely. The rates of exchange derived in their investigations were usually averaged for several amides in each peptide because more than one protected amide was present in almost every proteolytic fragment. An exception was provided by a small fragment containing Met 79, in which case an individual rate was assigned. The rate was in good agreement with the neutron diffraction results, with both techniques indicating that this particular amide was highly protected. Other assignments, particularly in the S peptide region, were not in good agreement. Tritium exchange indicated that amides 7 and 8, located in the middle part of the C-terminal α helix, were protected; the neutron data showed the protection of one end of the helix (amides 11–13). Nevertheless, the chemical methods provided a point of departure for the neutron diffraction experiments described previously. Recent results of the studies of kinetics of hydrogen exchange in the S peptide of RNase S using NMR (Kuwajima and Baldwin, 1983a, 1983b) are also in good agreement with the conclusions based on the neutron diffraction data.

8. CONCLUSIONS

Although our knowledge of the structure and mechanism of action of RNase is by no means complete, the gaps are rapidly being filled. High resolution structures of the complexes of RNase A with inhibitors are under investigation in at least three laboratories, and it can be expected that the detailed description of the active site during different stages of catalysis will become available in the near future. The structure of native

RNase S is being refined by least-squares, and these results will become invaluable for the interpretation of the structures of semisynthetic RNases. The interest in the semisynthetic structures originates chiefly from our need to understand the role of different amino acid side chains in catalysis, and because it is possible to attach artificial S peptides to the natural S protein, the active site of the enzyme can be appropriately modified. RNase S is unique in that the active site is divided between two separate chains that can be easily separated and reattached. This property has been exploited in many studies and will certainly be found useful in the future.

REFERENCES

Alber, T., Gilbert, W. A., Ponzi, D. R., and Petsko, G. A. (1983). *Ciba Symp.* **93,** 4–24.

Allewell, N. M., Mitsui, Y., and Wyckoff, H. W. (1973). *J. Biol. Chem.* **248,** 5291–5298.

Allewell, N. M., and Wyckoff, H. W. (1971). *J. Biol. Chem.* **246,** 4657–4663.

Antonov, I. V., Gurevich, A. Z., Dudkin, S. M., Karpeisky, M. Y., Sakharovsky, V. G. and Yakovlev, G. I. (1978). *Eur. J. Biochem.* **87,** 45–54.

Arus, C., Paolillo, L., Llorens, R., Napolitano, R., Pares, X., and Cuchillo, C. M. (1981). *Biochem. Biophys. Acta* **660,** 117–127.

Avey, H. P., Boles, M. O., Carlisle, C. H., Evans, S. A., Morris, S. J., Palmer, R. A., Woolhouse, B. A., and Shall, S. (1967). *Nature (London)* **213,** 557–562.

Barnard, E. A. (1969). *Nature (London)* **221,** 340–344.

Bello, J., and Harker, D. (1961). *Nature (London)* **192,** 756.

Bello, J., and Nowoswiat, E. F. (1965). *Biochim. Biophys. Acta* **105,** 325–332.

Bernstein, F. C., Koetzle, T. F., Williams, G. J. B., Meyer, E. F., Brice, M. D., Rodgers, J. R., Kennard, O., Shimanouchi, T., and Tasumi, M. (1977). *J. Mol. Biol.* **112,** 535–542.

Blackburn, P., and Moore, S. (1982). In *The Enzymes,* P. D. Boyer, Ed., 3rd Ed., Vol. 15. Academic Press, New York, p. 317.

Borkakoti, N. (1983a). *FEBS Lett.* **162,** 367–373.

Borkakoti, N. (1983b), *Eur. J. Biochem.* **132,** 89–94.

Borkakoti, N., Moss, D. A., and Palmer, R. A. (1982). *Acta Crystallogr.* **B38,** 2210–2217.

Borkakoti, N., Moss, D. A., Stanford, M. J., and Palmer, R. A. (1984). *J. Mol. Biol.,* in press.

Brayer, G. D., and McPherson, A. (1982). *J. Biol. Chem.* **257**, 3359–3361.

Breslow, R. (1982). *Science* **218**, 532–537.

Carlisle, C. H., Palmer, R. A., Mazumdar, S. K., Gorinsky, B. A., and Yeates, D. G. R. (1974). *J. Mol. Biol.* **85**, 1–18.

Carlson, W. D. (1976). Ph.D. Thesis, Yale University, New Haven.

Cohen, J. S., Hughes, L. J., and Wooten, J. B. (1983). *Mag. Res. Biol.* **2**, 130–247.

Cohen, J. S., and Shindo, H. (1975). *J. Biol. Chem.* **250**, 8874–8881.

Cohen, J. S., and Wlodawer, A. (1982). *Trends Biochem. Sci.* **7**, 389–391.

Deakyne, C. A., and Allen, L. C. (1979). *J. Am. Chem. Soc.* **101**, 3951–3959.

Dickerson, R. E., and Geis, I. (1969). *The Structure and Action of Proteins.* Harper and Row, New York, p. 80.

Doscher, M. S., Martin, P. D., and Edwards, B. F. P. (1983). *J. Mol. Biol.* **166**, 685–687.

Fankuchen, I. (1941). *J. Gen. Physiol.* **24**, 315–316.

Fletterick. R. J., and Wyckoff, H. W. (1975). *Acta Crystallogr.* **A31**, 698–700.

Fujinaga, M., Read, R. J., Sielecki, A., Ardelt, W., Laskowski, M., and James, M. N. G. (1982). *Proc. Natl. Acad. Sci. USA* **79**, 4868–4872.

Gilbert, W. A., Lord, R. L., Petsko G. A., and Thamann, T. J. (1982). *J. Raman Spectrosc.* **12**, 173–179.

Gilbert, W. A., and Petsko, G. A. (1984). *Biochemistry,* submitted.

Heinemann, U., and Saenger, W. (1982). *Nature (London)* **299**, 27–31.

Hendrickson, W. A., and Konnert, J. H. (1981). In *Biomolecular Structure, Conformation, Function and Evolution,* R. Srinivasan, Ed., Vol. 1. Pergamon, New York, p. 43.

Holmes, R. R., Deiters, J. A., and Gallucci, J. C. (1978). *J. Am. Chem. Soc.* **100**, 7393–7402.

Iwahashi, K., Nakamura, K., Mitsui, Y., Ohgi, K., and Irie, M. (1981). *J. Biochem. (Japan)* **90**, 1685–1690.

Jentoft, J. E., Gerken, T. A., Jentoft, N., and Dearborn, D. G. (1981). *J. Biol. Chem.* **256**, 231–236.

Karpeisky, M. Y., and Yakovlev, G. I. (1981). *Sov. Sci. Rev., Sect. D* **2**, 145–257.

Kartha, G. (1967). *Nature (London)* **214**, 234–235.

Kartha, G., Bello, J., and Harker, D. (1967). *Nature (London)* **213**, 862–865.

Kartha, G., Bello, J., and Harker, D. (1968). In *Structural Chemistry and Molecular Biology,* A. Rich and N. Davidson, Eds., Freeman, San Francisco, p. 29–37.

King, M. V., Magdoff, B. S., Adelman, M. B., and Harker, D. (1956). *Acta Crystallogr.* **9**, 460–465.

Krieger, M., Chambers, J. L., Christoph, G. G., Stroud, R. M., and Trus, B. L. (1974). *Acta Crystallogr.* **A30**, 740–748.

Kunitz, M. (1940). *J. Gen. Physiol.* **24**, 15–32.

Kuwajima, K., and Baldwin, R. L. (1983a). *J. Mol. Biol.* **169**, 281–297.

Kuwajima, K., and Baldwin, R. L. (1983b). *J. Mol. Biol.* **169**, 299–323.

Lewis, P. N., Momany, F. A., and Scheraga, H. A. (1973). *Biochem. Biophys. Acta* **303**, 211–229.

Lindquist, R. N., Lynn, J. L., and Lienhard, G. E. (1973). *J. Am. Chem. Soc.* **95**, 8762–8768.

Luzzati, V. (1952). *Acta Crystallogr.* **5**, 802–810.

Markley, J. L. (1975). *Biochemistry* **14**, 3546–3554.

Marsh, D. J., and Petsko, G. A. (1973). *J. Appl. Crystallogr.* **6**, 76–80.

Martin, P. D., Petsko, G. A., and Tsernoglou, D. (1976). *J. Mol. Biol.* **108**, 265–269.

Mauguen, Y., Hartley, R. W., Dodson, E. J., Dodson, G. G., Bricogne, G., Chothia, C., and Jack, A. (1982). *Nature (London)* **297**, 162–164.

Meadows, D. H., Jardetzky, O., Epand, R. M., Ruterjans, H. H., and Scheraga, A. (1968). *Proc. Natl. Acad. Sci. USA* **60**, 766–772.

Meadows, D. H., Markley, J. L., Cohen, J. S., and Jardetzky, O. (1967). *Proc. Natl. Acad. Sci. USA* **58**, 1307–1313.

Mitsui, Y., Urata, Y., Torii, K., and Irie, M. (1978). *Biochem. Biophys. Acta* **535**, 299–308.

Mitsui, Y., and Wyckoff, H. W. (1975). *J. Mol. Biol.* **94**, 17–31.

Nakamura, K. T., Iwahashi, K., Yamamoto, Y., Iitaka, Y., Yoshida, N., and Mitsui, Y. (1982). *Nature (London)* **299**, 564–566.

Narayana, S. V. L., Weininger, M. S., Heuss, K. L., and Argos, P. (1982). *J. Appl. Crystallogr.* **15**, 571–573.

Pandin, M., Padlan, E. A., DiBello, C., and Chaiken, I. M. (1976). *Proc. Natl. Acad. Sci. USA* **73**, 1844–1847.

Pavlovsky, A. G., Borisova, S. N., Broisov, V. V., Antonov, I. V., and Karpeisky, M. Y. (1978). *FEBS Lett.* **92**, 258–262.

Plummer, T. H., Tarentino, A., and Maley, F. (1968). *J. Biol. Chem.* **243**, 5158–5164.

Potts, J. T., Berger, A., Cooke, J., and Anfinsen, C. B. (1962). *J. Biol. Chem.* **237**, 1851–1855.

Powers, T. B. (1976). Ph.D. Thesis, Department of Molecular Biophysics and Biochemistry, Yale University, New Haven.

Prince, E., Wlodawer, A., and Santoro, A. (1978). *J. Appl. Crystallogr.* **11**, 173–178.

Richards, F. M., and Vithayathil, P. J. (1959). *J. Biol. Chem.* **234**, 1459–1465.

Richards, F. M., and Wyckoff, H. W. (1971). In *The Enzymes*, P. D., Boyer, Ed., 3rd ed. Vol. 4. Academic Press, New York, pp. 647–806.

Richards, F. M., and Wyckoff, H. W. (1973). In *Atlas of Molecular Structures in Biology—1: Ribonuclease S'*, D. C. Phillips and F. M. Richards, Eds., Clarendon, Oxford.

Richards, F. M., Wyckoff, H. W., and Allewell, N. (1970). In *The Neurosciences*. F. O. Schmitt, Ed., Rockefeller University Press, New York, p. 901.

Richards, F. M., Wyckoff, H. W., Carlson, W. D., Allewell, N. M., Lee, B., and Mitsui, Y. (1971). *Cold Spring Harbor Symp. Quant. Biol.* **36**, 35–43.

Roberts, G. C. K., Dennis, E. A., Meadows, D. H., Cohen, J. S., and Jardetzky, O. (1969). *Proc. Natl. Acad. Sci. USA* **62**, 1151–1158.

Rosa, J. J., and Richards, F. M. (1979). *J. Mol. Biol.* **133**, 399–416.

Rosa, J. J., and Richards, F. M. (1981). *J. Mol. Biol.* **145**, 835–851.

Rubin, B., Carperos, V., and Kezar, E. (1982). *J. Biol. Chem.* **257**, 8896–8897.

Rübsamen, H., Khandker, R., and Witzel, H. (1974). *Hoppe-Seyler's Z. Physiol. Chem.* **355**, 687–708.

Sasaki, D. M., Martin, P. M., Doscher, M. S., and Tsernoglou, D. (1979). *J. Mol. Biol.* **135**, 301–304.

Saunders, M., Wishnia, A., and Kirkwood, J. (1957). *J. Am. Chem. Soc.* **79**, 3289–3290.

Schreier, A. A., and Baldwin, R. L. (1976). *J. Mol. Biol.* **105**, 409–426.

Shindo, H., Hayes, M. B., and Cohen, J. S. (1976). *J. Biol. Chem.* **251**, 2644–2647.

Sjölin, L., and Wlodawer, A. (1981). *Acta Crystallogr.* **A37**, 594–604.

Smyth, D. G., Stein W. H., and Moore, S. (1963). *J. Biol. Chem.* **238**, 227–234.

Taylor, H. C., Richardson, D. C., Richardson, J. S., Wlodawer, A., Komoriya, A., and Chaiken, I. M. (1981). *J. Mol. Biol.* **149**, 313–317.

Timchenko, A. A., Ptitsyn, O. B., Dolgikh, D. A., and Fedorov, B. A. (1978). *FEBS Lett.* **88**, 105–108.

Torii, K., Urata, Y., Iitaka, Y., Sawada, F., and Mitsui, Y. (1978). *J. Biochem. (Japan)* **83**, 1239–1247.

Umeyama, H., Nakagawa, S., and Fuji, T. (1979). *Chem. Pharm. Bull.* **27**, 974–980.

Usher, D. A., Erenrich, E. S., and Eckstein, F. (1972). *Proc. Natl. Acad. Sci. USA* **69**, 115–118.

Usher, D. A., Richardson, D. I., and Eckstein, F. (1970). *Nature (London)* **228**, 663–665.

Valle, G., Zanotti, G., Filippi, B., and Del Pra, A. (1977). *Biopolymers* **16**, 1371–1376.

Wlodawer, A. (1980). *Acta Crystallogr.* **B36**, 1826–1831.

Wlodawer, A. (1982). *Progr. Biophys. Mol. Biol.* **40,** 115–159.

Wlodawer, A., Bott, R., and Sjölin, L. (1982a) *J. Biol. Chem.* **257,** 1325–1332.

Wlodawer, A., and Hendrickson, W. H. (1982). *Acta Crystallogr.* **A38,** 239–247.

Wlodawer, A., Miller, M., and Sjölin, L. (1983). *Proc. Natl. Acad. Sci. USA* **80,** 3628–3631.

Wlodawer, A., and Sjölin, L. (1981). *Proc. Natl. Acad. Sci. USA* **78,** 2853–2855.

Wlodawer, A., and Sjölin, L. (1982a). *Nucl. Instrum. Methods.* **201,** 117–122.

Wlodawer, A., and Sjölin, L. (1982b). *Proc. Natl. Acad. Sci. USA* **79,** 1418–1422.

Wlodawer, A., and Sjölin, L. (1983). *Biochemistry* **22,** 2720–2728.

Wlodawer, A., Sjölin, L., and Santoro, A. (1982b). *J. Appl. Cryst.* **15,** 79–81.

Wodak, S. Y., Liu, M. Y., and Wyckoff, H. W. (1977). *J. Mol. Biol.* **116,** 855–875.

Wyckoff, H. W., Carlson, W., and Wodak, S. (1977). In *Nucleic Acid–Protein Recognition*, H. J. Vogel, Ed., Academic Press, New York, pp. 569–580.

Wyckoff, H. W., Doscher, M., Tsernoglou, D., Inagami, T., Johnson, L. N., Hardman, K. D., Allewell, N. M., Kelly, D. M., and Richards, F. M. (1967a). *J. Mol. Biol.* **27,** 563–578.

Wyckoff, H. W., Hardman, K. D., Allewell, N. M., Inagami, T., Johnson, L. N., and Richards, F. M. (1967b). *J. Biol. Chem.* **242,** 3784–3788.

Wyckoff, H. W., Hardman, K. D., Allewell, N. M., Inagami, T., Tsernoglou, D., Johnson, L. N., and Richards, F. M. (1967c). *J. Biol. Chem.* **242,** 3749–3753.

Wyckoff, H. W., Tsernoglou, D., Hanson, A. W., Knox, J. R., Lee, B., and Richards, F. M. (1970). *J. Biol. Chem.* **245,** 305–328.

Aminoacyl-tRNA Synthetases

DAVID M. BLOW
PETER BRICK
Blackett Laboratory
Imperial College of Science and Technology
London, England

CONTENTS

1. INTRODUCTION

Life depends on the translation of genetic information into functional structure in a reliable way. In practice, the assembly of proteins is virtually error-free. Aminoacyl-tRNA synthetases* carry out a step in the translation of the nucleic acid message, which is a crucial step because

---

* Aminoacyl-tRNA synthetase is abbreviated aaRS. Individual aaRS are identified by the three-letter abbreviation for amino acids, as in TyrRS. In the case of TyrRS, the enzyme from *B. stearothermophilus* is referred to unless otherwise indicated.

442

it mediates the association of a specific nucleic acid (tRNA) with a specific amino acid. These enzymes therefore require an accurate mechanism for the recognition of a specific amino acid (difficult because amino acids are small and not very different from each other) and also an accurate protein-mediated mechanism for nucleic acid recognition.

Aminoacyl-tRNA synthetases form a group of enzymes of very varied properties (Schimmel and Söll, 1979). In *E. coli* they range from monomeric to tetrameric; there are also examples of $\alpha_2\beta_2$ structures. The individual polypeptide chains range from 37,000 dalton (TrpRS) to 110,000 dalton (IleRS). Two zinc atoms are essential for activity of the dimeric MetRS molecule (Posorske et al., 1979). Other aaRS are not known to have strongly bound cations.

In prokaryotes there are normally only 20 aaRS, and each has to be able to recognize all the isoaccepting tRNA molecules. In eukaryotes, mitochondria and chloroplasts have at least some different synthetases; some of these are needed to implement differences in the genetic code in these organelles (Barrell et al., 1979; de Bruijn, 1983).

The protein synthesis mechanisms function with amazing fidelity. Loftfield and Vanderjagt (1972) have shown that the incorporation of valine in place of isoleucine occurs at a level less than 3×10^{-4}, and there are no data to suggest a higher error rate in other cases (Popp et al., 1976; Edelmann and Gallant, 1977).

The majority of aaRS function through a mechanism involving an aminoacyl adenylate intermediate formed by aminoacylation of ATP with elimination of pyrophosphate. Three of the synthetases (ArgRS, GlnRS, and GluRS) require tRNA for the aminoacyl adenylate formation (Söll and Schimmel, 1974). The formation of the aminoacyl adenylate does not always show a high selectivity for the cognate amino acid. Baldwin and Berg (1966) discovered that IleRS will catalyze formation of a valyl adenylate in the presence of valine and ATP. In this case, addition of $tRNA^{Ile}$ produces no valyl-tRNA. Instead, the tRNA has the effect of initiating hydrolysis of valyl adenylate. Fersht and Kaethner (1976) showed that in the case of ValRS there is a transient formation of threonyl-$tRNA^{Val}$, which is rapidly hydrolyzed. The hydrolysis is catalyzed by ValRS at a distinct second binding site. In comparison with the process by which we try to reduce errors in our publications, these mechanisms have been referred to as "proofreading" or "editing" mechanisms.

It now begins to appear that proofreading properties are found only in those enzymes where it is not possible to achieve sufficient specificity in a single recognition step. The level of fidelity reported by Loftfield (3 × 10^{-4}) implies a difference of interaction energy ΔW of 5.0 kcal/mole at 37°C (Pauling, 1957) in a single recognition step, such that the favored ligand is preferred over other candidates that are present at a similar concentration. The rejection of phenylalanine by TyrRS would require a slightly higher ΔW, as phenylalanine is normally a more abundant amino acid than tyrosine, but Fersht et al. (1980) have shown that adequate selectivity is shown in the aminoacylation step of TyrRS (preferential activation rate of 10^5 for tyrosine over phenylalanine). It seems likely that all the aaRS having a single type of small subunit (Asn, Asp, Cys, His, Lys, Trp, Tyr at least) can perform adequately without an editing function.

2. CRYSTALLIZATION OF AMINOACYL-tRNA SYNTHETASES

The first report of crystallization of an aminoacyl tRNA synthetase was the crystallization of LysRS from yeast (Rymo et al., 1970). The crystals were trigonal with one molecule (100,000 daltons) in the crystal asymmetric unit, but the diffraction pattern only extended to a resolution of 3.5 to 4.0 Å. The crystals were very radiation-sensitive, being limited to a lifetime of 4 to 5 hr in the beam from a rotating anode generator, even though the crystals were grown and maintained at 4°C.

This appeared to be a rather general problem. In unpublished experiments, we obtained small crystals of TrpRS and TyrRS from *E. coli* that were also very sensitive to radiation and appeared unsuitable for diffraction analysis. On the other hand, a MetRS fragment, which has lost some 200 amino acids from the C-terminus, gives well-ordered crystals that are reasonably stable to radiation (Waller et al., 1971). No crystals of intact MetRS have been reported.

A successful approach to the problem was to crystallize an aaRS from a thermophile, and Reid et al. (1973) obtained crystals of TyrRS from *B. stearothermophilus* that were satisfactorily ordered and much less sensitive to radiation. A further successful step was taken by Carter and Carter (1979), who showed that the crystal structure of TrpRS from *B. stearothermophilus* was considerably stabilized by crystallization in the presence of the amino acid tryptophan. The known unit cells of aaRS crystals are listed in Table 1.

Despite continued effort in many laboratories, there is only one authentic report of crystallization of an aaRS–tRNA complex (Giegé et al., 1980; Lorber et al., 1983). The conditions of crystallization were quite conventional: ammonium sulfate (48–53% saturated), magnesium chloride (5–15 mM), enzyme at 3–12 mg/ml concentration, with stoichiometry tRNA/synthetase between 1.8 and 2.2. Crystals were obtained both at 4°C and at room temperature, and the addition of spermine is not important. The crystals of AspRS–tRNA^{Asp} complex thus obtained have a large unit cell (I432, $a = b = c = 480$ Å) which is very unfavorable for X-ray diffraction studies. However, they have provided a strong stimulus for further attempts to crystallize analogous complexes in other systems.

3.   TYROSYL-tRNA SYNTHETASE

3.1.   Structure Determination

Starting from an isomorphous replacement study at a nominal 2.7 Å resolution (Irwin et al., 1976), a density modification technique was used to assist recognition of poorly resolved parts of the structure (Bhat and Blow, 1982). An account of the structure refined at 3.0 Å resolution has been published (Bhat et al., 1982). Figure 1 shows the main chain conformation. In further work, diffraction data to 2.1 Å have been used for further refinement of the structure (Bhat et al., 1984). A partial amino acid sequence was established by conventional methods, but the complete amino acid sequence was only determined when the gene was isolated and sequenced (Winter et al., 1983). The amino acid sequence of TyrRS from *E. coli* has been determined by similar techniques (Barker et al., 1982a).

3.2.   Domain Structure and Disorder

In this crystal structure, different parts of the TyrRS molecule show an unusual range in their degree of order. The peptide chain may be divided into three domains: the N-terminal domain (residues 1–220) is an α-β domain in which helices are folded around a central sheet of five parallel

Table 1. Crystals of Aminoacyl-tRNA Synthetases

Aminoacyl-tRNA Synthetase	Species	Molecule	Space group
Aspartyl-	Yeast	Dimer: $2 \times 60,000$	$P4_12_12$ $P4_12_12$
AspRS:tRNA^{Asp} complex Leucyl-[a]	Yeast	—	I432
Lysyl-	Yeast	Monomer: 100,000	$P3_121$
Methionyl-(fragment)	E. coli K12	Monomeric fragment 64,000	$P2_1$
Tryptophanyl- + tryptophan	B. stear.	Dimer: $2 \times 37,500$	P312
Tryptophanyl- + Trp-ATP	B. stear.	Dimer: $2 \times 37,500$	$P4_12_12$
Tyrosyl-	B. stear.	Dimer: $2 \times 45,000$	$P3_121$ $P6_122$

[a] The claim by Chirikjian et al. (1972) to have crystallized LeuRS has been withdrawn (Nikodem et al. 1978).

β strands and one short antiparallel strand; after a linking peptide, the α helical domain (residues 250–320) contains five α helices; then there remains a C-terminal domain of 99 amino acids whose structure is unknown because no resolved structure appears in the electron density map, though an "island" of 1500 Å^3 of unresolved high electron density is attributed to part of this domain (Bhat and Blow, 1982). If the lack of density is due to stochastic disorder, the Debye–Waller B factor must be at least 50 Å^2 throughout this domain. In the α-β domain, the β sheet appears to form a relatively rigid structure ($B \simeq 5$ Å^2) around which the domain is folded. The adjacent parts of several α helices also have a low B factor, but

Unit Cell Dimensions (A)		Monomers in Asymmetric Unit	Resolution Limit (A)	References
$a = 92$	$c = 185$	1	3	
$a = 89$	$c = 480$	2	4	Dietrich et al., 1980
$a = 354$		2 AspRS +2 tRNA^{Asp}	8	Giegé et al., 1980
				Lorber et al., 1983
$a = 118$	$c = 190$	1	4	Rymo et al., 1970
$a = 78$	$b = 46$			Waller et al.,
$c = 88$	$\beta = 109$	1	2.5	1971
				Zelwer et al., 1982
$a = 91$	$c = 93$	1	—	Carter and Carter, 1979
$a = 61$	$c = 234$	1	—	Coleman and Carter, 1983
$a = 63$	$c = 239$	1	2.1	Reid et al., 1973
				Bhat et al., 1982
$a = 78$	$c = 266$	1	3	P. Brick, unpublished

disorder increases toward the surface of the molecule. There are four segments of chain on the surface of the domain where B rises above 30 Å^2, and where the detailed interpretation of the chain conformation becomes uncertain. Two of these (residues 82–85 and 151–159) are lengths of chain that are close to each other and not far from the α amino group binding site of the substrate. Residues 232–235 in the surface-linking peptide between the α-β and the α domains have a very high B factor.

The α helical domain is adequately ordered to allow confident structural interpretation except for a loop between two helices (289–292) and in the last helix where the density fades and becomes uninterpretable.

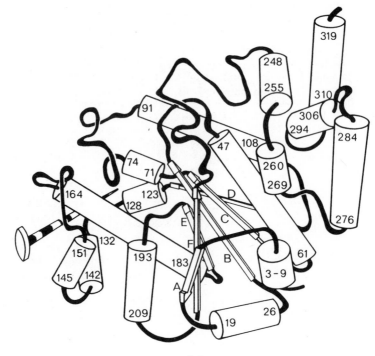

(a).

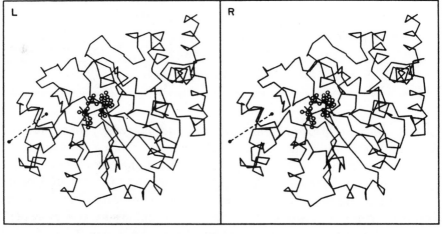

(b)

Figure 1. (*a*) Schematic secondary structure drawing for residues 1-320 of TyrRS. One monomer is shown, and the molecular twofold axis is indicated. (*b*) Corresponding stereoscopic view, showing α-carbon positions and the site of the inhibitor tyrosinyl adenylate (Bhat et al., 1982).

The position of the "island" density relative to the rest of the molecule is uncertain, as it is not possible to determine which "island" is associated with a particular molecule in the crystal.

3.3. Structure of α-β Domain

The five parallel strands of the sheet (strands B to F) shown in Figure 2 form a structure reminiscent of the Rossmann dinucleotide-binding fold (Rossmann et al., 1975). The order of strands in the sheet is the same as in the Rossmann fold, except that the sixth strand is absent: in its place is the short antiparallel strand (strand A), which is near the N-terminus. The six strands have the usual twist (Chothia, 1973). Each linkage between strands includes an α helix that lies against the sheet, running in the opposite sense to the strand; in every case the linkage through this α helix from one strand to the next is right-handed (Richardson, 1976; Sternberg and Thornton, 1976). From the center of the sheet, the linkage of strands B, C, and D leads in one direction and the linkage of strands E and F in the other. The same separation of C-terminal ends of strands is at the center of the sheet, between strands B and E, as has been discussed by Brändén (1981).

The structure is considerably more complex than the usual dinucleotide-binding fold, because there are several additional elements in the structure. The three strands B, C, and D form a unit that corresponds topologically to the "mononucleotide-binding fold" (Rao and Rossmann, 1973; Rossmann et al., 1977). However, the linkage between strands C and D includes a helix and a long section of extended chain before it joins the α helix, which lies against strand D. (This structure covers the C-terminal ends of strands C and D and would prevent the actual binding of a nucleotide in any way similar to the mononucleotide binding illustrated by Rossmann et al., 1977.) The α helical linkage between strands B and C is more direct, without any extended loops. Even here, it would be a mistake to imagine that there is close similarity with the structure observed in the nicotinamide adenine dinucleotide (NAD) binding domain of the dehydrogenases. In the dehydrogenases, there is a sharp change in direction of the polypeptide chain at the end of the first β strand, where it leads directly into the α helix with a linking unit of only three residues. In order to achieve this sharp change in direction, a glycine is required

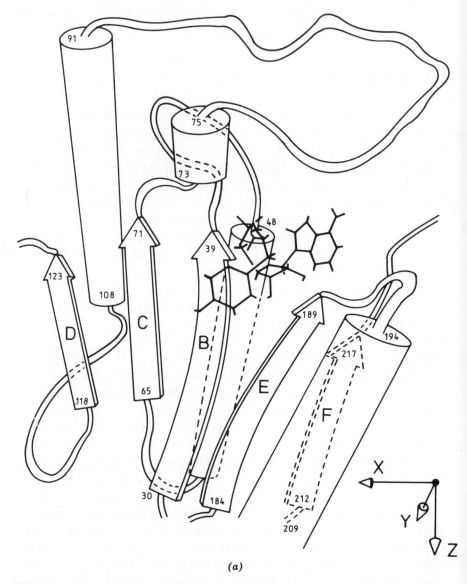

(a)

Figure 2. Schematic drawings showing the five strands of parallel β sheet in relation to the site of binding of tyrosyl adenylate.

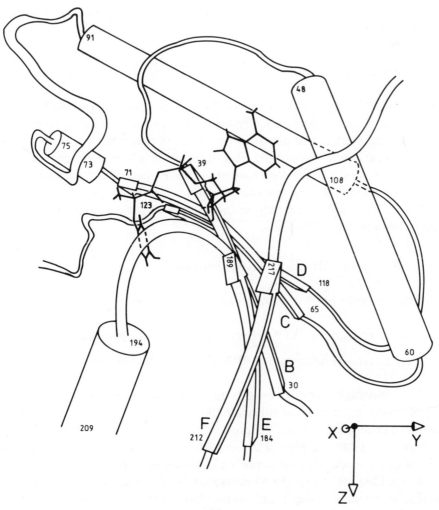

(b)

Figure 2. (Continued)

at the end of the first β strand (Rossmann et al., 1975). In TyrRS this change in direction is less abrupt, with eight residues between strand and helix, and there is no requirement for glycine.

There seems to be no evidence of any sequence homology between TyrRS and the dehydrogenases.

3.4. Dimerization

The contact between two chains of the dimeric TyrRS molecule forms a crystallographic twofold axis. Crystallographic results on this crystal form can give no evidence about any differences between the molecules, and in particular about "half of the sites" reactivity, which has been reported (Jakes and Fersht, 1975; Bosshard et al., 1978), because the crystal symmetry forces the two subunits to be identical. The dimer contact is made through a hydrophobic patch on the surface of the α-β subunit, mainly involving residues 128–136.

3.5. Substrate Binding Stoichiometry

TyrRS catalyzes both steps of a two-step reaction:

Step 1: tyrosine + ATP → tyrosyl adenylate + pyrophosphate

Step 2: tyrosyl adenylate + tRNA^{Tyr} → tyrosyl-tRNA^{Tyr} + AMP

Crystallographic studies have been made of a number of complexes of TyrRS with substrates and substrate-like molecules (Table 2). These studies have accurately defined the positions of tyrosine and adenosine in the tyrosyl adenylate complex of the enzyme during step 1, but they give no direct information about the binding of tRNA in step 2.

Monteilhet and Blow (1978) measured the dissociation constant of tyrosine in crystals, using a radioactive label and a filter assay. The dissociation constant in crystals in 2 M ammonium sulfate, measured as 20 μM, is in good agreement with the dissociation constant in solution measured from enzyme assays and equilibrium dialysis in low salt. Monteilhet and Blow also showed the binding of ATP to be about two orders of magnitude weaker, though the dissociation constant was too great to be measured. The only accurate measurement of the dissociation constant of ATP (K_m = 3.9 mM) (Fersht et al., 1975) comes from kinetic measurements, and this is in general agreement with the crystal observations (Table 3).

There is a fundamental disagreement between crystal and solution studies of the binding stoichiometry of tyrosine. Solution studies consistently report that in the absence of ATP, where step 1 of the reaction is prevented from taking place, only one tyrosine is bound per dimer. Because of symmetry, the crystallographic studies show tyrosine binding equally at both sites of the dimer. But the electron density indicates tyrosine binding at both sites with essentially full occupancy. This occupancy is measured by an electron count, where the density in a difference map is compared with electron density differences due to heavy atoms believed to bind with high occupancy. It is also measured by an independent method using ^{14}C-labeled tyrosine. The total quantity of tyrosine bound to a crystal is measured using the radioactive label, and the number of molecules in the crystal is estimated by direct measurement of its size. A further complication is that a chemical active site assay on dissolved crystals shows that the crystalline protein contains only about 80% of the theoretical number of binding sites. Though neither of these methods is capable of high precision, they give results that agree excellently. After allowing for the 80% factor mentioned above, the crystal studies show almost exactly two tyrosines bound per dimer.

These results suggest that in solution, some mechanism operates to prevent the binding of a tyrosine molecule to both sites of the dimer unless the adenylate is formed. This could, for instance, be caused by a conformational change. This mechanism does not operate in crystals, and the background level of the difference electron density map shows directly that some conformational changes occur.

Another discrepancy between crystal and solution studies, which may relate to the same effect, concerns substitution of a cysteine (now known to be Cys 35). Koch (1974), using 5,5'-dithiobis(2-nitrobenzoate) (DTNB) as cysteine reagent, found that only one cysteine could be blocked per dimer in solution. In crystals, methyl mercury and p-chloromercuribenzoate can be fully substituted at both sites (Irwin et al., 1976).

Crystallographic and solution studies agree that adenosine derivatives such as ATP are bound with a much lower affinity. Because of the lower

Table 2. Crystallographic Studies of Ligands Bound to TyrRS

Ligand	Concentration	Resolution	Binding Site[a]	Reference
Tyrosine	Saturated	2.7 Å	T	Monteilhet and Blow, 1978
ATP	2 mM	2.7 Å	(T, R?)	Monteilhet and Blow, 1978
Adenylyl (α, β methylene) diphosphate	3 mM	2.7 Å	(T, R?)	Monteilhet and Blow, 1978
AMP	4 mM	2.7 Å	(T, R?)	Monteilhet and Blow, 1978
Tyrosinyl adenylate	40 μM	2.7 Å	TRA	Monteilhet and Blow, 1978
Tyrosine + ATP[b] (bound as tyrosyl adenylate)	Saturated 12 mM	2.7 Å	TRA[c]	Rubin and Blow, 1981
Pyrophosphate	100 mM	4 Å(h01)	(T)	Rubin and Blow, 1981
Puromycin	Saturated	2.7 Å	T(R?)[c]	Rubin and Blow, 1981
CpC	2 mM	4 Å(h01)	(T)	Rubin and Blow, 1981
CpCp	10 mM	4 Å(h01)	(T)	Rubin and Blow, 1981
CpA	10 mM	4 Å(h01)	(T)	Rubin and Blow, 1981
ApC	Saturated	4 Å(h01)	(T)	Rubin and Blow, 1981
ApCpC	1 mM	4 Å(h01)	(T)	Rubin and Blow, 1981
Ammonium arsenate	60% saturated	2.7 Å	(T)[d]	Rubin and Blow, 1981
Tyrosinol + ATP	3 mM, 15 mM	2.7 Å	T(R?)	Monteilhet et al., 1984
Tyrosinol, AMP, pyrophosphate	3 mM, 10 mM, 7 mM	2.7 Å	T(R?)	Monteilhet et al., 1984
Tyrosinol, adenosine, pyrophosphate	3 mM, 2.5 mM, 5 mM	2.7 Å	T(R?)	Monteilhet et al., 1984

[a] T, tyrosine site. R, ribose site. A, adenosine site. R?, peak within 1 Å of ribose site. Parentheses indicate a partially occupied site.

[b] 1 mM CpCpA was also present in this experiment but was not believed to have any effect.

[c] Considerable background difference density.

[d] Also numerous sites in solvent-accessible regions.

affinity of binding, the stoichiometry of the interaction is harder to investigate.

3.6. Substrate Binding Conformation

Santi and Peña (1973) studied many inhibitors of *E. coli* TyrRS and found L-tyrosinyl adenylate (Figure 3) to be very tightly bound ($K_i = 2.10^8 M$). The electron density difference map for tyrosinyl adenylate (Monteilhet and Blow, 1978) has a clear and unambiguous interpretation (Figure 4). The conformation of the inhibitor in relation to selected nearby amino acid side chains is shown in Figure 5. Extremely similar difference density is found for the true intermediate tyrosyl adenylate (Figure 3) (Rubin and Blow, 1981), but this map shows greatly increased background density throughout the molecule (Figure 6), suggesting that minor structural changes occur to the enzyme when tyrosyl adenylate is bound but do not occur when tyrosinyl adenylate is formed. We conclude that the conformations of tyrosyl adenylate and tyrosinyl adenylate, when bound to the enzyme, are very similar. The minor structural changes have not yet been analyzed in detail. It is not obvious why they should occur with the true intermediate tyrosyl adenylate but not with the inhibitor tyrosinyl adenylate.

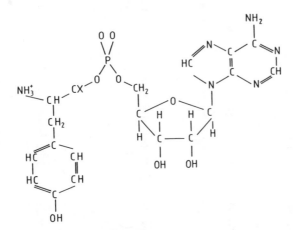

Figure 3. Tyrosyl adenylate ($X = O$); Tyrosinyl adenylate ($X = H_2$).

Table 3. Dissociation Constant and Binding Stoichiometry for Tyrosine and ATP

Medium[a]	Method of Analysis	Ligand[d]				Reference
		Tyrosine (alone)	ATP (alone)	Tyrosine (ATP present)	ATP (tyrosine present)	
Solution	Kinetic	—	—	2.3(—)	3900(1)	Fersht et al., 1980
Solution	Eq. dialysis	10(0.92)	—	5–7[b](1.6)	—	Fersht et al., 1975
Solution	Eq. gel. filtration	—(0.96)	>400(—)	—(1.73)	>200[b](1.7–1.9)	Fersht et al., 1975 Mulvey and Fersht, 1978
Crystals	Radioactive label	~10(1.50) (1.66)[c]	>1000(—)	—	—	Monteilhet and Blow, 1978 Rubin and Blow, 1981 Monteilhet et al., 1984
Crystals	Difference map	—(1.52)	—(0.2)[e]	—	—	Monteilhet and Blow, 1978

[a] Solution measurements generally in Tris-Cl buffer 144 mM (pH 7.78), 10 mM MgCl$_2$. Crystal measurements in 2 M ammonium sulfate, 100 mM Tris-Cl (pH 7.0), 10 mM MgCl$_2$.

[b] In presence of pyrophosphate.

[c] Tyrosinol.

[d] The entry in the table gives the dissociation constant of the ligand in μM followed by the stoichiometry per dimer in parenthesis.

[e] At ATP concentration 1–2 mM.

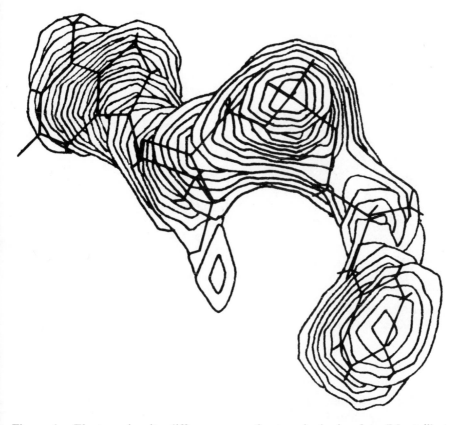

Figure 4. Electron density difference map for tyrosinyl adenylate (Monteilhet and Blow, 1978).

The difference maps for tyrosine and tyrosinol (Monteilhet and Blow, 1978, Monteilhet et al., 1984) are entirely consistent with the above interpretation, and suggest that the tyrosyl group is bound in exactly the same way before and after adenylation by ATP.

The most surprising discovery about the tyrosine binding site is that in the absence of tyrosine, a variety of other aromatic groups can occupy this site. These include not only p-methoxy phenylalanine (as in puromycin; see Rubin and Blow, 1981) but adenine, which evidently enters the site when ATP or AMP are bound in the absence of tyrosine (Monteilhet and Blow, 1978). Outside the tyrosine binding site, both puromycin and ATP appear to be disordered.

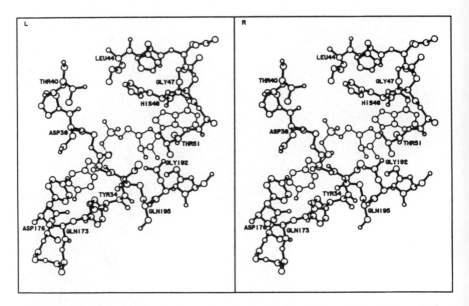

Figure 5.   The positions of selected amino acid side-chains of TyrRS in relation to the inhibitor tyrosinyl adenylate.

Detailed examination of the groove forming the tyrosine site suggests that although one dimension is accurately shaped for an aromatic group (van der Waals spacing of 3.4 Å), the perpendicular dimension is considerably wider than is needed for tyrosine or phenylalanine, making the acceptance of adenine more understandable. Callendar and Berg (1966) had noted that *E. coli* TyrRS would catalyze pyrophosphate exchange using 3-fluoro-L-tyrosine or 3-hydroxy-L-tyrosine, and B. D. Sykes (personal communication) informs us that 2- or 3-fluoro-L-tyrosine can be incorporated into proteins. Another unexpected feature of the groove is that it is far from hydrophobic. In addition to two polar groups at the bottom of the groove, which doubtless function to discriminate in favor of tyrosine against phenylalanine (Tyr 34 and Asp 176), the upper surface of the groove also has several amide and hydroxyl groups.

The interpretation of these observations awaits detailed analysis, but the following factors may be relevant. First, the function of TyrRS is to discriminate against other amino acids that occur in living cells in significant concentration, especially phenylalanine. Other molecules may be unimportant. Second, it is expected that the hydroxyl group of tyrosine

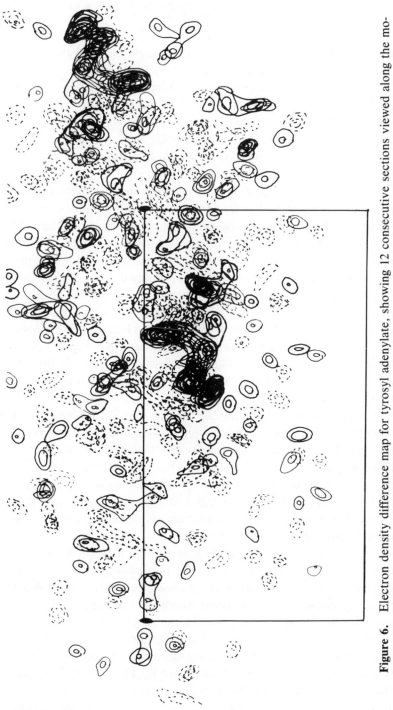

Figure 6. Electron density difference map for tyrosyl adenylate, showing 12 consecutive sections viewed along the molecular twofold axis (center right). The electron density representing two dyad-related molecules of tyrosyl adenylate can be seen (Rubin and Blow, 1981).

459

will be bound by polar groups, which will strongly bind a water molecule in the absence of tyrosine. This water molecule would become buried when phenylalanine is bound, and would hinder phenylalanine from occupying the same position as tyrosine. For rapid binding of tyrosine, a low energy route may be needed for this water molecule to be expelled. Because the tyrosine side chain is more polar than phenylalanine, some polar character of the binding pocket would improve the discrimination against phenylalanine.

A detailed comparison of difference electron density maps containing tyrosinol with corresponding maps for tyrosine (Monteilhet et al., 1984; Monteilhet and Blow, 1978) shows that the extra oxygen in the carboxylate density of tyrosine can be precisely identified. Two other peaks consistently seen in the tyrosine + ATP, tyrosinol + ATP, and tyrosinol + AMP + PP$_i$ maps cannot be definitely assigned. One of these is centered 0.9 Å from the center of the ribose ring as seen in the tyrosinyl adenylate derivative, and it may possibly represent a ribose ring at low occupancy. None of these studies, even with ATP at 15 mM, shows any significant density in the adenine site. It is evident that unless adenosine is covalently bound to a ligand like tyrosine, which is firmly bound in a defined orientation, the adenine ring is not sufficiently strongly bound to appear in difference maps.

In the difference maps showing the binding of ATP, tyrosinol + AMP + PP$_i$, and tyrosinol + ATP, an additional two or three peaks consistently appear. One of these may represent the position of the ribose moiety.

Lowe and Tansley (1984) have shown that the configuration at the α-phosphorus of ATP is inverted when tyrosine activation is catalyzed by TyrRS. They conclude that the enzyme-catalyzed reaction is an "in-line" displacement of pyrophosphate. Similar observations have been made for IleRS and MetRS (Lowe et al., 1983a, b).

The position of tyrosinyl adenylate is shown in relation to α-carbon positions for the enzyme in Figure 1. The phosphate group lies at the C-terminal end of central strands of the β sheet, between strands B and E. As previously noted, this is the point where the linkage of the strands of the sheet changes its sense, and the binding of a ribose-phosphate group in this position is exactly in line with the general binding properties of molecules containing a Rossmann fold, discussed by Bränden (1981). To this extent, the binding of tyrosyl adenylate by TyrRS bears comparison with the binding of NAD by the dehydrogenases.

If this structural analogy is pursued, it is found that the adenine of tyrosyl adenylate, which makes contact with side chains beyond the C-terminal end of strand F of the sheet, occupies a position somewhat analogous to the nicotinamide of NAD in the dehydrogenases. The tyrosine binding site is formed by a deep groove lying against strands B and C of the sheet, on the opposite side from the helix that links them. This position is not similar to any part of the NAD binding in dehydrogenases, which is all near the C-terminal ends of the strands of the sheet (Bhat et al., 1982).

In step 2 of the reaction, the enzyme and tyrosyl adenylate have to interact with the 3'-terminal adenine of a tRNA molecule. Following the theory that a "mononucleotide-binding fold" indeed has the function of binding a mononucleotide, one can hypothesize that this adenine could be near the C-terminal ends of strands B, C, and D of the sheet. As already discussed, this is clearly not possible in the TyrRS structure observed in crystals, because the long and partly disordered loop between strands C and D gets in the way.

One anticipates that tRNA binding involves positively charged groups on the enzyme. In competitive labeling experiments, Bosshard, Koch, and Hartley (1978) identified two or more lysine residues whose availability for substitution was affected by tRNA. These are now known to be lysines 225, 230, and 233, which lie on the disordered chain linking the α-β domain to the α helical domain. Other positively charged groups near the ends of strands B, C, and D are Lys 82, Lys 83, and Arg 86 on the disordered chain linking strands C and D. (In *E. coli*, Lys 83 is replaced by Ala, but residue 87 is Lys.) The other strongly disordered region in the α-β domain (residues 151–159) is remote from the mononucleotide-binding fold but carries two positive charges (Lys 152, Arg 157). (In *E. coli* there is a further lysine at 155.) The amino acid sequence of the disordered C-terminal domain, which cannot be identified crystallographically, has one dramatic concentration of positively charged groups. This is near the C-terminus of the molecule and has the sequence Arg-Arg-Gly-Lys-Lys-Lys at residues 411–416. (The last Lys is not conserved in *E. coli*.)

We have the working hypothesis that the binding of tRNA will cause a significant structural change of the enzyme near the active site. The occurrence of four lengths of disordered polypeptide chain, each carrying several positive charges, seems suggestive evidence that tRNA binding may be accompanied by a significant ordering of these parts of the structure.

Table 4. Sequence Homology of MetRS and TyrRS[a]

Enzyme	197			200										210									220		
MetRS (yeast)	Asn	Ile	Leu	Ile	Thr	Ser	Ala	Leu	Pro	Tyr	Val	Asn	Asn	Val	Pro	His	Leu	Gly	Asn	Ile	Gly	Ser	Val	Leu		
MetRS (E. coli)	Lys	Ile	Leu	Val	Thr	Cys	Ala	Leu	Pro	Tyr	Ala	Asn	Gly	Ser	Ile	His	Leu	Gly	His	His	Met	Leu	Glu	His	Ile	Gln
TyrRS (B. stear.)	Arg	Val	Thr	Leu	Tyr	Cys	Gly	Phe	Asp	Pro	Thr	Ala	Asp	Ser	Leu	His	Ile	Gly	His	Leu	Ala	Thr	Ile	Leu	Thr	
TyrRS (E. coli)	Pro	Ile	Ala	Leu	Tyr	Cys	Gly	Phe	Asp	Pro	Thr	Ala	Asp	Ser	Leu	His	Leu	Gly	His	Leu	Val	Pro	Leu	Leu	Cys	

Residue numbering: MetRS (yeast) 6 … 30; MetRS (E. coli) 10 … 40 … 50; TyrRS (B. stear.) 32. (Boxed residues: the Cys column at position 202 and the His-Leu-Gly motif beginning at position 212.)

[a] See Walter et al., 1983; Barker et al., 1982b; Winter et al., 1983; and Barker et al., 1982a.

3.7. Site-Directed Mutagensis and Tests of Structural Interpretation

Tyrosyl-tRNA synthetase is the first enzyme system in which it has proved possible to design site-directed mutagenesis experiments from crystallographic data and to interpret the results of enzyme kinetic experiments on the mutant enzyme in the light of structural data.

The first experiments involved the active site cysteine (Cys 35), which forms part of a conserved sequence in *B. stearothermophilus* and *E. coli* (Table 4). Blocking of this residue with DTNB or mercurials destroys activity. The structural results show the − SH group interacting with the 3′-hydroxyl of the ribose of the adenylate. Any hydrogen bond would be extremely weak, and no other chemical function appears likely. The effect of − SH blockers would be simply steric, preventing correct positioning of the ribose.

Cys 35 has been altered to serine, alanine, and glycine by site-directed mutagenesis (Winter et al., 1982; Wilkinson et al., 1983). All mutants show activity in both aminoacylation and pyrophosphate exchange. The mutants have a slower turnover in both reactions, and the K_M for ATP is significantly increased. These properties seem consistent with a slightly less favorable binding of the ribose.

A number of more drastic changes have been made to the enzyme in subsequent experiments (G. P. Winter and A. R. Fersht, personal communication). In one case the polypeptide chain is terminated at residue 320 (Waye et al., 1983). The enzyme so produced has full activity in pyrophosphate exchange (step 1) but no activity in aminoacylation of tRNA (steps 1 and 2). In another case, a single point mutation, changing Thr 51 to Pro 51, improves the affinity for ATP as measured by K_M, by a factor 100 (Wilkinson *et al.*, 1984).

4. METHIONYL-tRNA SYNTHETASE

4.1. Crystal Structure of Methionyl-tRNA Synthetase Fragment

Methionyl-tRNA synthetase incubated with trypsin was crystallized from 48–49% saturated ammonium sulfate in 0.1 *M* phosphate, pH 7.2 (Waller et al., 1971). The tryptic fragment is monomeric (M.W. 64,000) and retains full enzymatic activity. The monoclinic crystals (Table 1) contain one molecular fragment in the crystal asymmetric unit. Using $Pt(CN)_4^{2-}$ and $UO_2F_5^{3-}$ as isomorphous derivatives, data to approximately 3.9 Å resolution were collected on precession photographs (Monteilhet et al., 1974). These were used to compute a low resolution electron density map, which revealed a molecule roughly $90 \times 52 \times 44$ Å with a slightly bilobal shape (Zelwer et al., 1976). The resolution was subsequently extended to 2.5 Å resolution, using oscillation photographs and including a "double" $Pt(CN)_4^{2-} + UO_2F_5^{3-}$ derivative, $Sm(NO_3)_2$, and 8-bromo-ATP-Cd^{2+} as additional isomorphous derivatives (Risler et al., 1981; Zelwer et al., 1982). Unfortunately, Sm binds at the U site, which is also the main Cd site. The occupancy of the ATP analog is rather low, and apparently Cd binds weakly at many sites. The isomorphous replacement results were improved by use of a density modification procedure (Schevitz et al., 1981).

This map at 2.5 Å resolution revealed a structure that is described in terms of three domains: (1) at the N-terminus, an α-β domain consisting of five parallel β strands with linking α helices; (2) a domain with less ordered structure, which forms an insertion between the third and fourth strands of the β sheet; (3) a C-terminal domain, rich in α helices. The interpretation of the structure remains incomplete—the model structure contains 480 amino acid residues, while the fragment is known to contain 560 residues. Zelwer et al. (1982) originally had only fragments of the amino acid sequence available, but subsequently the complete sequence of the fragment has become known by sequencing the cloned gene (Barker et al., 1982b).

Comparison of the five-stranded β sheet of MetRS and the associated helices of the α-β domain shows a considerable similarity with five parallel strands of the α-β domain of TyrRS. The strands are assembled in the same order, and there is close positional similarity for some of the α helices linking the strands (Blow et al., 1983). There is a particularly close structural homology between the structures of the first parallel β strand (strand B in TyrRS) and the returning helix that leads to the next strand. The connection between this strand and helix has already been discussed as a case where the TyrRS α-β domain has obvious differences from the structures found in the dehydrogenases. In the two synthetases, the joining chain has exactly the same number of residues. The α-carbon positions of the β strand, helix, and joining residues can all be superimposed with an rms difference of 1.6 Å (Figure 7).

This structural homology corresponds precisely to a sequence homology already noted by Barker et al. (1982a; see Table 4). This aligns a cysteine (Cys 35 in TyrRS) and two histidines, as well as several more commonly occurring amino acids, in a length of 14 residues. This sequence alignment is the more difficult to interpret when it is realized that the cysteine has been shown not to be essential for activity in TyrRS (Wilkinson et al., 1983) Indeed, in the various sequences now available (see also Walter et al., 1983) only two amino acids are completely invariant, and it is partly the chance of comparing TyrRS (*B. stearothermophilus*) with MetRS (*E. coli*) that gave Barker and Winter a recognizable alignment. But the correspondence with the structural homology shown in Figure 7 makes the alignment significant.

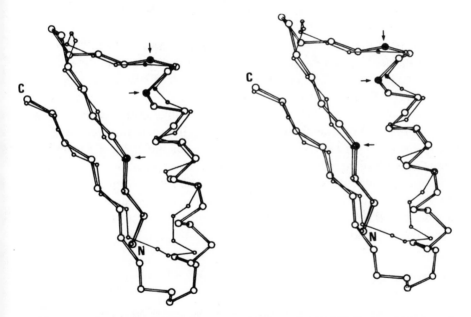

Figure 7. Superposition of α-carbon positions of MetRS (large circles) and TyrRS (small circles) in the region of closest structural homology. The diagram shows two β strands (strands B and C in TyrRS) and a linking helix in each structure (MetRS 8-52, TyrRS 32-71). The arrows mark the homologous cysteine and histidines indicated in Table 4 (Blow et al., 1983).

It is noteworthy that there is a similarly spaced occurrence of cysteine and histidine in the TrpRS sequence (Winter and Hartley, 1977).

4.2. Active Site

The fragment of MetRS retains the firmly bound Zn^{2+} ion, which is known to be essential for activity (Porsorske et al., 1979). The site has not yet been defined: it may possibly be the major U and Sm site.

The binding of 8-bromo-ATP, soaked into the crystals at 10 mM, is very weak. As with TyrRS and other synthetases, K_M for ATP is high. The major binding site is near the center of the C-terminal end of the β sheet, and this may be the site of adenylation. It is very far from the major U site.

5. TRYPTOPHANYL-tRNA SYNTHETASE

TrpRS from *B. stearothermophilus* has been crystallized in several different forms. Crystals of different habit (and very probably different unit cells) can be obtained according to the presence or absence of tryptophan and of ATP (Carter and Carter, 1979).

In the presence of excess ATP the crystalline TrpRS evidently catalyzes a reaction analogous to Step 2, in which a second molecule of ATP takes the place of tRNA^{Trp}, with formation of the 3' (or 2') tryptophan ester of ATP (Coleman and Carter, 1983). The formation of this ester in solution had already been noted (Weiss et al, 1959; Joseph and Muench, 1971). This ester is chemically analogous to the terminal nucleotide of aminoacyl-tRNA, and the structure should be helpful in understanding the mode of binding of aminoacyl-tRNA to the synthetase.

REFERENCES

Baldwin, A. N., and Berg, P. (1966) *J. Biol. Chem.* **241**, 839–845.

Barker, D. G., Bruton, C. J., and Winter G. (1982a). *FEBS Lett.* **150**, 419–423.

Barker, D., Ebel, J. P., Jakes, R., and Bruton, C. J. (1982b). *Eur. J. Biochem.* **127**, 449–457.

Barrell, B. G., Bankier, A. T., and Drouin, A. J. (1979). *Nature* **282**, 189–194.

Bhat, T. N., and Blow, D. M. (1982). *Acta Crystallogr.* **A38**, 22–29.

Bhat, T. N., Blow, D. M., and Brick, P. (1984). In preparation.

Bhat, T. N., Blow, D. M., Brick, P., and Nyborg, J. (1982). *J. Mol. Biol.* **158**, 699–709.

Blow, D. M., Bhat, T. N., Metcalfe, A., Risler, J. L., and Brunie, S. (1983). *J. Mol. Biol.* **171**, 571–576.

Bosshard, H. R., Koch, G. L. E., and Hartley, B. S. (1978). *J. Mol. Biol.* **119**, 377–389.

Brändén, C. I. (1981). *Quart. Rev. Biophys.* **13**, 317–339.

Bruton, C. J., and Cox, L. A. M. (1979). *Eur. J. Biochem.* **100**, 301–308.

Calendar, R., and Berg, P. (1966). *Biochem.* **5**, 1690–1695.

Carter, C. W., and Carter, C. W. (1979). *J. Biol. Chem.* **254**, 12,219–12,223.

Chirikjian, J. G., Wright, H. T., and Fresco, J. R. (1972). *Proc. Nat. Acad. Sci. USA* **69**, 1638–1641.

Chothia, C. (1973). *J. Mol. Biol.* **75**, 295–302.

Coleman, D. E., and Carter, C. W. (1983). Personal Communication.

de Bruijn, M. H. L. (1983). *Nature* **304**, 234–241.

Dessen, P., Blanquet, S., Zaccai, G., and Jacrot, B. (1978). *J. Mol. Biol.* **126**, 293–313.

Dietrich, A., Giegé, R., Comarmond, M. B., Thierry, J. C., Moras, D. (1980). *J. Mol. Biol.* **138**, 129–135.

Edelmann, P., and Gallant, J. (1977). *Cell* **10**, 131–137.

Fersht, A. R., and Kaethner, M. M. (1976). *Biochemistry* **15**, 3342–3346.

Fersht, A. R., Mulvey, R. S., and Koch, G. L. E. (1975). *Biochemistry* **14**, 13–18.

Fersht, A. R., Shindler, J. S., and Tsui, W-C. (1980). *Biochemistry* **19**, 5520–5524.

Giegé, R., Lorber, B., Ebel, J., Moras, D., and Thierry, J. C. (1980). *C. R. Acad. Sci.* **291**, 393–396.

Irwin, M. J., Nyborg, J., Reid, B. R., and Blow, D. M. (1976). *J. Mol. Biol.* **105**, 577–586.

Jakes, R., and Fersht, A. R. (1975). *Biochemistry* **14**, 3344–3350.

Joseph, D. R., and Muench, K. H. (1971). *J. Biol. Chem.* **246**, 7610–7615.

Koch, G. L. E. (1974). *Biochemistry* **13**, 2307–2312.

Loftfield, R. B., and Vanderjagt, D. (1972). *Biochem. J.* **128**, 1353–1356.

Lorber, B., Giegé, R., Ebel, J. P., Berthet, C., Thierry, J. C., and Moras, D. (1983). *J. Biol. Chem.* **258**, 8429–8435.

Lowe, G., Sproat, B. S., and Tansley, G. (1983a). *Eur. J. Biochem.* **130**, 341–345.

Lowe, G., Sproat, B. S., Tansley, G., and Cullis, P. M. (1983b). *Biochem.* **22**, 1229–1236.

Lowe, G., and Tansley, G. (1984). *Tetrahedron* **40**, 113–117.

Monteilhet, C., and Blow, D. M. (1978). *J. Mol. Biol.* **122**, 407–417.

Monteilhet, C., Blow, D. M., and Brick, P. (1984). *J. Mol. Biol.* **173**, 477–486.

Monteilhet, C., Zelwer, C., and Risler, J. L. (1974). *FEBS Lett.* **46**, 101–105.

Mulvey, R. S., and Fersht, A. R. (1977). *Biochemistry* **16**, 4005–4013.

Nikodem, V., Johnson, R. C., and Fresco, J. R. (1978). *Abstracts—Cold Spring Harbor Meeting on tRNA*, p. 21.

Pauling, L. (1957). *Festschrift Arthur Stoll,* p. 597, Birkhauser, Basel.

Popp, R. A., Bailiff, E. G., Hirsch, G. P., and Conrad, R. A. (1976). *Interdiscipl. Top. Gerontol.* **9**, 209–218.

Posorske, L. H., Cohn, M., Yanagisawa, N., and Auld, D. S. (1979). *Biochim. Biophys. Acta* **576**, 128–133.

Rao, S. T., and Rossmann, M. G. (1973). *J. Mol. Biol.* **76**, 241–256.

Reid, B. R., Koch, G. L. E., Boulanger, Y., Hartley, B. S., and Blow, D. M. (1973). *J. Mol. Biol.* **80**, 199–201.

Richardson, J. S. (1976). *Proc. Nat. Acad. Sci. USA* **73**, 2619–2623.

Risler, J. L., Zelwer, C., and Brunie, S. (1981). *Nature* **292**, 384–386.

Rossmann, M. G., Garavito, R. M., and Eventoff, W. (1977). In *Pyridine Nucleotide-dependent dehydrogenases,* H. Sund, Ed., De Gruyter, Berlin, pp. 3–30.

Rossmann, M. G., Liljas, A., Brändén, C. I., and Banaszak, L. J. (1975). In *The Enzymes,* P. D. Boyer, Ed., Vol. 9. Academic Press, New York, pp. 61–102.

Rubin, J., and Blow, D. M. (1981). *J. Mol. Biol.* **145**, 489–500.

Rymo, L., Lagerkvist, U., and Wonacott, A. (1970). *J. Biol. Chem.* **245**, 4308–4316.

Santi, D. V., and Peña, V. A. (1973). *J. Med. Chem.* **16**, 273–280.

Schevitz, R. W., Podjarny, A. D., Zwick, M., Hughes, J. J., and Sigler, P. B. (1981). *Acta Crystallogr.* **A37**, 669–677.

Schimmel, P., and Söll, D. (1979). *Ann. Rev. Biochem.* **48**, 601–648.

Söll, D., and Schimmel, P. R. (1974), In *The Enzymes,* P. D. Boyer, Ed., Vol. 10. Academic Press, New York, pp. 489–538.

Sternberg, M. J. E., and Thornton, J. M. (1976). *J. Mol. Biol.* **105**, 367–382.

Waller, J. P., Risler, J. L., Monteilhet, C., and Zelwer, C. (1971). *FEBS Lett.* **16**, 186–188.

Walter, P., Gangloff, J., Bonnet, J., Boulanger, Y., Ebel, J. P., and Fasiolo, F. (1983). *PNAS* **80**, 2437–2441.

Waye, M. M. Y., Winter, G., Wilkinson, A. J., and Fersht, A. R. (1983). *EMBO J.,* **2**, 1827–1829.

Weiss, S. B., Zachau, H. G., and Lipmann, F. (1959). *Arch. Biochem. Biophys.* **83,** 101–114.

Wilkinson, A. J., Fersht, A. R., Blow, D. M., and Winter, G. (1983). *Biochemistry,* **22,** 3581–3586.

Winter, G., Fersht, A. R., Wilkinson, A. J., Zoller, M., and Smith, M. (1982). *Nature* **299,** 756–758.

Winter, G., and Hartley, B. S. (1977). *FEBS Lett.* **80,** 340–342.

Winter, G., Koch, G. L. E., Hartley, B. S., and Barker, D. G. (1983). *Eur. J. Biochem.* **132,** 383–387.

Zelwer, C., Risler, J. L., and Brunie, S. (1982). *J. Mol. Biol.* **155,** 63–81.

Zelwer, C., Risler, J. L., and Monteilhet, C. (1976), *J. Mol. Biol.* **102,** 93–101.

Helix Comparison Tables

The following tables A1 through C11 give a comparison, in uniform format, of helix parameters for most of the double helical oligomers discussed in the Chapters 1, 2 and 3 on A, B, and Z-DNA respectively. This comparison was produced with the aid of the HELIX program written by John Rosenberg, and the ROLL, CYLIN, and TORAN programs written by Richard Dickerson. The roll and tilt tables, Series A, contain parameters related to base plane orientation. The cylindrical tables, Series B, list parameters describing the orientation of C1' and P atoms about a best overall helix axis. The torsion angle tables, Series C, give the six main-chain and the glycosyl torsion angles for each nucleotide. Within each series, the helices tabulated are:

1. A-helical ^{I}CCGG/^{I}CCGG, with two tetramers stacked so as to form a continuous octameric helix.
2. A-helical GG^{Br}UA^{Br}UACC.
3. A-helical GGCCGGCC.
4. A-helical r(GCG)d(TATACGC).
5. B-helical CGCGAATTCGCG at room temperature.
6. B-helical CGCGAATTCGCG, 16K.
7. B-helical CGCGAATTCGCG complexed with *cis*-dichlorodiaminoplatinum(II).
8. B-helical CGCGAATT^{Br}CGCG with bent axis, MPD20.

9. B-helical CGCGAATT^{Br}CGCG with straight axis, MPD7.
10. Z-helical CGCG.
11. Z-helical CGCGCG.
    Coordinates of helices 1 and 5 through 10 from R. E. Dickerson;
    those of helix 2 from O. Kennard, Z. Shakked and D. Rabinovich;
    those of helices 3, 4 and 11 from A. Wang and A. Rich.

SERIES A. ROLL AND TILT ANGLE TABLES

Individual roll and tilt angles Φ_R and Φ_T (PHI/R and PHI/T in the tables) are the components of the base plane normal in the direction of the minor groove and the long axis of the base pair respectively. Angles Θ_R and Θ_T (THET/R and THET/T) measure the corresponding components of the change in base normal orientation from one base pair to the next, and hence measure the angles between base planes or best planes through base pairs. Θ_R is positive if the roll angle between two successive base pairs opens toward the minor groove and negative if the angle opens toward the major groove. Θ_T is positive if the angle between base planes opens outward toward strand 1 of the double helix. Analytical expressions for Φ_R, Φ_T, Θ_R, and Θ_T are to be found in Fratini et al.* Θ values for the step between base pair j and base pair $j + 1$ are listed in row j, with zeros entered in the final row.

INCLIN is the inclination angle that the C6(pyrimidine)–C8(purine) vector along one base pair makes with a plane perpendicular to the helix axis. It is approximately equal to $-\Phi_T$.

PRTW, propeller twist, is the dihedral angle between normals to the two base planes of a pair, viewed along the long axis of the base pair, and is positive for clockwise rotation of the nearer base. Propeller twist is approximately equal to the difference between Φ_R for strand 2 and strand 1. (Note: These tables list only magnitudes of propeller twist; the occasional negative values can be detected by comparison with the Φ_R differences.)

BUCKLE is the angle that would remain between base normals in a base pair after the propeller twist has been rotated back to zero.

SLIDE, DISP, and SLIP indicate shifts of base pairs relative to the helix axis or to neighboring base pairs. DISP is the displacement of the

---

* Fratini. (1982). *J. Biol. Chem.* **24**, 14,686–14,707.

C6–C8 vector of a base pair from the helix axis, viewed in projection down that axis. It has an average value of $+4.0$ Å for the A helices, -0.2 Å for the B helices, and 0.5–1.5 Å for Z-DNA. Displacement can be regarded as one of the prime attributes distinguishing the A helix from the B. SLIP is the shift of the midpoint of the C6–C8 vector along the C6–C8 direction, once again viewed in projection down the helix axis. SLIP is zero if a perpendicular from the helix axis to the projected C6–C8 vector strikes the vector at its midpoint, and is positive if the midpoint is displaced toward strand 1 of the double helix. SLIP appears to be a relatively uninformative quantity in A and B helices, but shows a striking alternation in Z-DNA. C6/C8 is the length of the C6–C8 vector in projection down the helix axis.

SLIDE measures the *relative* displacement of two successive base pairs along the direction of their long axes. It is independent of helix axis, and depends in fact only on the coordinates of the C6 and C8 atoms on two adjacent base pairs. If these atoms on strands 1 and 2 of the helix at base pair 1 are denoted by 1,1 and 1,2 and those at the adjacent base pair 2 are labeled 2,1 and 2,2 and if the components of vectors $\mathbf{m}$ and $\mathbf{M}$ are given by:

$$\mathbf{m}_x = \tfrac{1}{2}(x_{1,1} - x_{1,2} + x_{2,1} - x_{2,2}), \qquad \text{etc. for } \mathbf{m}_y \text{ and } \mathbf{m}_z$$

$$\mathbf{M}_x = \tfrac{1}{2}(x_{1,1} + x_{1,2} - x_{2,1} - x_{2,2}), \qquad \text{etc. for } \mathbf{M}_y \text{ and } \mathbf{M}_z$$

then the quantity SLIDE is calculated from:

$$\text{SLIDE} = \frac{\mathbf{M}_x\mathbf{m}_x + \mathbf{M}_y\mathbf{m}_y + \mathbf{M}_z\mathbf{m}_z}{(\mathbf{m}_x^2 + \mathbf{m}_y^2 + \mathbf{m}_z^2)^{1/2}}$$

(Vector $\mathbf{M}$ connects the midpoints of the C6–C8 lines in the two base pairs. If for each strand of helix, a line is drawn connecting C6 or C8 atoms on the adjacent base pairs for that strand, vector $\mathbf{m}$ connects the midpoints of these lines and provides an average value for the C6–C8 base pair long axis vector.)

SERIES B. CYLINDRICAL PARAMETERS

In the phosphate backbone half of each table, R, PHI, and Z are cylindrical coordinates of the phosphorus atoms about a best overall helix axis. The *d* is the length of the vector between successive phosphorus atoms along one strand of the double helix, *q* is its projection on a plane perpendicular to the helix axis, and *h* is its projection along the axis, or the rise per helix step. Π (PI) is the local pitch angle: $h/d = \sin(\Pi)$.

In the "rotation and rise" half of each table, *s5′* is the helical rotation semiangle from P past O5′ to C1′ in the 5′-to-3′ direction, and *s3′* is the semiangle from C1′ past O3′ to P. An *s5′* angle plus its following *s3′* yields helical rotation angle *r*, the angle from one phosphorus to the next. An *s3′* plus the following *s5′* yields helix twist angle *t*, the angle from one C1′ atom to the next. Distance *h*(C1′) is the vertical rise along the helix axis between C1′ atoms, analogous to *h* measured between phosphorus atoms in the upper part of the table.

Angle t_g (TG) and rise h_g (HG) are the global twist and rise: t_g is the change in orientation of C1′–C1′ vectors of two successive base pairs viewed in projection down the helix axis, and h_g is the mean of the rise between successive C1′ atoms on the two ends of the base pair. Hence t_g and h_g are properties of the double helix, whereas all other quantities in this table pertain to individual helix strands. Because strand 1 descends the *z* axis and rotates φ in a negative direction, quantities, *h*, Π, *s5′*, *s3′*, *r*, *t* and *h*(C1′) are negative for strand 1 and positive for strand 2. Quantities *d*, *q*, t_g, and h_g are magnitudes.

SLIDE, DISP, SLIP, and C1/C1 are defined as in the roll angle tables, except that C1′ sugar atoms are used as markers instead of C6 and C8 atoms of bases.

SERIES C. TORSION ANGLES

The Series C tables give main-chain and glycosyl torsion angles in the convention recommended in 1982 by the IUPAC-IUB Joint Commission on Biochemical Nomenclature*:

Main chain: P—α—O5′—β—C5′—γ—C4′—δ—C3′—ϵ—O3′—ζ—P

Pyrimidines: O4′——C1′—x—N1——C2

Purines: O4′——C1′—x—N9——C4

* *Eur. J. Biochem.* **131**, 9–15 (1983).

The zero torsion angle is at the fully eclipsed position of the outside atoms, and positive angles occur with clockwise rotation of the more distant bond. Angles are tabulated between −180° and +180°.

Table A1. Roll and Tilt in A-Helical ¹CCGG/¹CCGG

STRAND	1 ROLL	AND TILT	ANGLES		STRAND	2 ROLL	AND TILT	ANGLES
PHI/R	PHI/T	THET/R	THET/T		PHI/R	PHI/T	THET/R	THET/T
-2.91	25.01	4.05	-6.63		8.73	12.65	7.99	-9.95
-10.49	13.20	18.78	2.18		10.56	8.24	9.21	-0.62
2.02	12.48	3.41	3.21		14.34	13.83	7.27	-4.18
-3.61	15.41	2.63	6.95		11.38	18.08	2.63	-6.95
-11.38	18.08	7.27	4.18		3.61	15.41	3.41	-3.21
-14.34	13.83	9.21	0.62		-2.02	12.48	18.78	-2.18
-10.56	8.24	7.99	9.95		10.49	13.20	4.05	6.63
-8.73	12.65	0.00	0.00		2.91	25.01	0.00	0.00

BEST PLANE	THROUGH	BOTH BASES								
PHI/R	PHI/T	THET/R	THET/T	INCLIN	PR TW	BUCKLE	SLIDE	DISP	SLIP	C6/C8
5.19	11.11	2.55	-1.15	-13.96	16.09	9.93	1.15	3.68	-0.14	9.83
0.69	11.76	12.58	-1.11	-12.08	21.41	-2.31	1.66	3.11	-0.01	9.83
7.29	12.22	4.91	0.47	-13.01	12.40	0.71	1.24	3.37	-0.71	9.63
3.03	16.28	3.92	0.00	-16.91	15.13	0.04	0.11	3.68	-0.55	9.61
-3.03	16.28	4.91	-0.47	-16.91	15.13	0.04	1.24	3.68	0.55	9.61
-7.29	12.22	12.58	1.11	-13.01	12.40	0.71	1.66	3.37	0.71	9.63
-0.69	11.76	2.55	1.15	-12.08	21.41	-2.31	1.15	3.11	0.01	9.83
-5.19	11.11	0.00	0.00	-13.96	16.09	9.93	0.00	3.68	0.14	9.83

Table A2. Roll and Tilt in A-Helical GGbrUAbrUACC

STRAND	1 ROLL	AND TILT	ANGLES		STRAND	2 ROLL	AND TILT	ANGLES
PHI/R	PHI/T	THET/R	THET/T		PHI/R	PHI/T	THET/R	THET/T
3.67	6.80	2.32	1.02		7.70	13.58	12.00	1.76
1.53	9.29	0.24	4.97		10.60	20.53	5.70	-6.26
-5.44	13.32	10.01	0.20		4.07	18.52	10.24	-8.06
-1.30	11.92	3.36	9.04		7.08	12.83	2.06	-4.62
-7.36	18.85	11.94	1.19		1.77	10.73	12.03	-4.61
-4.16	17.26	7.63	5.52		9.23	8.91	0.06	-3.03
-7.67	19.57	16.41	-2.64		3.47	9.72	3.65	-2.63
-0.02	15.09	0.00	0.00		2.38	8.68	0.00	0.00

BEST PLANE	THROUGH	BOTH BASES								
PHI/R	PHI/T	THET/R	THET/T	INCLIN	PR TW	BUCKLE	SLIDE	DISP	SLIP	C6/C8
5.66	8.89	6.11	0.56	-9.45	6.41	1.59	1.80	3.70	0.21	9.67
5.92	12.64	0.91	0.20	-13.61	12.37	2.85	1.17	3.45	-0.31	9.49
-1.75	14.24	12.39	-0.53	-15.04	10.27	1.86	1.32	3.92	-0.06	9.43
4.00	14.33	0.65	0.51	-13.93	8.67	-3.00	0.94	3.98	-0.35	9.47
-4.45	14.73	14.98	-1.77	-14.69	10.84	0.34	1.54	3.79	0.18	9.43
3.91	12.94	2.20	-1.33	-13.38	14.61	-0.39	1.38	4.17	-0.20	9.50
-1.63	12.28	8.58	-0.26	-13.27	13.35	2.91	1.94	3.55	-0.03	9.59
0.70	12.03	0.00	0.00	-12.04	5.12	-0.30	0.00	3.98	-0.73	9.75

Table A3. Roll and Tilt in A-Helical GGCCGGCC

STRAND 1 ROLL AND TILT ANGLES				STRAND 2 ROLL AND TILT ANGLES			
PHI/R	PHI/T	THET/R	THET/T	PHI/R	PHI/T	THET/R	THET/T
0.84	18.60	11.76	-4.35	1.26	14.55	14.28	-2.97
2.15	15.00	2.50	5.54	6.71	14.22	2.61	-11.74
-4.10	20.49	7.46	-10.65	4.66	4.94	-1.46	12.03
-6.19	4.31	14.92	13.12	-5.84	17.65	14.92	-13.12
5.84	17.65	-1.46	-12.03	6.19	4.31	7.46	10.65
-4.66	4.94	2.61	11.74	4.10	20.49	2.50	-5.54
-6.71	14.22	14.28	2.97	-2.15	15.00	11.76	4.35
-1.26	14.55	0.00	0.00	-0.84	18.60	0.00	0.00

BEST PLANE THROUGH BOTH BASES

PHI/R	PHI/T	THET/R	THET/T	INCLIN	PR TW	BUCKLE	SLIDE	DISP	SLIP	C6/C8
-0.60	12.24	11.25	-1.47	-13.47	5.27	6.23	0.88	3.61	-0.43	9.50
3.27	11.69	2.48	-3.81	-12.74	5.32	3.84	1.25	3.96	0.34	9.57
0.73	8.74	2.17	3.42	-10.25	14.50	4.99	1.19	3.51	0.20	9.60
-4.81	10.77	12.42	0.00	-11.03	9.43	0.02	2.15	3.11	1.03	9.48
4.81	10.77	2.17	-3.43	-11.03	9.43	0.02	1.19	3.11	-1.03	9.48
-0.73	8.74	2.48	3.81	-10.25	14.50	4.99	1.25	3.51	-0.20	9.60
-3.27	11.69	11.25	1.47	-12.74	5.32	3.84	0.88	3.96	-0.34	9.57
0.60	12.24	0.00	0.00	-13.47	5.27	6.23	0.00	3.61	0.43	9.50

Table A4. Roll and Tilt in A-Helical r(GCG)d(TATACGC)

STRAND 1 ROLL AND TILT ANGLES				STRAND 2 ROLL AND TILT ANGLES			
PHI/R	PHI/T	THET/R	THET/T	PHI/R	PHI/T	THET/R	THET/T
6.23	15.20	2.04	3.85	14.84	10.43	2.44	-2.58
-3.61	20.29	6.73	-1.55	8.05	15.63	11.24	3.77
-6.71	15.72	7.81	8.77	8.17	25.13	4.67	-10.40
-9.18	20.24	17.33	2.27	0.77	16.40	20.75	-1.65
-1.59	19.72	10.00	6.45	13.06	18.22	5.80	-5.30
-4.29	24.56	16.33	1.42	7.51	18.81	14.19	-2.82
0.44	24.56	4.09	0.01	12.33	20.50	-0.03	-9.06
-8.95	21.88	10.72	3.41	1.84	15.06	12.16	0.33
-10.47	19.36	8.86	5.06	4.41	17.59	-3.23	1.11
-13.13	17.04	0.00	0.00	-10.22	16.75	0.00	0.00

BEST PLANE THROUGH BOTH BASES

PHI/R	PHI/T	THET/R	THET/T	INCLIN	PR TW	BUCKLE	SLIDE	DISP	SLIP	C6/C8
9.68	13.15	3.34	1.52	-13.64	9.20	-1.11	1.62	4.47	-0.37	9.60
2.18	18.91	8.91	-0.34	-19.03	12.17	-1.91	1.85	3.78	-0.16	9.19
0.60	19.67	5.05	0.84	-20.31	16.35	0.58	1.06	4.47	-0.72	9.35
-5.54	18.98	22.31	0.36	-19.14	10.40	-1.39	1.24	5.11	-0.18	9.31
7.17	19.73	5.30	-0.25	-19.83	14.69	-1.19	1.48	5.24	0.13	9.29
0.13	21.62	17.63	-0.44	-21.46	12.47	0.71	1.36	4.44	0.37	9.06
7.25	22.53	1.11	-2.59	-22.95	12.15	-0.28	1.34	4.15	-0.06	8.92
-4.26	20.61	12.70	1.96	-20.40	12.15	-3.58	1.42	4.44	-0.17	9.02
-3.62	20.32	4.02	1.60	-20.53	15.17	-3.69	1.07	4.86	0.17	9.09
-11.55	16.93	0.00	0.00	-17.37	2.93	-0.20	0.00	4.55	0.94	9.21

Table A5. Roll and Tilt in Native B-Helical CGCGAATTCGCG

STRAND 1 ROLL AND TILT ANGLES			
PHI/R	PHI/T	THET/R	THET/T
-4.40	13.47	7.46	-1.51
-4.18	8.86	-2.70	2.31
-12.26	5.40	4.27	12.48
-11.65	12.46	2.61	0.12
-14.42	4.06	1.90	3.66
-13.41	-1.30	-1.68	6.29
-13.85	-2.76	1.16	7.14
-10.59	-3.11	-1.00	0.96
-7.26	-8.68	8.19	4.94
4.47	-4.24	-18.23	-2.61
-9.71	-9.04	7.35	9.22
1.27	-1.86	0.00	0.00

STRAND 2 ROLL AND TILT ANGLES			
PHI/R	PHI/T	THET/R	THET/T
8.65	11.71	6.19	-3.65
6.65	13.20	-6.05	-0.67
-8.62	11.66	10.95	-4.30
-1.34	4.47	3.69	-5.89
1.79	-1.41	0.18	-4.83
4.10	-4.53	-3.55	-2.51
3.16	-4.92	0.76	-1.10
6.50	-2.97	1.31	-1.58
8.66	0.93	-0.76	-10.95
9.15	-5.78	-5.99	-5.94
7.17	-5.91	-12.16	-3.12
-1.09	-7.59	0.00	0.00

BEST PLANE THROUGH BOTH BASES

PHI/R	PHI/T	THET/R	THET/T	INCLIN	PR TW	BUCKLE	SLIDE	DISP	SLIP	C6/C8
2.74	13.13	7.09	-1.71	-13.78	13.19	-2.18	-0.26	0.89	0.43	9.49
1.36	12.77	-3.16	0.33	-12.71	11.50	-3.24	-0.47	0.84	0.48	9.53
-10.11	9.90	5.02	0.86	-9.86	5.92	-2.33	-0.66	0.28	0.62	9.63
-8.69	6.17	5.77	-1.20	-7.41	11.94	2.28	-0.04	0.83	0.97	9.49
-4.81	0.68	0.91	-1.00	-1.27	16.68	0.14	0.40	0.58	1.21	9.76
-3.05	-2.98	-6.04	1.57	3.03	17.66	0.15	0.62	0.15	1.09	9.93
-7.17	-4.28	0.30	2.57	4.33	17.08	0.84	0.31	-0.31	0.62	9.82
-3.79	-5.11	2.67	0.14	4.43	17.19	2.68	0.10	-0.35	0.36	9.76
2.91	-5.30	1.71	-0.44	4.00	17.32	0.17	-0.81	-0.55	0.17	9.73
6.88	-3.24	-9.82	-2.59	3.64	5.20	-2.80	-0.65	-1.42	0.77	9.75
-0.93	-3.75	-0.98	0.84	3.96	17.78	-7.13	-0.17	-0.98	0.81	9.74
0.23	-3.12	0.00	0.00	3.64	4.93	-2.16	0.00	-2.05	0.29	9.91

Table A6. Roll and Tilt in B-Helical CGCGAATTCGCG, 16K

STRAND 1 ROLL AND TILT ANGLES			
PHI/R	PHI/T	THET/R	THET/T
-9.38	13.61	11.54	1.60
-5.49	10.31	-1.92	-0.16
-12.53	3.78	10.00	9.62
-5.40	9.47	0.25	3.08
-10.91	7.14	-1.40	2.02
-15.13	0.87	-4.41	3.89
-18.45	-5.13	-0.23	8.48
-14.68	-6.91	1.78	-6.88
-2.85	-21.05	5.72	16.50
8.60	-2.16	-25.37	-8.72
-12.44	-12.53	12.48	8.71
6.12	-5.03	0.00	0.00

STRAND 2 ROLL AND TILT ANGLES			
PHI/R	PHI/T	THET/R	THET/T
13.21	9.00	-5.54	-7.34
2.06	6.38	-3.14	3.42
-6.49	8.44	14.44	-4.54
5.40	3.56	-0.31	-4.46
3.30	1.82	6.01	-3.32
8.39	2.25	-3.17	-10.45
5.35	-4.29	0.34	-3.34
8.26	-3.36	-1.77	-1.66
7.33	0.58	0.28	-17.09
11.02	-12.17	-11.85	-6.93
7.41	-12.56	-23.99	2.51
-9.56	-11.17	0.00	0.00

BEST PLANE THROUGH BOTH BASES

PHI/R	PHI/T	THET/R	THET/T	INCLIN	PR TW	BUCKLE	SLIDE	DISP	SLIP	C6/C8
2.65	13.38	4.66	-1.22	-14.02	23.08	-4.87	-0.41	0.63	0.45	9.43
-1.35	12.61	-0.61	-0.17	-11.79	9.39	-6.83	-0.59	0.53	0.44	9.55
-9.43	8.60	10.67	-0.60	-8.34	7.48	-4.20	-0.58	0.13	0.65	9.61
-1.78	5.35	1.88	-1.27	-6.12	11.60	0.84	0.00	0.59	0.94	9.63
-2.48	2.57	1.86	-2.98	-3.11	14.86	2.81	0.33	0.24	0.93	9.73
-0.78	-1.57	-9.79	-1.23	0.69	23.76	4.48	0.64	0.13	0.64	9.88
-8.86	-5.52	0.04	5.10	5.45	23.83	1.17	0.32	0.04	0.21	9.70
-5.46	-4.81	4.88	-1.23	4.75	23.10	-0.98	0.14	0.10	0.22	9.75
3.93	-6.85	2.17	-1.73	7.12	18.86	-6.31	-0.84	-0.24	0.36	9.53
9.35	-5.15	-15.69	-4.49	5.72	7.83	-3.06	-0.72	-1.38	1.29	9.64
-2.71	-7.25	-2.58	1.00	7.55	21.16	-10.23	-0.16	-1.18	1.55	9.53
-0.71	-7.37	0.00	0.00	8.27	16.29	0.18	0.00	-2.89	0.96	9.68

Table A7. Roll and Tilt in B-Helical CGCGAATTCGCG Cisplatin Complex

STRAND 1 ROLL AND TILT ANGLES				STRAND 2 ROLL AND TILT ANGLES			
PHI/R	PHI/T	THET/R	THET/T	PHI/R	PHI/T	THET/R	THET/T
-0.64	14.98	6.61	3.80	14.16	15.32	10.93	-4.27
-5.06	16.82	0.87	-4.28	12.63	20.48	-16.95	-8.00
-11.72	6.53	-3.70	15.27	-14.77	9.89	13.29	-6.91
-21.33	13.88	2.07	-0.95	-3.02	-1.57	-0.63	-1.40
-23.33	-1.24	2.21	5.64	-1.73	-4.56	6.92	2.62
-16.27	-10.09	-0.09	8.12	7.25	0.41	2.19	-12.64
-11.36	-8.90	-0.05	13.01	11.64	-7.59	4.77	-1.38
-8.26	-1.29	2.96	-9.03	19.13	0.14	-8.03	-2.37
0.58	-14.08	6.49	9.69	7.12	7.79	8.19	-18.99
9.95	-0.97	-19.26	-7.70	16.72	-4.95	-3.49	-15.65
-7.19	-7.72	9.19	1.58	17.78	-8.05	-24.50	-0.53
7.10	-5.91	0.00	0.00	-3.43	-4.20	0.00	0.00

BEST PLANE THROUGH BOTH BASES										
PHI/R	PHI/T	THET/R	THET/T	INCLIN	PR TW	BUCKLE	SLIDE	DISP	SLIP	C6/C8
7.52	16.06	5.24	-3.00	-16.62	14.91	-2.42	-0.17	0.96	1.36	9.49
1.36	15.97	-5.94	-0.98	-17.42	18.05	2.98	-0.95	0.52	0.95	9.30
-13.71	10.28	3.59	3.97	-9.61	4.23	-2.28	-0.72	-0.17	1.15	9.71
-14.55	7.25	5.34	-1.26	-8.19	21.63	-3.47	0.00	0.40	1.30	9.78
-10.50	-2.11	5.16	-0.42	2.15	21.74	-1.67	0.08	0.06	1.28	9.73
-1.41	-7.07	-4.29	0.64	6.05	24.76	2.12	0.50	-0.80	1.16	9.79
-2.06	-7.39	0.16	1.65	7.40	23.06	-2.00	0.46	-0.72	0.65	9.87
2.05	-5.65	0.75	-0.05	3.84	27.78	6.29	-0.06	-0.66	0.12	9.85
6.04	-2.61	4.86	-4.35	2.14	16.81	-2.03	-1.07	-1.28	-0.41	9.67
12.63	-2.18	-7.81	-4.52	2.17	7.46	-1.70	-0.67	-2.10	0.06	10.09
4.90	0.06	-2.50	-5.78	0.79	26.99	-14.24	-0.62	-0.80	-0.26	9.92
3.44	-3.25	0.00	0.00	3.96	10.73	-2.22	0.00	-2.32	-0.50	10.02

Table A8. Roll and Tilt in B-Helical CGCGAATTbrCGCG, MPD20

STRAND 1 ROLL AND TILT ANGLES				STRAND 2 ROLL AND TILT ANGLES			
PHI/R	PHI/T	THET/R	THET/T	PHI/R	PHI/T	THET/R	THET/T
1.79	8.54	-2.36	-3.20	3.50	8.24	11.80	6.20
-5.25	4.06	-4.18	4.15	7.17	18.15	-8.28	-5.37
-12.11	2.22	2.84	11.10	-11.45	10.45	15.48	-3.05
-11.58	7.63	-4.83	5.60	0.63	4.41	2.68	-6.76
-20.56	3.09	6.81	4.50	2.58	-1.62	3.82	-9.19
-13.31	-3.11	-3.41	7.52	10.00	-6.83	-0.66	-4.73
-14.67	-3.45	6.15	7.64	12.72	-4.98	-2.43	-1.31
-6.10	-2.22	-7.55	-6.02	11.26	1.38	-1.36	-1.72
-8.36	-13.93	6.74	12.96	7.12	6.09	-4.34	-18.81
2.27	-1.92	-18.37	-4.64	4.13	-10.29	-2.90	0.08
-12.26	-10.78	4.04	12.17	7.35	-5.55	-8.91	4.37
-3.21	-3.17	0.00	0.00	-0.62	1.34	0.00	0.00

BEST PLANE THROUGH BOTH BASES										
PHI/R	PHI/T	THET/R	THET/T	INCLIN	PR TW	BUCKLE	SLIDE	DISP	SLIP	C6/C8
2.70	8.51	3.53	-1.18	-8.48	1.73	-0.17	-0.15	0.14	0.34	9.82
0.57	8.22	-6.17	1.32	-9.51	16.14	3.30	-0.55	0.17	0.05	9.49
-10.98	5.74	5.85	2.38	-6.12	5.87	0.69	-0.61	-0.18	0.06	9.71
-7.24	3.55	1.98	1.21	-4.77	12.58	2.61	0.05	0.56	0.43	9.67
-6.44	0.35	5.49	-1.51	-0.82	23.39	-0.12	0.54	0.70	0.62	9.71
0.35	-3.02	-5.46	0.74	3.48	23.57	-3.14	0.67	0.65	0.59	10.05
-3.52	-3.00	2.36	0.52	3.77	27.42	-1.10	0.28	0.31	0.39	9.84
1.00	-3.31	-1.24	-0.76	2.37	17.77	3.85	0.22	0.30	0.49	9.73
1.96	-3.06	0.38	-0.11	2.74	21.09	-2.40	-0.80	-0.13	0.47	9.85
3.58	-1.76	-7.23	-4.35	2.92	7.68	-6.37	-0.69	-1.16	1.18	9.88
-1.33	-5.48	-3.33	4.01	5.15	20.46	-6.19	-0.43	-0.92	1.34	9.82
-2.17	-2.43	0.00	0.00	1.90	4.34	1.96	0.00	-2.20	0.99	9.82

Table A9.   Roll and Tilt in B-Helical CGCGAATTbrCGCG, MPD7

STRAND 1 ROLL AND TILT ANGLES				STRAND 2 ROLL AND TILT ANGLES			
PHI/R	PHI/T	THET/R	THET/T	PHI/R	PHI/T	THET/R	THET/T
-8.13	-0.47	3.52	5.37	5.88	3.24	-5.24	-19.46
-4.44	1.08	-5.26	3.63	3.10	-13.90	-12.15	16.58
-10.31	-0.52	1.82	13.48	-5.69	2.70	3.93	-10.94
-10.98	7.92	-4.98	4.59	0.67	-9.67	-5.96	-0.01
-20.07	2.14	5.45	0.78	0.87	-9.16	1.91	-4.65
-11.69	-8.46	-2.18	4.31	9.73	-10.41	-1.38	-2.75
-8.86	-9.93	-0.37	5.06	13.35	-6.73	3.08	-1.29
-3.72	-8.31	-6.54	3.86	18.38	1.05	-6.63	-0.43
-4.54	-7.53	8.43	7.30	6.50	10.27	0.29	-18.60
5.71	0.41	-22.85	5.52	5.99	-5.56	-6.34	-1.16
-19.25	1.47	11.79	1.57	2.65	-3.29	-36.13	-8.73
-5.93	-4.17	0.00	0.00	-27.30	-19.07	0.00	0.00

BEST PLANE THROUGH BOTH BASES										
PHI/R	PHI/T	THET/R	THET/T	INCLIN	PR TW	BUCKLE	SLIDE	DISP	SLIP	C6/C8
-0.66	3.74	2.86	-0.48	-3.46	14.55	-4.09	-0.26	0.16	0.31	9.88
0.20	3.00	-5.51	2.97	-0.74	16.48	-14.30	-0.41	0.04	0.49	9.69
-8.02	3.26	2.23	0.93	-2.96	5.77	-3.68	-0.61	-0.99	0.45	9.70
-6.74	0.54	-2.09	-0.13	-0.93	17.22	-3.59	0.36	-0.16	0.64	9.93
-7.56	-4.55	4.12	-0.27	3.69	22.39	0.14	0.46	0.12	0.33	9.95
1.06	-7.33	-5.04	-0.38	7.77	21.62	-3.61	0.43	-0.09	0.18	9.69
0.09	-7.59	2.06	1.99	7.70	22.35	-1.56	0.35	-0.16	0.12	9.91
5.68	-3.91	-3.52	-0.21	3.52	23.05	-0.42	0.06	-0.49	-0.10	9.73
3.64	-0.50	3.06	-0.34	-0.77	16.78	1.21	-0.56	-0.41	-0.43	9.85
6.52	1.65	-12.19	-2.86	-0.35	5.92	-5.85	-0.73	-1.01	-0.24	9.93
-6.70	-1.31	-11.19	-1.00	-0.03	22.20	-1.88	-0.01	0.07	0.16	9.81
-15.62	-8.60	0.00	0.00	10.84	24.50	-2.92	0.00	0.03	0.62	9.75

Table A10.   Roll and Tilt in Z-Helical CGCG

STRAND 1 ROLL AND TILT ANGLES				STRAND 2 ROLL AND TILT ANGLES			
PHI/R	PHI/T	THET/R	THET/T	PHI/R	PHI/T	THET/R	THET/T
7.02	12.22	7.56	12.83	7.14	5.01	7.04	-2.10
0.02	-1.05	2.12	-13.41	1.40	6.54	6.57	2.17
2.45	12.42	15.63	11.12	-1.23	4.10	6.48	-0.29
-11.33	2.06	0.00	0.00	-6.65	5.26	0.00	0.00

BEST PLANE THROUGH BOTH BASES										
PHI/R	PHI/T	THET/R	THET/T	INCLIN	PR TW	BUCKLE	SLIDE	DISP	SLIP	C6/C8
6.54	4.69	8.55	-1.34	-5.68	6.62	5.96	-5.47	4.06	-2.15	9.75
-0.84	5.83	3.31	0.49	-4.66	6.17	-3.83	0.52	3.85	2.19	9.85
0.33	5.51	11.11	2.53	-6.25	7.51	4.04	-5.37	3.93	-2.08	9.84
-9.68	3.94	0.00	0.00	-3.48	5.19	0.46	0.00	4.06	2.08	9.82

Table A11.   Roll and Tilt in Z-Helical CGCGCG

STRAND 1 ROLL AND TILT ANGLES				STRAND 2 ROLL AND TILT ANGLES			
PHI/R	PHI/T	THET/R	THET/T	PHI/R	PHI/T	THET/R	THET/T
1.54	8.57	5.31	2.86	2.04	8.04	3.07	-4.18
-2.78	5.95	6.72	-2.20	0.61	12.19	5.12	3.38
-2.53	10.83	0.10	6.94	3.49	6.54	6.27	-5.88
-1.42	4.14	0.82	-4.79	-1.31	12.31	4.47	2.32
3.74	8.36	-1.49	-3.93	3.75	8.53	3.03	1.69
6.91	11.42	0.00	0.00	2.16	6.10	0.00	0.00

BEST PLANE THROUGH BOTH BASES										
PHI/R	PHI/T	THET/R	THET/T	INCLIN	PR TW	BUCKLE	SLIDE	DISP	SLIP	C6/C8
2.14	9.26	5.09	0.93	-9.19	1.39	-1.75	-5.29	3.19	-2.40	9.73
-1.66	8.44	4.45	0.00	-8.42	5.63	1.31	0.75	3.12	1.87	9.78
1.22	8.64	4.95	-0.30	-8.51	6.74	0.36	-5.38	3.04	-2.42	9.77
-2.34	9.03	2.08	0.16	-8.30	5.78	-0.15	0.59	3.02	1.95	9.77
3.67	8.22	0.21	-2.43	-8.24	0.34	0.45	-5.07	2.97	-2.25	9.83
5.04	9.84	0.00	0.00	-10.03	6.32	-2.46	0.00	2.63	1.68	9.74

479

Table B1. Cylindrical Parameters for A-Helical ¹CCGG/¹CCGG

PHOSPHATE BACKBONE TABLE, 5" TO 3" DIRECTION IN EACH STRAND

R	PHI	Z	D	Q	H	PI
STRAND	1					
8.94	-165.51	13.86	5.50	5.00	-2.29	-24.64
8.78	161.75	11.57	6.43	5.76	-2.86	-26.42
9.19	124.44	8.71	**	**	**	**
**	**	**	**	**	**	**
9.80	56.83	2.32	5.69	5.27	-2.15	-22.19
10.09	26.12	0.17	6.60	6.10	-2.53	-22.56
9.57	-9.89	-2.37	0.00	0.00	0.00	0.00
STRAND	2					
8.94	-122.32	5.60	5.50	5.00	2.29	24.64
8.78	-89.58	7.89	6.43	5.76	2.86	26.42
9.19	-52.27	10.75	**	**	**	**
**	**	**	**	**	**	**
9.80	15.35	17.14	5.69	5.27	2.15	22.19
10.09	46.05	19.29	6.60	6.10	2.53	22.56
9.57	82.07	21.83	0.00	0.00	0.00	0.00

ROTATION AND RISE TABLE, 5" TO 3" DIRECTION

S5"	S3	R(P)	T(C1")	TG	H(C1")	HG	SLIDE	DISP	SLIP	C1/C1
STRAND	1									
0.00	-1.82	0.00	-32.51	34.95	-2.364	2.835	2.82	6.13	-0.20	11.07
-30.69	-2.05	-32.74	-36.51	31.00	-3.013	2.894	3.21	5.57	-0.05	10.67
-34.46	-2.85	-37.31	-36.72	36.49	-2.979	2.591	3.00	5.85	-0.68	10.50
-33.87	**	**	-28.11	33.49	-3.259	3.259	1.77	6.15	-0.50	10.52
**	-5.63	**	-34.20	36.49	-2.203	2.591	3.00	6.15	0.50	10.52
-28.57	-2.14	-30.70	-32.38	31.00	-2.776	2.894	3.21	5.85	0.68	10.50
-30.25	-5.77	-36.02	-35.90	34.95	-3.306	2.835	2.82	5.57	0.05	10.67
-30.13	0.00	0.00	0.00	0.00	0.000	0.000	0.00	6.13	0.20	11.07
STRAND	2									
0.00	1.82	0.00	32.51	34.95	2.364	2.835	0.00	0.00	0.00	0.00
30.69	2.05	32.74	36.51	31.00	3.013	2.894	0.00	0.00	0.00	0.00
34.46	2.85	37.31	36.72	36.49	2.979	2.591	0.00	0.00	0.00	0.00
33.87	**	**	28.11	33.49	3.259	3.259	0.00	0.00	0.00	0.00
**	5.63	**	34.20	36.49	2.203	2.591	0.00	0.00	0.00	0.00
28.57	2.14	30.70	32.38	31.00	2.776	2.894	0.00	0.00	0.00	0.00
30.25	5.77	36.02	35.90	34.95	3.306	2.835	0.00	0.00	0.00	0.00
30.13	0.00	0.00	0.00	0.00	0.000	0.000	0.00	0.00	0.00	0.00

Starred values are missing because the two stacked tetramers are not bridged by phosphates at the center of the octamer helix.

Table B2. Cylindrical Parameters for A-Helical GGbrUAbrUACC

PHOSPHATE BACKBONE TABLE, 5' TO 3' DIRECTION IN EACH STRAND

R	PHI	Z	D	G	H	PI
STRAND	1					
8.85	22.75	17.49	6.02	5.35	-2.76	-27.26
9.06	-11.98	14.73	5.74	4.86	-3.07	-32.26
8.97	-43.23	11.67	5.99	5.08	-3.18	-32.05
9.32	-75.40	8.49	5.96	5.13	-3.02	-30.50
9.01	-107.88	5.47	5.80	5.02	-2.90	-30.00
9.27	-139.74	2.57	6.01	5.33	-2.77	-27.46
9.15	-173.40	-0.21	0.00	0.00	0.00	0.00
STRAND	2					
9.08	75.96	8.44	6.21	5.71	2.42	22.99
9.44	111.83	10.86	5.59	4.89	2.71	29.03
9.69	141.43	13.57	6.24	5.42	3.10	29.75
9.63	173.99	16.67	5.94	4.93	3.31	33.89
9.52	-156.20	19.98	6.03	5.17	3.10	30.92
9.54	-124.73	23.08	6.37	5.88	2.46	22.72
9.44	-88.68	25.54	0.00	0.00	0.00	0.00

ROTATION AND RISE TABLE, 5' TO 3' DIRECTION

S5'	S3'	R(P)	T(C1')	TG	H(C1')	HG	SLIDE	DISP	SLIP	C1/C1
STRAND	1									
0.00	-1.61	0.00	-35.58	32.47	-2.955	2.747	3.45	6.15	0.35	10.70
-33.96	-0.77	-34.73	-35.47	33.90	-3.134	2.879	2.89	5.95	-0.20	10.29
-34.70	3.45	-31.25	-30.39	28.98	-3.102	3.114	2.58	6.41	-0.14	10.34
-33.84	1.66	-32.18	-31.50	34.08	-3.024	3.059	2.55	6.46	-0.28	10.20
-33.17	0.69	-32.48	-32.43	30.10	-2.739	2.967	2.96	6.30	0.16	10.00
-33.12	1.26	-31.86	-30.85	33.39	-2.861	2.755	3.17	6.67	-0.10	10.26
-32.11	-1.55	-33.65	-36.42	30.45	-3.178	2.947	3.27	6.04	-0.19	10.35
-34.87	0.00	0.00	0.00	0.00	0.000	0.000	0.00	6.47	-0.91	10.41
STRAND	2									
0.00	1.21	0.00	32.27	30.45	2.717	2.947	0.00	0.00	0.00	0.00
31.06	4.81	35.87	36.89	33.39	2.650	2.755	0.00	0.00	0.00	0.00
32.08	-2.49	29.59	30.62	30.10	3.194	2.967	0.00	0.00	0.00	0.00
33.11	-0.55	32.56	31.83	34.08	3.094	3.059	0.00	0.00	0.00	0.00
32.38	-2.57	29.81	29.17	28.98	3.126	3.114	0.00	0.00	0.00	0.00
31.74	-0.27	31.47	31.58	33.90	2.624	2.879	0.00	0.00	0.00	0.00
31.85	4.21	36.05	35.31	32.47	2.539	2.747	0.00	0.00	0.00	0.00
31.10	0.00	0.00	0.00	0.00	0.000	0.000	0.00	0.00	0.00	0.00

Table B3. Cylindrical Parameters for A-Helical GGCCGGCC

PHOSPHATE BACKBONE TABLE, 5' TO 3' DIRECTION IN EACH STRAND

R	PHI	Z	D	G	H	PI
STRAND	1					
9.93	63.45	5.23	6.19	5.25	-3.27	-31.90
9.90	32.73	1.96	6.38	5.96	-2.28	-20.98
10.27	-1.55	-0.32	5.67	4.74	-3.11	-33.23
9.95	-28.60	-3.43	6.90	5.94	-3.52	-30.70
8.41	-65.10	-6.95	5.69	5.13	-2.45	-25.53
9.19	-98.63	-9.40	6.07	5.03	-3.39	-34.02
9.34	-130.10	-12.80	0.00	0.00	0.00	0.00
STRAND	2					
9.93	116.55	-5.23	6.19	5.25	3.27	31.90
9.90	147.27	-1.96	6.38	5.96	2.28	20.98
10.27	-178.45	0.32	5.67	4.74	3.11	33.23
9.95	-151.40	3.43	6.90	5.94	3.52	30.70
8.41	-114.90	6.95	5.69	5.13	2.45	25.53
9.19	-81.37	9.40	6.07	5.03	3.39	34.02
9.34	-49.90	12.80	0.00	0.00	0.00	0.00

ROTATION AND RISE TABLE, 5' TO 3' DIRECTION

S5'	S3'	R(P)	T(C1')	TG	H(C1')	HG	SLIDE	DISP	SLIP	C1/C1
STRAND	1									
0.00	-0.93	0.00	-34.05	36.61	-3.040	3.251	2.40	6.14	-0.34	10.31
-33.13	2.41	-30.72	-26.60	27.47	-2.824	3.112	2.68	6.48	0.43	10.28
-29.01	-5.27	-34.28	-37.47	44.14	-2.431	2.159	3.08	6.01	0.10	10.45
-32.20	5.15	-27.05	-26.67	16.12	-3.991	3.991	2.61	5.55	0.96	10.48
-31.82	-4.69	-36.51	-41.32	44.14	-1.888	2.159	3.08	5.55	-0.96	10.48
-36.63	3.11	-33.52	-31.91	27.47	-3.401	3.112	2.68	6.01	-0.10	10.45
-35.02	3.54	-31.48	-30.83	36.61	-3.463	3.251	2.40	6.48	-0.43	10.28
-34.37	0.00	0.00	0.00	0.00	0.000	0.000	0.00	6.14	0.34	10.31
STRAND	2									
0.00	0.93	0.00	34.05	36.61	3.040	3.251	0.00	0.00	0.00	0.00
33.13	-2.41	30.72	26.60	27.47	2.824	3.112	0.00	0.00	0.00	0.00
29.01	-5.27	34.28	37.47	44.14	2.431	2.159	0.00	0.00	0.00	0.00
32.20	-5.15	27.05	26.67	16.12	3.991	3.991	0.00	0.00	0.00	0.00
31.82	4.69	36.51	41.32	44.14	1.888	2.159	0.00	0.00	0.00	0.00
36.63	-3.11	33.52	31.91	27.47	3.401	3.112	0.00	0.00	0.00	0.00
35.02	-3.54	31.48	30.83	36.61	3.463	3.251	0.00	0.00	0.00	0.00
34.37	0.00	0.00	0.00	0.00	0.000	0.000	0.00	0.00	0.00	0.00

Table B4. Cylindrical Parameters for A-Helical r(GCG)d(TATACGC)

PHOSPHATE BACKBONE TABLE, 5' TO 3' DIRECTION IN EACH STRAND

R	PHI	Z	D	Q	H	PI
STRAND	1					
8.41	131.96	7.04	5.62	5.17	-2.19	-22.98
8.29	95.90	4.84	5.53	5.07	-2.21	-23.55
9.16	62.58	2.63	5.87	5.19	-2.74	-27.84
9.72	30.81	-0.11	6.13	5.26	-3.16	-31.00
9.71	-0.59	-3.27	5.54	4.89	-2.60	-27.96
9.48	-30.10	-5.87	5.85	5.24	-2.61	-26.44
8.59	-63.34	-8.47	6.16	5.73	-2.26	-21.50
9.60	-99.52	-10.73	5.62	5.21	-2.11	-22.05
10.04	-130.21	-12.84	0.00	0.00	0.00	0.00
STRAND	2					
9.32	130.76	-3.46	5.72	5.02	2.75	28.72
8.90	162.62	-0.71	5.75	5.16	2.54	26.25
8.85	-163.59	1.83	5.54	5.02	2.35	25.12
9.18	-131.35	4.18	5.98	5.53	2.28	22.40
9.58	-97.16	6.46	5.65	5.03	2.58	27.15
10.19	-67.91	9.04	5.85	4.92	3.18	32.87
9.84	-39.56	12.21	6.08	5.44	2.72	26.58
9.55	-7.04	14.93	6.29	6.00	1.88	17.40
9.37	29.91	16.82	0.00	0.00	0.00	0.00

ROTATION AND RISE TABLE, 5' TO 3' DIRECTION

S5'	S3'	R(P)	T(C1')	TG	H(C1')	HG	SLIDE	DISP	SLIP	C1/C1
STRAND	1									
0.00	-3.18	0.00	-36.84	37.92	-2.627	2.296	3.54	6.93	-0.33	10.64
-33.66	-2.41	-36.07	-37.25	32.62	-2.243	2.153	3.25	6.26	-0.30	9.99
-34.85	1.53	-33.32	-32.14	32.75	-2.783	2.763	2.68	6.95	-0.67	10.11
-33.66	1.89	-31.77	-28.65	29.96	-2.993	3.062	2.45	7.60	-0.23	10.27
-30.54	-0.86	-31.40	-31.33	33.15	-2.596	2.514	3.06	7.69	0.23	9.99
-30.47	0.96	-29.52	-29.24	29.47	-2.837	2.632	2.60	6.95	0.36	9.37
-30.20	-3.04	-33.24	-34.33	32.07	-2.091	2.281	3.09	6.62	0.08	9.53
-31.28	-4.90	-36.18	-33.40	35.84	-2.323	2.373	2.86	6.91	-0.30	9.91
-28.50	-2.19	-30.69	-30.72	35.75	-2.653	2.641	2.76	7.34	0.33	10.14
-28.53	0.00	0.00	0.00	0.00	0.000	0.000	0.00	6.98	1.03	10.25
STRAND	2									
0.00	1.27	0.00	33.25	35.75	2.629	2.641	0.00	0.00	0.00	0.00
31.97	-0.11	31.86	31.45	35.84	2.423	2.373	0.00	0.00	0.00	0.00
31.56	2.22	33.79	34.00	32.07	2.471	2.281	0.00	0.00	0.00	0.00
31.78	0.46	32.24	32.87	29.47	2.427	2.632	0.00	0.00	0.00	0.00
32.41	1.78	34.19	33.27	33.15	2.432	2.514	0.00	0.00	0.00	0.00
31.49	-2.24	29.25	26.52	29.96	3.131	3.062	0.00	0.00	0.00	0.00
28.76	-0.41	28.35	28.50	32.75	2.743	2.763	0.00	0.00	0.00	0.00
28.91	3.61	32.51	31.85	32.62	2.064	2.153	0.00	0.00	0.00	0.00
28.25	8.71	36.95	38.99	37.92	1.965	2.296	0.00	0.00	0.00	0.00
30.28	0.00	0.00	0.00	0.00	0.000	0.000	0.00	0.00	0.00	0.00

Table B5. Cylindrical Parameters for Native B-Helical CGCGAATTCGCG

PHOSPHATE BACKBONE TABLE, 5' TO 3' DIRECTION IN EACH STRAND

R	PHI	Z	D	Q	H	PI
STRAND	1					
9.80	-36.34	29.04	6.64	5.73	-3.36	-30.37
10.08	-69.81	25.68	6.47	5.71	-3.05	-28.13
10.04	-102.78	22.63	6.83	6.25	-2.75	-23.75
10.45	-138.24	19.88	6.88	6.62	-1.88	-15.83
10.59	-174.91	18.00	6.90	6.28	-2.84	-24.31
10.26	150.05	15.16	6.29	5.42	-3.19	-30.46
9.68	118.68	11.97	6.87	6.01	-3.31	-28.86
9.37	81.94	8.66	6.70	6.20	-2.55	-22.40
9.55	43.70	6.11	6.55	4.15	-5.07	-50.72
9.39	18.41	1.04	7.05	6.76	-1.99	-16.40
9.73	-22.95	-0.95	0.00	0.00	0.00	0.00
STRAND	2					
8.34	174.29	-3.64	6.63	5.74	3.30	29.91
8.57	-146.02	-0.34	6.45	4.93	4.16	40.17
8.42	-112.28	3.83	7.12	6.29	3.34	27.94
8.70	-69.20	7.16	6.77	6.34	2.38	20.56
8.86	-26.88	9.54	6.71	5.58	3.73	33.74
8.60	10.35	13.27	6.70	5.61	3.67	33.23
8.47	48.67	16.94	6.70	5.62	3.65	32.96
8.57	87.20	20.59	6.17	4.58	4.13	42.05
9.00	117.31	24.72	6.60	4.73	4.60	44.23
8.71	148.23	29.32	6.68	5.58	3.66	33.27
8.93	-174.90	32.99	0.00	0.00	0.00	0.00

ROTATION AND RISE TABLE, 5' TO 3' DIRECTION

S5'	S3'	R(P)	T(C1')	TQ	H(C1')	HQ	SLIDE	DISP	SLIP	C1/C1
STRAND	1									
0.00	-11.02	0.00	-38.61	39.36	-3.021	3.005	1.53	3.31	0.46	10.55
-27.59	-5.89	-33.47	-33.58	38.09	-2.773	2.870	1.35	3.28	0.57	10.60
-27.70	-5.27	-32.96	-30.45	27.74	-3.618	4.069	0.44	2.70	0.64	10.45
-25.18	-10.28	-35.46	-33.93	36.43	-2.267	3.011	1.87	3.28	1.16	10.32
-23.65	-13.03	-36.68	-34.32	37.15	-2.893	3.312	2.19	2.99	1.24	10.42
-21.29	-13.75	-35.04	-31.46	32.05	-3.171	3.246	2.55	2.58	1.19	10.34
-17.71	-13.65	-31.37	-35.76	35.88	-3.305	3.356	2.10	2.15	0.43	10.09
-22.10	-14.65	-36.75	-39.23	40.88	-3.101	3.051	2.04	2.10	0.20	10.33
-24.58	-13.66	-38.24	-20.45	30.39	-4.436	4.073	0.58	1.89	0.06	10.45
-6.80	-18.49	-25.29	-43.70	40.27	-2.604	3.005	1.35	1.03	0.81	10.39
-25.21	-16.15	-41.36	-27.11	36.54	-4.264	3.765	1.25	1.41	0.69	10.66
-10.96	0.00	0.00	0.00	0.00	0.000	0.000	0.00	0.38	0.45	10.86
STRAND	2									
0.00	20.04	0.00	48.99	36.54	3.267	3.765	0.00	0.00	0.00	0.00
28.95	10.74	39.69	36.61	40.27	3.406	3.005	0.00	0.00	0.00	0.00
25.87	7.87	33.74	37.27	30.39	3.709	4.073	0.00	0.00	0.00	0.00
29.39	13.69	43.08	43.68	40.88	3.000	3.051	0.00	0.00	0.00	0.00
29.99	12.32	42.31	37.99	35.88	3.407	3.356	0.00	0.00	0.00	0.00
25.67	11.56	37.23	40.03	32.05	3.321	3.246	0.00	0.00	0.00	0.00
28.46	9.86	38.33	41.20	37.15	3.731	3.312	0.00	0.00	0.00	0.00
31.33	7.19	38.53	38.72	36.43	3.754	3.011	0.00	0.00	0.00	0.00
31.53	-1.42	30.11	18.97	27.74	4.521	4.069	0.00	0.00	0.00	0.00
20.39	10.52	30.91	42.35	38.09	2.968	2.870	0.00	0.00	0.00	0.00
31.83	5.04	36.87	39.07	39.36	2.989	3.005	0.00	0.00	0.00	0.00
34.03	0.00	0.00	0.00	0.00	0.000	0.000	0.00	0.00	0.00	0.00

Table B6. Cylindrical Parameters for B-Helical CGCGAATTCGCG, 16K

PHOSPHATE BACKBONE TABLE, 5' TO 3' DIRECTION IN EACH STRAND

R	PHI	Z	D	Q	H	PI
STRAND	1					
9.77	-35.09	29.36	6.52	5.59	-3.36	-31.03
9.92	-68.05	25.99	6.61	5.62	-3.48	-31.77
10.03	-100.76	22.51	6.88	6.37	-2.60	-22.25
10.47	-136.86	19.91	6.90	6.62	-1.93	-16.29
10.19	-174.21	17.97	6.70	6.12	-2.74	-24.09
9.78	150.16	15.24	6.48	5.93	-2.61	-23.77
9.51	114.40	12.63	6.74	5.90	-3.26	-28.91
9.46	78.18	9.37	6.73	6.21	-2.59	-22.61
9.88	40.78	6.78	6.35	3.99	-4.94	-51.08
10.06	17.72	1.84	6.83	6.42	-2.33	-19.96
10.36	-18.91	-0.49	0.00	0.00	0.00	0.00
STRAND	2					
7.50	170.03	-3.04	6.32	5.60	2.93	27.63
7.87	-147.29	-0.10	6.40	4.99	4.00	38.68
7.93	-110.43	3.89	7.20	6.15	3.75	31.36
8.78	-67.64	7.64	6.81	6.41	2.29	19.69
9.26	-26.14	9.93	6.73	5.75	3.50	31.33
9.04	10.44	13.43	6.52	5.39	3.67	34.23
8.81	45.53	17.10	6.92	5.96	3.52	30.60
8.39	85.99	20.62	6.30	5.10	3.70	35.97
9.03	119.77	24.33	6.27	4.16	4.69	48.43
8.67	146.87	29.02	6.75	5.63	3.72	33.50
9.01	-176.08	32.74	0.00	0.00	0.00	0.00

ROTATION AND RISE TABLE, 5' TO 3' DIRECTION

S5'	S3'	R(P)	T(C1')	TQ	H(C1')	HG	SLIDE	DISP	SLIP	C1/C1
STRAND	1									
0.00	-11.24	0.00	-37.85	39.14	-3.237	3.119	1.45	3.05	0.45	10.44
-26.61	-6.34	-32.96	-33.70	36.80	-2.578	2.845	1.15	2.96	0.50	10.67
-27.35	-5.36	-32.71	-30.82	28.54	-3.733	4.086	0.60	2.57	0.68	10.28
-25.46	-10.63	-36.09	-35.02	37.28	-2.487	2.942	1.97	3.08	1.14	10.28
-24.39	-12.96	-37.35	-36.36	36.41	-2.686	3.131	2.21	2.70	1.00	10.20
-23.40	-12.23	-35.63	-35.31	32.51	-2.639	3.001	2.57	2.55	0.69	10.18
-23.08	-12.68	-35.76	-36.26	36.40	-3.338	3.267	2.09	2.46	-0.02	9.87
-23.58	-12.65	-36.22	-36.22	40.44	-3.076	3.008	2.05	2.56	0.11	10.07
-23.57	-13.83	-37.40	-17.88	30.95	-4.519	4.344	0.49	2.22	0.26	10.32
-4.05	-19.01	-23.06	-40.33	39.06	-2.426	2.894	1.27	1.05	1.38	10.29
-21.33	-15.31	-36.64	-22.05	36.63	-4.394	3.665	1.09	1.22	1.41	10.44
-6.74	0.00	0.00	0.00	0.00	0.000	0.000	0.00	-0.48	1.12	10.91
STRAND	2									
0.00	26.07	0.00	60.69	36.63	2.936	3.665	0.00	0.00	0.00	0.00
34.62	8.06	42.68	36.91	39.06	3.362	2.894	0.00	0.00	0.00	0.00
28.84	8.02	36.87	39.68	30.95	4.169	4.344	0.00	0.00	0.00	0.00
31.65	11.14	42.79	43.54	40.44	2.939	3.008	0.00	0.00	0.00	0.00
32.40	9.09	41.50	35.39	36.40	3.196	3.267	0.00	0.00	0.00	0.00
26.29	10.28	36.57	36.19	32.51	3.363	3.001	0.00	0.00	0.00	0.00
25.91	9.19	35.10	39.60	36.41	3.577	3.131	0.00	0.00	0.00	0.00
30.41	10.04	40.45	41.53	37.28	3.398	2.942	0.00	0.00	0.00	0.00
31.50	2.29	33.78	20.88	28.54	4.440	4.086	0.00	0.00	0.00	0.00
18.59	8.51	27.10	38.36	36.80	3.112	2.845	0.00	0.00	0.00	0.00
29.85	7.20	37.05	40.22	39.14	3.001	3.119	0.00	0.00	0.00	0.00
33.02	0.00	0.00	0.00	0.00	0.000	0.000	0.00	0.00	0.00	0.00

Table B7. Cylindrical Parameters for B-Helical Cisplatin Complex

PHOSPHATE BACKBONE TABLE, 5' TO 3' DIRECTION IN EACH STRAND

R	PHI	Z	D	Q	H	PI
STRAND	1					
10.43	-33.41	28.09	6.53	5.44	-3.61	-33.59
10.59	-63.39	24.48	6.88	6.17	-3.04	-26.19
10.23	-97.83	21.44	6.72	6.19	-2.60	-22.78
10.87	-131.79	18.84	7.10	6.96	-1.39	-11.31
10.57	-169.64	17.45	6.64	5.97	-2.92	-26.07
9.73	156.51	14.53	6.73	5.81	-3.39	-30.23
9.55	121.42	11.14	6.65	5.63	-3.53	-32.10
8.96	86.18	7.61	6.26	5.42	-3.14	-30.07
9.05	51.17	4.47	6.65	4.32	-5.06	-49.51
8.70	23.11	-0.58	7.02	6.65	-2.25	-18.70
8.94	-21.15	-2.83	0.00	0.00	0.00	0.00
STRAND	2					
8.81	175.72	-4.17	6.53	5.70	3.19	29.28
9.31	-147.79	-0.98	6.26	5.05	3.69	36.11
8.96	-115.75	2.71	7.27	6.67	2.90	23.51
8.22	-70.32	5.61	6.90	6.47	2.40	20.32
8.79	-25.74	8.01	6.33	5.30	3.46	33.13
8.66	9.64	11.47	6.70	5.24	4.17	38.53
8.08	45.91	15.64	6.71	5.62	3.66	33.07
8.36	85.87	19.30	6.44	4.76	4.34	42.38
8.57	118.48	23.64	6.38	4.16	4.84	49.29
7.76	147.47	28.48	6.80	5.93	3.33	29.30
8.55	-170.25	31.81	0.00	0.00	0.00	0.00

ROTATION AND RISE TABLE, 5' TO 3' DIRECTION

S5'	S3'	R(P)	T(C1')	TG	H(C1')	HG	SLIDE	DISP	SLIP	C1/C1
STRAND	1									
0.00	-11.59	0.00	-34.76	40.12	-3.117	3.271	1.33	3.45	1.16	10.39
-23.17	-6.81	-29.98	-32.00	35.81	-2.688	2.738	1.09	2.85	1.30	10.76
-25.19	-9.24	-34.43	-33.73	30.69	-3.483	4.065	0.72	2.29	1.26	10.55
-24.49	-9.48	-33.96	-32.22	34.14	-1.796	2.836	1.63	2.78	1.39	10.76
-22.74	-15.11	-37.85	-35.42	42.80	-2.838	3.184	2.27	2.48	1.42	10.50
-20.31	-13.54	-33.85	-32.86	28.52	-3.424	3.551	2.03	1.53	1.13	11.00
-19.33	-15.77	-35.09	-34.90	33.18	-3.607	3.355	2.15	1.77	0.64	10.01
-19.13	-16.11	-35.24	-37.59	43.21	-3.253	2.983	2.13	1.82	0.18	10.16
-21.48	-13.53	-35.01	-18.17	27.84	-4.567	4.244	0.12	1.15	-0.42	10.44
-4.64	-23.42	-28.06	-53.81	41.53	-2.820	3.090	1.29	0.37	0.27	10.80
-30.39	-13.87	-44.26	-24.59	38.20	-4.274	3.575	0.83	1.59	-0.08	11.18
-10.72	0.00	0.00	0.00	0.00	0.000	0.000	0.00	0.22	-0.14	10.54
STRAND	2									
0.00	22.83	0.00	51.52	38.20	2.876	3.575	0.00	0.00	0.00	0.00
28.69	7.80	36.49	30.06	41.53	3.360	3.090	0.00	0.00	0.00	0.00
22.27	9.76	32.03	35.18	27.84	3.921	4.244	0.00	0.00	0.00	0.00
25.42	20.02	45.44	52.08	43.21	2.713	2.983	0.00	0.00	0.00	0.00
32.06	12.52	44.58	34.83	33.18	3.103	3.355	0.00	0.00	0.00	0.00
22.31	13.06	35.37	25.82	28.52	3.678	3.551	0.00	0.00	0.00	0.00
12.76	23.51	36.27	56.45	42.80	3.531	3.184	0.00	0.00	0.00	0.00
32.94	7.03	39.97	36.07	34.14	3.876	2.836	0.00	0.00	0.00	0.00
29.04	3.57	32.60	25.49	30.69	4.647	4.065	0.00	0.00	0.00	0.00
21.92	7.07	28.99	41.00	35.81	2.788	2.738	0.00	0.00	0.00	0.00
33.92	8.36	42.28	45.75	40.12	3.425	3.271	0.00	0.00	0.00	0.00
37.39	0.00	0.00	0.00	0.00	0.000	0.000	0.00	0.00	0.00	0.00

Table B8. Cylindrical Parameters for B-Helical CGCGAATTbrCGCG, MPD20

PHOSPHATE BACKBONE TABLE, 5' TO 3' DIRECTION IN EACH STRAND

R	PHI	Z	D	G	H	PI
STRAND	1					
9.51	-23.80	30.46	6.54	5.58	-3.41	-31.47
9.79	-57.37	27.05	6.77	5.74	-3.58	-31.95
9.28	-92.31	23.47	7.00	6.28	-3.08	-26.15
9.51	-131.32	20.38	6.95	6.76	-1.62	-13.50
10.05	-171.62	18.76	6.88	6.26	-2.86	-24.59
10.10	152.19	15.90	6.43	5.41	-3.49	-32.83
9.82	120.73	12.41	6.94	6.13	-3.25	-27.94
9.87	84.45	9.16	5.95	5.13	-3.02	-30.47
10.20	54.88	6.14	6.83	5.15	-4.48	-41.04
9.71	25.02	1.65	6.98	6.77	-1.71	-14.20
10.27	-14.45	-0.06	0.00	0.00	0.00	0.00
STRAND	2					
7.64	179.43	-2.82	6.81	5.82	3.54	31.34
7.96	-136.81	0.73	6.18	4.69	4.02	40.54
8.05	-102.72	4.74	7.07	6.12	3.55	30.11
8.48	-59.38	8.29	6.80	6.41	2.27	19.50
9.03	-16.60	10.56	6.48	5.65	3.17	29.27
9.32	19.23	13.73	6.78	5.85	3.43	30.42
9.29	55.86	17.16	6.67	5.71	3.43	30.99
9.49	91.28	20.59	6.38	4.51	4.51	45.02
9.60	118.58	25.10	6.58	4.96	4.33	41.14
8.97	149.32	29.43	6.69	5.60	3.66	33.13
8.74	-173.82	33.09	0.00	0.00	0.00	0.00

ROTATION AND RISE TABLE, 5' TO 3' DIRECTION

S5'	S3'	R(P)	T(C1')	TG	H(C1')	HG	SLIDE	DISP	SLIP	C1/C1
STRAND	1									
0.00	-11.36	0.00	-39.67	39.57	-2.949	3.051	1.43	2.61	0.35	10.73
-28.31	-5.25	-33.57	-33.63	37.40	-3.200	3.121	1.31	2.56	0.30	10.55
-28.38	-6.56	-34.94	-36.14	29.68	-3.582	4.059	0.65	2.13	0.12	10.94
-29.58	-9.44	-39.02	-37.45	36.11	-2.412	3.075	2.00	2.99	0.50	10.58
-28.01	-12.28	-40.29	-36.83	37.70	-2.887	3.290	2.29	3.11	0.52	10.33
-24.55	-11.64	-36.20	-32.02	32.21	-3.251	3.346	2.58	3.07	0.70	10.25
-20.38	-11.08	-31.46	-35.61	35.94	-3.517	3.240	2.29	2.72	0.34	9.73
-24.53	-11.76	-36.29	-34.69	38.99	-2.966	2.859	2.10	2.76	0.30	10.09
-22.93	-6.63	-29.56	-18.68	29.99	-4.480	4.257	0.48	2.32	0.28	10.40
-12.05	-17.81	-29.86	-43.00	42.11	-2.202	2.729	1.32	1.32	1.21	10.39
-25.19	-14.27	-39.47	-27.80	38.25	-4.426	3.826	1.20	1.50	1.26	10.91
-13.52	0.00	0.00	0.00	0.00	0.000	0.000	0.00	0.24	1.09	10.86
STRAND	2									
0.00	22.09	0.00	54.73	38.25	3.226	3.826	0.00	0.00	0.00	0.00
32.64	11.11	43.76	40.83	42.11	3.256	2.729	0.00	0.00	0.00	0.00
29.72	4.38	34.09	36.87	29.99	4.033	4.257	0.00	0.00	0.00	0.00
32.49	10.85	43.34	43.87	38.99	2.753	2.859	0.00	0.00	0.00	0.00
33.02	9.76	42.78	36.81	35.94	2.963	3.240	0.00	0.00	0.00	0.00
27.05	8.78	35.83	35.96	32.21	3.441	3.346	0.00	0.00	0.00	0.00
27.17	9.46	36.63	36.69	37.70	3.693	3.290	0.00	0.00	0.00	0.00
27.23	8.19	35.41	34.39	36.11	3.737	3.075	0.00	0.00	0.00	0.00
26.20	1.11	27.31	19.36	29.68	4.535	4.059	0.00	0.00	0.00	0.00
18.26	12.48	30.74	42.87	37.40	3.041	3.121	0.00	0.00	0.00	0.00
30.39	6.48	36.87	39.86	39.57	3.153	3.051	0.00	0.00	0.00	0.00
33.38	0.00	0.00	0.00	0.00	0.000	0.000	0.00	0.00	0.00	0.00

Table B9. Cylindrical Parameters for B-Helical CGCGAATTbrCGCG, MPD7

PHOSPHATE BACKBONE TABLE, 5' TO 3' DIRECTION IN EACH STRAND

R	PHI	Z	D	Q	H	PI
STRAND	1					
9.83	138.75	0.00	6.79	5.26	-4.31	-39.33
10.06	108.13	-4.31	6.60	5.28	-3.96	-36.86
9.31	76.78	-8.27	7.10	6.66	-2.47	-20.38
9.36	35.01	-10.74	6.96	6.85	-1.25	-10.38
9.35	-7.92	-11.99	6.60	5.67	-3.38	-30.78
9.55	-42.84	-15.37	6.62	5.42	-3.80	-35.04
9.54	-75.81	-19.17	6.68	5.33	-4.03	-37.09
8.82	-109.23	-23.20	6.46	5.74	-2.95	-27.15
8.93	-147.00	-26.14	6.59	4.53	-4.78	-46.55
8.86	-176.51	-30.93	6.65	5.78	-3.29	-29.66
9.42	146.81	-34.22	0.00	0.00	0.00	0.00
STRAND	2					
9.02	-7.00	-34.36	6.74	5.53	3.85	34.85
9.49	27.65	-30.51	6.24	4.71	4.10	41.05
9.21	56.77	-26.41	7.15	6.67	2.59	21.25
9.17	99.33	-23.81	6.64	6.44	1.62	14.09
9.24	140.28	-22.20	6.66	5.93	3.03	27.08
9.41	177.35	-19.17	6.77	5.60	3.79	34.10
9.20	-147.63	-15.37	6.91	5.76	3.82	33.57
9.06	-110.88	-11.55	6.40	5.10	3.87	37.21
9.30	-78.66	-7.68	6.67	5.11	4.28	39.98
8.29	-45.51	-3.39	6.92	6.53	2.30	19.44
9.46	-3.00	-1.09	0.00	0.00	0.00	0.00

ROTATION AND RISE TABLE, 5' TO 3' DIRECTION

S5'	S3'	R(P)	T(C1')	TQ	H(C1')	HG	SLIDE	DISP	SLIP	C1/C1
STRAND	1									
0.00	-12.66	0.00	-35.10	36.62	-3.609	3.672	1.53	2.60	0.43	10.67
-22.44	-8.17	-30.61	-30.69	39.15	-3.076	2.799	1.66	2.49	0.65	10.55
-22.52	-8.83	-31.35	-36.48	30.26	-3.696	4.021	0.73	1.43	0.33	10.56
-27.64	-14.12	-41.77	-41.19	37.48	-2.155	2.869	2.24	2.28	0.66	10.87
-27.06	-15.87	-42.93	-39.79	40.22	-3.024	3.344	2.32	2.49	0.28	10.50
-23.92	-11.00	-34.92	-31.24	31.24	-3.365	3.340	2.30	2.35	0.35	9.93
-20.24	-12.73	-32.97	-31.06	33.55	-3.662	3.319	2.26	2.28	0.11	10.12
-18.33	-15.10	-33.43	-42.89	42.05	-3.366	2.742	2.00	1.91	-0.26	10.31
-27.79	-9.97	-37.76	-22.26	28.57	-4.622	4.354	0.65	1.98	-0.55	10.85
-12.29	-17.23	-29.52	-48.77	41.62	-2.584	2.892	1.23	1.46	-0.20	10.67
-31.54	-5.14	-36.68	-33.37	35.07	-3.485	3.728	1.29	2.41	0.07	11.21
-28.23	0.00	0.00	0.00	0.00	0.000	0.000	0.00	2.32	0.55	10.80
STRAND	2									
0.00	7.60	0.00	33.01	35.07	3.971	3.728	0.00	0.00	0.00	0.00
25.42	9.23	34.65	32.91	41.62	3.200	2.892	0.00	0.00	0.00	0.00
23.68	5.44	29.12	32.17	28.57	4.086	4.354	0.00	0.00	0.00	0.00
26.72	15.83	42.55	43.09	42.05	2.118	2.742	0.00	0.00	0.00	0.00
27.26	13.70	40.96	38.85	33.55	2.977	3.319	0.00	0.00	0.00	0.00
25.15	11.92	37.07	33.50	31.24	3.315	3.340	0.00	0.00	0.00	0.00
21.58	13.44	35.02	39.81	40.22	3.665	3.344	0.00	0.00	0.00	0.00
26.37	10.38	36.75	36.49	37.48	3.582	2.869	0.00	0.00	0.00	0.00
26.12	6.10	32.22	20.84	30.26	4.347	4.021	0.00	0.00	0.00	0.00
14.74	18.41	33.16	51.30	39.15	2.523	2.799	0.00	0.00	0.00	0.00
32.88	9.62	42.51	36.29	36.62	3.736	3.672	0.00	0.00	0.00	0.00
26.66	0.00	0.00	0.00	0.00	0.000	0.000	0.00	0.00	0.00	0.00

Table B10. Cylindrical Parameters for Z-Helical CGCG

PHOSPHATE BACKBONE TABLE, 5' TO 3' DIRECTION IN EACH STRAND

R	PHI	Z	D	G	H	PI
STRAND	1					
5.93	43.83	12.58	7.07	2.83	-6.47	-66.38
8.07	59.19	6.10	6.62	6.39	-1.73	-15.16
5.25	111.44	4.37	0.00	0.00	0.00	0.00
STRAND	2					
5.91	-105.10	0.34	7.47	2.85	6.90	67.57
8.37	-116.89	7.25	6.72	6.63	1.08	9.29
5.85	-168.85	8.33	0.00	0.00	0.00	0.00

ROTATION AND RISE TABLE, 5' TO 3' DIRECTION

S5'	S3'	R(P)	T(C1')	TG	H(C1')	HG	SLIDE	DISP	SLIP	C1/C1
STRAND	1									
0.00	48.47	0.00	30.13	12.90	-3.947	4.016	-5.12	1.61	-2.25	10.70
-18.34	33.70	15.36	29.47	44.91	-3.738	3.603	2.65	1.36	2.23	10.66
-4.22	56.47	52.25	27.81	14.20	-3.851	4.014	-5.12	1.49	-2.23	10.66
-28.67	0.00	0.00	0.00	0.00	0.000	0.000	0.00	1.60	2.19	10.56
STRAND	2									
0.00	-47.23	0.00	-30.37	14.20	4.177	4.014	0.00	0.00	0.00	0.00
16.87	-28.65	-11.79	-32.30	44.91	3.468	3.603	0.00	0.00	0.00	0.00
-3.64	-48.32	-51.96	-24.68	12.90	4.084	4.016	0.00	0.00	0.00	0.00
23.64	0.00	0.00	0.00	0.00	0.000	0.000	0.00	0.00	0.00	0.00

Table B11. Cylindrical Parameters for Z-Helical CGCGCG

PHOSPHATE BACKBONE TABLE, 5' TO 3' DIRECTION IN EACH STRAND

R	PHI	Z	D	G	H	PI
STRAND	1					
6.21	157.84	3.65	6.37	1.89	-6.08	-72.73
7.10	172.26	-2.43	5.90	5.56	-1.99	-19.67
5.91	-138.21	-4.41	7.17	1.92	-6.91	-74.46
7.83	-138.57	-11.32	6.50	6.50	0.14	1.27
6.72	-86.19	-11.18	0.00	0.00	0.00	0.00
STRAND	2					
7.08	50.82	-15.31	6.27	2.00	5.94	71.36
7.58	35.61	-9.37	6.02	5.74	1.82	17.60
6.63	-11.41	-7.55	6.26	1.89	5.97	72.48
7.80	-23.24	-1.58	5.81	5.56	1.69	16.92
6.74	-67.47	0.11	0.00	0.00	0.00	0.00

ROTATION AND RISE TABLE, 5' TO 3' DIRECTION

S5'	S3'	R(P)	T(C1')	TG	H(C1')	HG	SLIDE	DISP	SLIP	C1/C1
STRAND	1									
0.00	41.94	0.00	17.11	7.59	-4.059	4.069	-5.16	0.73	-2.52	10.67
-24.83	39.25	14.42	43.34	50.22	-3.795	3.676	2.94	0.64	2.00	10.68
4.09	45.44	49.53	15.71	8.16	-3.852	3.990	-5.34	0.57	-2.60	10.62
-29.72	29.37	-0.36	47.44	52.44	-3.339	3.365	2.92	0.57	2.09	10.70
18.08	34.30	52.38	17.52	9.74	-4.365	4.330	-4.81	0.50	-2.33	10.76
-16.78	0.00	0.00	0.00	0.00	0.000	0.000	0.00	0.20	1.85	10.78
STRAND	2									
0.00	-30.95	0.00	-9.19	9.74	4.295	4.330	0.00	0.00	0.00	0.00
21.75	-36.96	-15.21	-46.31	52.44	3.392	3.365	0.00	0.00	0.00	0.00
-9.35	-37.67	-47.02	-13.89	8.16	4.128	3.990	0.00	0.00	0.00	0.00
23.78	-35.61	-11.83	-43.42	50.22	3.557	3.676	0.00	0.00	0.00	0.00
-7.81	-36.42	-44.23	-13.20	7.59	4.080	4.069	0.00	0.00	0.00	0.00
23.22	0.00	0.00	0.00	0.00	0.000	0.000	0.00	0.00	0.00	0.00

Table C1. Torsion Angles for A-Helical ^{I}CCGG/^{I}CCGG

STRAND 1

ALPHA	BETA	GAMMA	DELTA	EPSILON	ZETA	CHI
0.0	0.0	142.8	90.1	−168.4	−58.2	−149.4
−68.4	−169.6	51.0	79.8	−162.1	−68.3	−159.5
−80.3	−178.0	72.5	87.2	−159.7	−66.5	−163.9
−74.9	−175.8	55.3	80.4	0.0	0.0	−152.1

STRAND 2

ALPHA	BETA	GAMMA	DELTA	EPSILON	ZETA	CHI
0.0	0.0	60.0	75.9	−156.7	−68.8	−168.9
−70.1	173.6	56.7	76.7	−150.8	−65.5	−162.3
−73.1	166.1	70.0	72.6	−170.1	−71.7	−170.7
−72.7	−173.5	78.8	150.4	0.0	0.0	−127.3

Table C2. Torsion Angles for A-Helical GGbrUAbrUACC

STRAND 1

ALPHA	BETA	GAMMA	DELTA	EPSILON	ZETA	CHI
0.0	0.0	23.8	88.7	−144.0	−60.6	−175.9
−95.1	−177.3	80.1	81.3	−146.5	−77.8	−166.8
−72.0	178.2	54.9	83.1	−154.5	−75.8	−156.5
−64.8	178.2	59.9	87.6	−153.0	−75.5	−162.1
−65.3	172.3	47.3	87.7	−158.9	−81.6	−151.7
−58.0	174.2	48.3	87.5	−147.1	−77.8	−154.0
−53.9	164.1	48.0	88.1	−148.7	−78.6	−156.2
−53.6	177.3	30.9	86.7	0.0	0.0	−154.0

STRAND 2

ALPHA	BETA	GAMMA	DELTA	EPSILON	ZETA	CHI
0.0	0.0	38.4	86.9	−146.4	−78.9	−176.7
−72.7	166.9	74.7	87.7	−146.1	−78.5	−176.3
−64.7	−177.2	46.2	88.9	−150.2	−79.3	−158.7
−45.1	162.6	53.0	83.2	−162.0	−75.8	−158.3
−67.0	173.5	44.4	90.9	−161.4	−77.1	−156.3
−57.5	167.6	45.3	90.9	−155.6	−84.1	−154.7
−50.9	158.5	49.9	86.2	−165.5	−73.3	−152.0
−64.6	172.2	66.7	88.0	0.0	0.0	−160.2

Table C3. Torsion Angles for A-Helical GGCCGGCC

STRAND 1

ALPHA	BETA	GAMMA	DELTA	EPSILON	ZETA	CHI
0.0	0.0	92.7	123.8	−158.4	−74.0	−142.2
−77.4	−159.0	30.8	120.4	−162.5	−99.3	−144.9
−19.3	160.1	30.3	115.5	172.0	−72.6	−132.8
−130.9	−171.3	94.8	79.9	−127.8	−76.3	−166.3
−41.9	170.7	54.3	116.3	175.5	−101.5	−140.7
−56.8	−167.1	40.1	85.1	−164.3	−51.9	−160.5
−91.7	−175.1	68.4	70.3	−176.5	−47.3	−148.7
−105.8	−165.4	71.1	76.0	0.0	0.0	−153.1

STRAND 2

ALPHA	BETA	GAMMA	DELTA	EPSILON	ZETA	CHI
0.0	0.0	92.7	123.8	−158.4	−74.0	−142.2
−77.4	−159.0	30.8	120.4	−162.5	−99.3	−144.9
−19.3	160.1	30.3	115.5	172.0	−72.6	−132.8
−130.9	−171.3	94.8	79.9	−127.8	−76.3	−166.3
−41.9	170.7	54.3	116.3	175.5	−101.5	−140.7
−56.8	−167.1	40.1	85.1	−164.3	−51.9	−160.5
−91.7	−175.1	68.4	70.3	−176.5	−47.3	−148.7
−105.8	−165.4	71.1	76.0	0.0	0.0	−153.1

490

Table C4.   Torsion Angles for A-Helical r(GCG)d(TATACGC)

STRAND   1

ALPHA	BETA	GAMMA	DELTA	EPSILON	ZETA	CHI
0.0	0.0	58.2	85.5	-119.8	-95.4	-179.4
-28.2	156.0	24.3	76.5	-167.1	-55.2	-157.6
-89.0	-174.6	71.5	85.7	-133.8	-77.3	-163.3
-38.3	-172.7	21.2	94.8	-163.8	-66.8	-145.6
-93.9	-166.4	66.1	84.8	-163.3	-61.4	-159.0
-77.1	-178.1	59.4	74.8	-149.5	-79.8	-154.7
-58.4	157.0	50.8	77.9	-168.5	-80.7	-157.5
-86.3	-173.9	78.8	82.9	-148.9	-84.5	-169.4
-74.1	173.8	50.2	75.5	-163.4	-72.3	-155.9
-74.8	-170.0	60.8	77.1	0.0	0.0	-154.3

STRAND   2

ALPHA	BETA	GAMMA	DELTA	EPSILON	ZETA	CHI
0.0	0.0	111.5	106.0	-150.6	-62.8	-170.1
-124.4	-162.9	78.3	73.6	-143.2	-84.2	-165.9
-54.7	164.2	52.0	81.4	-130.8	-92.5	-172.6
-41.7	159.2	46.2	70.4	-160.0	-48.5	-157.7
-85.4	168.7	69.2	86.5	-142.3	-78.1	-157.7
-52.8	172.7	39.2	72.9	-168.9	-54.7	-153.3
-96.1	178.8	65.2	89.3	-141.0	-108.4	-164.1
-26.8	143.6	38.5	79.2	-172.1	-54.4	-158.0
-121.8	173.3	102.1	75.0	-135.5	-93.8	176.1
-14.9	175.1	7.8	93.8	0.0	0.0	-149.9

Table C5.   Torsion Angles for Native CGCGAATTCGCG

STRAND   1

ALPHA	BETA	GAMMA	DELTA	EPSILON	ZETA	CHI
0.0	0.0	174.2	156.8	-141.3	-143.9	-104.9
-65.6	169.8	40.1	128.1	174.2	-97.7	-110.5
-62.6	171.8	58.8	98.4	-176.6	-87.6	-135.1
-62.8	-179.9	57.2	155.7	-155.3	-152.5	-93.4
-43.0	142.8	52.4	119.6	179.9	-92.2	-126.2
-73.3	179.7	66.0	121.1	173.7	-88.5	-122.1
-56.6	-179.2	52.2	98.9	173.6	-85.9	-127.3
-59.2	173.3	64.1	108.9	170.6	-89.4	-125.7
-58.5	-179.5	60.4	128.7	-156.9	-94.1	-119.5
-67.2	169.1	47.2	142.9	-103.3	150.2	-89.6
-73.9	139.3	56.3	135.6	-161.8	-89.6	-125.1
-81.5	175.7	57.2	110.6	0.0	0.0	-112.0

STRAND   2

ALPHA	BETA	GAMMA	DELTA	EPSILON	ZETA	CHI
0.0	0.0	55.9	136.7	-158.7	-124.9	-127.6
-51.3	163.9	49.0	121.9	177.7	-93.0	-116.4
-63.0	168.9	60.4	85.7	174.8	-85.5	-133.8
-69.2	171.1	73.2	135.9	174.2	-98.4	-114.8
-56.6	-169.5	53.8	146.6	177.0	-97.2	-106.5
-57.0	-173.6	47.7	130.2	174.4	-101.3	-108.3
-58.3	173.6	60.1	109.2	178.8	-88.3	-131.2
-58.6	179.5	55.3	122.4	178.6	-94.5	-120.4
-59.1	-175.4	45.0	110.3	-176.7	-86.4	-114.3
-66.8	179.2	50.1	149.7	-100.1	171.6	-88.4
-72.3	138.5	44.6	112.8	-174.4	-96.8	-125.4
-65.0	170.6	46.7	78.7	0.0	0.0	-135.1

Table C6. Torsion Angles for CGCGAATTCGCG, 16K

STRAND 1

ALPHA	BETA	GAMMA	DELTA	EPSILON	ZETA	CHI
0.0	0.0	-170.5	156.1	-140.2	-136.1	-107.4
-41.0	167.1	20.5	142.9	-177.2	-113.1	-95.2
-70.7	159.8	72.5	88.4	-163.6	-93.3	-143.6
-39.5	167.4	47.2	142.2	-172.3	-129.9	-101.1
-15.8	150.1	32.1	126.9	179.0	-93.3	-110.5
-59.3	-173.0	44.3	126.1	162.6	-81.0	-105.8
-66.6	-176.9	63.2	103.7	-176.0	-87.8	-131.7
-48.0	175.8	46.3	117.8	168.9	-92.2	-112.6
-67.4	178.8	63.6	136.6	-130.8	-123.6	-129.2
-41.2	143.2	34.8	130.1	-101.6	144.5	-84.8
-67.5	132.9	44.9	137.7	-151.9	-99.4	-118.7
-91.9	165.7	69.7	107.1	0.0	0.0	-116.1

STRAND 2

ALPHA	BETA	GAMMA	DELTA	EPSILON	ZETA	CHI
0.0	0.0	63.0	144.4	-145.2	-156.4	-131.6
-21.1	135.0	29.3	121.0	-171.9	-99.5	-123.1
-53.6	151.3	68.8	80.8	-166.2	-90.3	-148.1
-73.5	171.5	80.7	136.3	176.1	-105.9	-110.2
-29.8	-172.2	28.0	144.4	172.2	-91.5	-105.0
-39.5	-176.0	31.2	132.3	168.2	-92.0	-106.9
-57.5	-179.0	51.8	118.8	-174.6	-102.5	-111.9
-31.2	176.3	32.4	130.5	175.1	-96.8	-112.4
-61.7	-171.9	49.6	115.4	-173.6	-86.1	-110.4
-46.9	177.9	33.1	145.3	-126.7	-178.0	-84.2
-57.3	117.1	63.9	99.3	-167.1	-87.3	-142.9
-82.1	168.3	56.8	72.1	0.0	0.0	-141.8

Table C7. Torsion Angles, CGCGAATTCGCG Plus Cisplatin

STRAND 1

ALPHA	BETA	GAMMA	DELTA	EPSILON	ZETA	CHI
0.0	0.0	-155.8	147.0	-165.3	-123.0	-93.8
-74.6	174.3	56.0	122.4	178.1	-103.3	-97.9
-67.3	165.3	71.2	103.6	-175.7	-94.0	-136.7
-61.0	-165.8	53.9	138.4	178.5	-123.0	-86.5
-33.5	141.9	58.6	110.5	178.9	-85.0	-139.1
-66.4	-169.8	48.2	148.4	176.5	-124.0	-101.0
-52.5	164.1	67.2	109.0	179.8	-95.7	-119.3
-46.7	168.2	58.2	101.4	163.9	-80.6	-115.0
-62.2	174.3	62.2	105.1	-166.2	-81.5	-132.0
-67.1	-178.4	49.9	147.6	-106.8	147.5	-76.0
-64.7	153.3	41.9	145.5	166.9	-77.1	-102.4
-101.2	-153.6	66.7	133.8	0.0	0.0	-95.1

STRAND 2

ALPHA	BETA	GAMMA	DELTA	EPSILON	ZETA	CHI
0.0	0.0	-49.9	136.5	-122.6	-173.5	-107.5
-58.8	152.3	33.6	125.1	172.1	-86.0	-111.5
-55.2	168.2	61.5	77.4	177.4	-71.8	-126.7
-63.3	176.0	73.7	154.2	164.0	-113.9	-97.0
-69.0	-159.8	63.6	142.4	160.9	-84.4	-88.5
-51.6	179.4	48.3	104.0	175.4	-76.5	-115.2
-64.1	168.9	66.2	90.5	179.3	-81.0	-79.0
-59.6	178.4	56.4	110.8	175.8	-82.4	-132.9
-69.5	-179.6	58.5	114.8	-169.5	-101.1	-117.7
-52.5	169.9	48.2	131.4	-154.5	-158.3	-96.9
-44.7	109.7	80.5	88.0	-141.4	-78.4	-140.3
-75.8	163.7	53.4	63.1	0.0	0.0	-135.8

492

Table C8. Torsion Angles for CGCGAATTbrCGCG, MPD20

STRAND 1

ALPHA	BETA	GAMMA	DELTA	EPSILON	ZETA	CHI
0.0	0.0	175.8	170.5	-165.3	-136.4	-90.9
-57.0	166.8	47.1	113.5	178.5	-85.4	-121.1
-61.6	173.3	58.3	107.9	-179.8	-94.8	-133.6
-65.1	-175.8	59.8	147.5	171.6	-113.1	-105.4
-56.8	173.9	64.4	136.7	-175.9	-99.6	-116.6
-66.3	177.7	61.0	126.3	175.0	-89.4	-118.8
-59.9	-179.1	55.3	95.8	173.9	-80.2	-129.6
-60.9	177.2	65.1	102.6	164.3	-76.7	-126.8
-59.6	179.9	49.3	99.2	-171.7	-76.9	-132.9
-65.0	173.7	61.8	149.5	-115.8	159.7	-94.3
-65.1	147.2	46.5	149.2	-173.2	-88.9	-111.8
-90.3	-167.3	54.9	100.2	0.0	0.0	-107.6

STRAND 2

ALPHA	BETA	GAMMA	DELTA	EPSILON	ZETA	CHI
0.0	0.0	38.3	154.7	-155.9	-147.4	-94.8
-50.8	158.0	47.2	136.4	-178.0	-110.6	-110.0
-53.4	164.6	51.9	82.8	172.1	-79.4	-140.6
-58.0	-180.0	62.7	146.1	166.1	-107.1	-100.2
-51.6	-174.6	54.4	140.4	173.6	-94.2	-107.9
-61.7	176.0	60.2	100.7	179.6	-78.0	-133.7
-61.1	169.0	67.7	100.3	-175.8	-83.3	-140.9
-53.6	175.1	57.0	109.0	-177.6	-84.3	-127.9
-60.3	-179.9	42.3	85.6	156.8	-49.4	-123.0
-92.8	-162.1	49.0	155.9	-74.0	157.9	-90.1
-92.2	150.2	42.7	123.5	-176.4	-108.4	-108.3
-67.9	175.8	51.1	85.7	0.0	0.0	-132.8

Table C9. Torsion Angles for CGCGAATTbrCGCG, MPD7

STRAND 1

ALPHA	BETA	GAMMA	DELTA	EPSILON	ZETA	CHI
0.0	0.0	160.7	142.3	-165.5	-110.7	-115.0
-69.5	178.0	52.3	121.8	175.1	-99.1	-105.3
-60.4	162.5	63.6	78.1	172.9	-78.5	-143.0
-67.2	-177.6	70.1	153.1	170.4	-115.2	-98.6
-56.4	-177.1	61.2	146.0	173.8	-97.1	-102.1
-65.1	-168.8	50.3	125.2	173.5	-98.8	-106.0
-51.8	168.6	59.9	101.1	-180.0	-84.7	-128.3
-55.1	167.2	60.8	90.6	161.2	-78.1	-138.9
-56.2	-171.0	51.1	140.2	-167.0	-103.3	-110.8
-72.3	175.3	57.1	141.4	-91.1	150.6	-94.6
-75.5	156.3	26.6	138.8	173.7	-109.5	-104.9
-50.1	173.2	59.8	106.2	0.0	0.0	-130.2

STRAND 2

ALPHA	BETA	GAMMA	DELTA	EPSILON	ZETA	CHI
0.0	0.0	54.7	134.8	-98.4	175.1	-141.6
-58.7	131.8	39.6	147.6	-132.5	-159.5	-113.0
-48.0	129.3	46.5	70.7	175.3	-77.3	-147.1
-54.4	169.6	69.4	158.6	-168.1	-143.3	-99.9
-40.5	160.6	45.2	132.1	176.3	-93.0	-111.5
-60.1	-177.9	57.8	114.4	-179.4	-86.9	-120.1
-55.3	166.6	63.2	86.6	175.0	-74.5	-143.8
-68.0	178.7	61.6	113.3	163.7	-82.8	-127.7
-70.9	-168.7	54.4	92.3	155.3	-44.8	-128.1
-77.3	-170.5	44.3	155.4	-109.4	169.5	-91.0
-79.6	140.4	62.7	110.9	-155.7	-77.7	-154.9
-71.6	165.1	46.1	72.8	0.0	0.0	-127.8

493

Table C10. Torsion Angles for Z-Helical CGCG

STRAND 1

ALPHA	BETA	GAMMA	DELTA	EPSILON	ZETA	CHI
0.0	0.0	-121.7	137.3	-78.3	77.6	-167.4
85.7	169.8	164.6	115.8	-120.0	-57.3	64.1
178.0	-154.5	86.7	137.2	-98.6	87.0	-177.5
40.8	-166.5	-160.5	103.0	0.0	0.0	64.2

STRAND 2

ALPHA	BETA	GAMMA	DELTA	EPSILON	ZETA	CHI
0.0	0.0	26.6	139.6	-92.3	71.6	-146.0
76.2	174.7	179.0	120.0	-179.2	24.4	74.7
174.8	-178.3	39.1	148.0	-71.4	47.2	-149.8
95.9	163.0	-176.1	147.7	0.0	0.0	77.2

Table C11. Torsion Angles for Z-Helical CGCGCG

STRAND 1

ALPHA	BETA	GAMMA	DELTA	EPSILON	ZETA	CHI
0.0	0.0	47.5	144.3	-92.5	80.7	-151.4
63.4	-173.6	174.1	94.0	-22.0	-69.9	57.2
-98.5	-100.4	50.2	152.1	-102.1	75.5	-157.6
70.1	-172.0	177.6	95.0	-178.8	65.0	52.5
169.3	167.0	42.8	142.0	-93.4	74.2	-145.2
74.0	178.0	179.6	148.7	0.0	0.0	79.0

STRAND 2

ALPHA	BETA	GAMMA	DELTA	EPSILON	ZETA	CHI
0.0	0.0	53.1	146.8	-89.3	77.6	-142.2
61.4	-173.1	174.6	95.2	-21.3	-74.5	69.5
-99.7	-112.5	54.7	148.9	-98.1	79.9	-159.9
64.2	178.8	179.0	102.7	-112.4	-69.8	64.8
-147.5	-123.7	50.2	143.3	-97.5	71.8	-156.1
78.4	-176.0	-174.0	148.8	0.0	0.0	79.4

Index